Cooperative Cognitive Radio Networks

CRC Press
Taylor & Francis Group
Boca Raton   London   New York

CRC Press is an imprint of the
Taylor & Francis Group, an **informa** business

Cooperative Cognitive Radio Networks

The Complete Spectrum Cycle

Mohamed Ibnkahla

QUEEN'S UNIVERSITY, KINGSTON, ONTARIO, CANADA

CRC Press
Taylor & Francis Group
Boca Raton London New York

CRC Press is an imprint of the
Taylor & Francis Group, an **informa** business

CRC Press
Taylor & Francis Group
6000 Broken Sound Parkway NW, Suite 300
Boca Raton, FL 33487-2742

First issued in paperback 2017

© 2015 by Taylor & Francis Group, LLC
CRC Press is an imprint of Taylor & Francis Group, an Informa business

No claim to original U.S. Government works

ISBN-13: 978-1-4665-7078-8 (hbk)
ISBN-13: 978-1-138-89278-1 (pbk)

Visit the Taylor & Francis Web site at
http://www.taylorandfrancis.com

and the CRC Press Web site at
http://www.crcpress.com

Contents

Preface

Cognitive radio is expected to be one of the enabling technologies for future generations of wireless communications systems. The cognitive concept allows cognitive radio users (or secondary users) to have opportunistic access to the radio spectrum when licenced users (or primary users) are not present or when the interference caused by secondary users' access is below a given threshold. This can result in a substantial improvement of the system's spectral efficiency. Continuous sensing and careful monitoring of the spectrum are required in order to allow secondary user access without harming licensed users. The radio spectrum cycle includes sensing, learning, planning, decision-making, and actions. The set of actions made by cognitive radio users can take different forms such as spectrum access, packet routing, handover, and resource allocation. The actions depend usually on the environmental information acquired during the sensing and learning processes. This information may include the status of the cognitive radio network and the primary network, quality of service, physical channel conditions, trading rules, economic indicators, social impact, environmental impact, etc.

In the past few years, research on cognitive radio has been scattered over a large number of conference and journal papers. These publications have treated the cognitive radio cycle selectively by focusing only on specific techniques or technologies rather than covering the spectrum cycle as a whole.

This book fills this gap by covering the entire spectrum cycle. It presents the state of the art of this field and proposes a unifying view of the different approaches and methodologies throughout the spectrum cycle. This book will be a benchmark that sets the foundations of cognitive communications that cover all layers of the protocol stack.

ORGANIZATION OF THE BOOK

Chapter 1 gives an overview of cognitive radio systems and explains the different components of the spectrum cycle, which will be covered throughout the book.

Chapters 2 through 5 present physical layer issues. The spectrum cycle aspects covered in these chapters include spectrum sensing, learning, and decision. Chapter 2 investigates the major spectrum sensing techniques in both single-band and multi-band frameworks with an in-depth discussion and comparison between the different approaches. Chapter 3 presents the basic principles of cooperative spectrum sensing. It also discusses advanced combining and cooperative diversity techniques for spectrum sensing and acquisition. Chapter 4 covers cooperative spectrum acquisition in the cases of spectrum interweave and underlay, where interference is considered as a design parameter. Chapter 5 discusses the different trade-offs that need to be taken into account during the design and operating phases of spectrum sensing and acquisition techniques and how the performance can be affected, for example, in terms of delay, throughput, or power efficiency.

Chapters 6 through 9 are devoted to link layer and medium access layer design issues. Learning, decision, access, and spectrum evacuation are the main spectrum cycle aspects covered in these chapters. Chapter 6 addresses the spectrum mobility problem. Various hand-off techniques and strategies are presented along with in-depth comparisons. Chapter 7 covers medium access protocols in the context of cognitive radio networks. In particular, it addresses cooperative and noncooperative media access control layer protocols that can be used in the absence of a common control channel. The challenges facing their design and the special requirements that need to be met are also covered in this chapter. Chapter 8 is devoted to cognitive radio ad hoc networks. It addresses the spectrum sharing and fairness problem and the basics of scaling laws. Special focus is given to local control schemes and fairness protocols for spectrum sharing. Chapter 9 investigates medium access protocols in cognitive radio ad hoc networks where mobility and primary exclusive regions are considered. Comparisons between the different protocols are provided, and several examples are given for illustration.

Chapter 10 covers the network layer and routing in cognitive radio networks. The aspects of the spectrum cycle covered in this chapter include learning, decision-making, and routing. A classification of routing protocols and their challenges are presented. In particular, the chapter investigates distributed and centralized approaches, mobility issues, as well as highly dynamic networks.

Chapter 11 presents an economic framework for cognitive radio networks based mainly on game theory. The spectrum cycle mechanisms covered in this chapter include learning, decision-making, and actions. Different strategies and market models are investigated including fixed-price markets, single auctions, and double auctions, where the physical layer parameters and interference levels are considered as part of the economic framework.

Chapter 12 gives an overview of security issues in cognitive radio networks. The spectrum cycle concepts covered in this chapter include learning, decision-making, and classification, which result in specific actions regarding other cognitive radio users in the network. The chapter covers trust as a metric used to determine if users in the network are good or bad. Several types of attacks are investigated such as route disruption, jamming, and primary user emulation attacks. Furthermore, this chapter provides an in-depth discussion on how these attacks affect the network and what measures can be taken to mitigate their risks.

REQUIRED BACKGROUND

The book addresses the main challenges and techniques in the cognitive radio spectrum cycle. The reader needs to have a basic knowledge of wireless communications systems. It is recommended (but not required) to read the chapters in the given order. The chapters are conceptually dependent and have been built with minimal interdependence in terms of mathematical derivations or protocol design.

BOOK FEATURES

- More than 400 references, listed at the end of each chapter.
- More than 170 figures and 15 tables.
- All aspects of the cognitive spectrum cycle are covered with several illustrative examples.
- Chapters are written in a tutorial style, making them easy to understand.
- Each chapter includes an in-depth survey of the state of the art of the corresponding topic.
- Step-by-step analysis of the different algorithms and systems supported by extensive computer simulations and illustrations are provided.

Mohamed Ibnkahla
Queen's University
Kingston, Ontario, Canada

MATLAB® is a registered trademark of The MathWorks, Inc. For product information, please contact:

The MathWorks, Inc.
3 Apple Hill Drive
Natick, MA 01760-2098 USA
Tel: 508-647-7000
Fax: 508-647-7001
E-mail: info@mathworks.com
Web: www.mathworks.com

Acknowledgments

I am deeply indebted to many people who helped me throughout the multiple phases of this project. My PhD and MSc students have contributed to the writing and simulation/experimental results of the chapters: MSc student G. Hattab (Chapters 1 through 3 and 5), PhD student A. Abu Alkheir (Chapters 3 and 4), PhD student P. Hu (Chapters 8 and 9), MSc student J. Mack (Chapter 10), MSc student A. Bloor (Chapter 11), and MSc student J. Spencer (Chapter 12). G. Hattab, Dr. W. Ejaz, and Dr. S. Aslam spent a lot of time reviewing the different versions of the chapters and I am very grateful to them.

I thank all the organizations and companies that have supported my research during the past 10 years. This includes the Natural Sciences and Engineering Research Council of Canada (NSERC), the Ontario Research Fund (ORF) Wisense Project, and NSERC DIVA Network.

I am thankful to my family for their encouragement, support, and love during the different phases of this project, in particular my wife, Houda, my son, Yasinn, and my daughters, Beyan and Noha.

Author

Mohamed Ibnkahla is a full professor in the Department of Electrical and Computer Engineering, Queen's University, Kingston, Ontario, Canada. He received his PhD and Habilitation à Diriger des Recherches from the Institut National Polytechnique de Toulouse, Toulouse, France, in 1996 and 1998, respectively. Dr. Ibnkahla led several national and international projects in the field of wireless communications, wireless sensor networks, and cognitive radio technology. His research interests include cognitive radio networks, cognitive networking, wireless sensor networks, and their applications in the e-Society, neural networks and adaptive signal processing. He has authored 5 books, over 50 journal papers and book chapters, and more than 100 conference papers. He has given a significant number of tutorials and invited talks in various international conferences and workshops. Dr. Ibnkahla serves as the editor of a number of international journals and book series and is a registered professional engineer of the province of Ontario, Canada.

1 Introduction to Cognitive Radio

1.1 INTRODUCTION

Spectrum scarcity is one of the biggest challenges that are being faced by modern communication systems. The efficient use of available licensed spectrum bands is becoming a predominant issue due to the ever-increasing demand of the radio spectrum as well as the inefficient utilization incurred by the rigid and static spectrum allocation set by the telecommunications regulating authorities. Cognitive radio (CR) has emerged as one of the most promising paradigms to address spectrum scarcity and underutilization problems. The basic idea of CR is that unlicensed users (or secondary users [SUs]) are allowed to use the spectrum when licensed users (or primary users [PUs]) are not present or when the interference caused by SUs is below a given threshold. Thus, by introducing cognition in the network, the spectrum utilization can be enhanced.

Cognition implies that the radio learns from its environment and adapts its operating parameters based on the acquired knowledge. Thus, the radio can actively exploit the possible empty bands in the licensed spectrum in order to make use of them. The network's capacity to learn, adapt its parameters, and make decisions is basically what makes the system cognitive and agile.

This chapter provides the basic concepts of cognitive radio networks (CRNs). Section 1.2 describes the cognitive cycle. Section 1.3 discusses the functions of the CR framework. Section 1.4 presents the paradigms of CR access. Finally, Section 1.5 gives an overview of the book.

1.2 COGNITIVE RADIO FRAMEWORK

The CR framework is based on interactions among different entities within the network as well as with the surrounding environment. Each entity has a cognitive presence in the network, replacing purely centralized control or predefined rules. The aim is to better exploit resources and manage behavior based on each entity's adaptation to the conditions of the network and its surrounding environment.

Traditionally, cross-layer design has been used to optimize different protocols by allowing the nodes to use relevant information when making decisions in the quest to improve the network's performance. However, this process has its limitations as it cannot optimize multiple goals and learn from its surrounding environment. Therefore, there is a need for a new evolved and intelligent technique to enable learning and planning. Cognition has been introduced as the process of learning

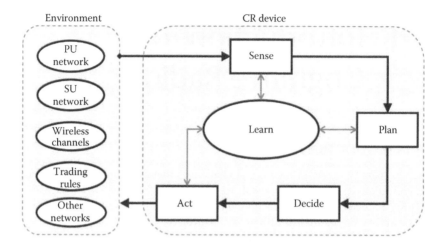

FIGURE 1.1   The CR cycle.

through perception, reasoning, and knowledge. CRNs are defined as self-aware, self-organizing, and adaptive networks that perform intelligent adaptations. These adaptations occur when the nodes in the network carry out observations regarding the state of the network, when the nodes share information between each other that is beyond the scope of the layered architecture, and when nodes learn and reason before making and carrying out optimized decisions.

Cognition has rapidly gained importance in the research fields as it enables more intelligent and better optimized networks. The cognitive cycle presented in [1] summarizes the major steps in the CR process (Figure 1.1). The cognitive loop enables intelligent adaptations through learning and sharing of information between the different entities in the network. There are six major elements that constitute the CR cycle:

- *Environment*: This includes the network and the surrounding environment such as physical channels, other users, devices, networks, and any objects that may affect the network conditions (e.g., weather conditions, obstacles, economic indicators, and trading rules).
- *Sense*: The different CR entities are capable of sensing and monitoring the environment such as the spectrum bands, interference levels, physical propagation channel parameters, and location of PUs and SUs.
- *Plan*: SUs make plans and assess these plans before taking decisions.
- *Decide*: Decision is based on knowledge and learning where the decision-making process optimizes the systems' resources.
- *Act*: The SU will act on the environment based on the decisions made. This can be through a sequence of actions (e.g., access to the media, routing of packets, allocation of resources, and modification of transmission schemes).

- *Learn*: Learning is the central element in the cognitive cycle where the nodes are equipped with a knowledge base and a learning tool that allows them to keep track of all information related to the network and environment conditions. This allows the system to learn from current and previous actions, predict future behaviors, and intelligently use them in the planning and decision-making processes.

1.2.1 Definition of Cognitive Radio

A CR system can be defined in [2] as follows:

> A radio system employing technology that allows the system to obtain knowledge of its operational and geographical environment, established policies and its internal state; to dynamically and autonomously adjust its operational parameters and protocols according to its obtained knowledge in order to achieve predefined objectives, and to learn from the results obtained.

The main feature of the CR paradigm is dynamic spectrum access, as opposed to conventional wireless paradigms, which abide by a static spectrum allocation policy. In its broadest sense, dynamic spectrum allocation is classified into three categories [3]:

- *Dynamic exclusive use model*: This method follows the current spectrum regulation policy, where licensed spectrum is used exclusively by the incumbent user. However, the owner may sell, share, or trade the spectrum property rights with other parties.
- *Spectrum commons model*: This model is developed on the foundation of unlicensed industrial, scientific, and medical (ISM) band success such as WiFi. This allows open sharing of spectrum among all users given that these users abide by several predetermined standards.
- *Hierarchical access model*: It is based on hierarchical access structure with PUs and SUs. The idea behind this model is that an SU may utilize the unoccupied channels as long as it is not causing harmful interference to PUs. This model can further be classified into stand-alone (i.e., it does not share network information) and cooperative (i.e., it involves spectrum information sharing to further optimize spectrum access).

Even though these access models are essentially different, all of them strive to accomplish the same objective, that is, enhance the spectrum utilization, and this is one of the several promises and potentials envisioned by the CR paradigm. There are also other advantages of CRNs:

- *Improve radio link performance*: CRNs can improve radio link performance by optimizing resource allocation to SUs (e.g., channel, power, rate, modulation scheme, and coding scheme).

- *Limit interference*: CRNs can help to reduce the impact of interference through dynamic spectrum access and adaptive resource allocation.
- *Balance traffic*: CRNs can help PU networks to offload traffic from densely occupied bands to other unoccupied bands. For instance, if a cellular network is experiencing high loads, then with the help of CRs, the network can opportunistically offload some of its traffic to other available bands.
- *Assist PUs*: CRs can cooperate with the PUs to help relay information from a PU transmitter to a PU receiver.

1.3 FUNCTIONS OF COGNITIVE RADIO FRAMEWORK

Conventional wireless paradigms are characterized by static spectrum allocation policies, where governmental agencies assign spectrum to licensed holders on a long-term basis and for large geographical region. CRNs are envisioned to change this trend by enabling coexistence of SUs with PUs via heterogeneous wireless architectures and dynamic spectrum access techniques. Therefore, each SU in the CRN must reliably perform the following tasks:

- Determine which portions (channels) of the spectrum are available.
- Select the best available channel.
- Coordinate access to this channel with other users.
- Vacate the channel when a PU is detected or when the interference level exceeds a predefined threshold.
- Cooperate with other users to improve network or user efficiency (optional).

From these tasks, one can deduce that the SU must have the following characteristics [4]:

- *Cognitive capabilities*: The CR must be aware of the surrounding environment to intelligently adapt its parameters in order to provide reliable communications and enhance spectrum utilization.
- *Reconfigurability*: The CR must appropriately adjust its operating parameters such as the transmit power, modulation technique, and routing scheme to enhance the spectrum utilization and limit the interference with PU networks.

1.3.1 TRANSCEIVER ARCHITECTURE

A typical CR transceiver consists of two key elements: the radio front end and the baseband processing unit [5] as shown in Figure 1.2. In the radio front end, the CR must have a wideband architecture to be able to monitor a large part of the spectrum. For instance, the wideband antenna is tuned to capture PU's transmissions over multiple bands. Then, the sensed bands must be processed to determine if there are any available opportunities. If there is an available channel, the user can

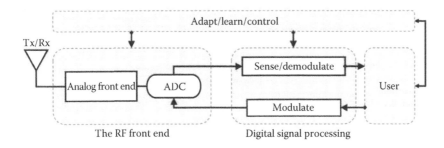

FIGURE 1.2   Block diagram of a typical CR transceiver.

tune its parameters (e.g., transmission frequency, modulation scheme, and so on), and transmit over that channel.

1.3.2   CRITICAL DESIGN CHALLENGES

The CR framework imposes unique challenges at all levels of abstraction due to the coexistence of different types of users with different quality of service (QoS) requirements. For instance, SUs must avoid interference with PUs, and to guarantee this, SUs must reliably detect PU signals, including very weak signals. This is a difficult procedure due to the random nature of the wireless channel. In addition, the SU must immediately vacate the channel when the PU returns, and this mandates the SU to perform periodic sensing of the channel. That is, data transmission interruptions are difficult to avoid, and thus, QoS support for SUs may not be guaranteed. Alternatively, the SU must seamlessly hand over from one channel to another. Conventional handoff techniques may not work well in the CR context. In CRNs, the handoff process is usually initiated when the PU returns, and this is challenging due to the unpredictable behavior of PUs. In addition to the aforementioned challenges, there are several design issues that must be explored such as detecting weak signals in large dynamic range, carrier generation, multiband sensing, and reconfiguration [5,6].

1.3.3   FUNCTIONS OF THE SPECTRUM MANAGEMENT PROCESS

Due to the unique challenges imposed on the CR framework, an appropriate management process is of paramount importance to address them. Broadly speaking, a comprehensive spectrum management process consists of four major elements (Figure 1.3) [5]:

1. *Spectrum sensing and awareness*: The SU must be aware of the spectrum occupancy, and this can be done through local spectrum sensing to determine if the sensed channels are vacant or occupied by PUs [7–10].

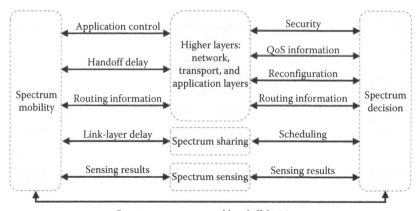

FIGURE 1.3   Spectrum management framework.

2. *Spectrum decision*: The SU must make a decision whether to access a channel or not. Such important decision is not merely based on the spectrum sensing results, but also the QoS requirements, and other internal and external policies.

3. *Spectrum sharing*: The SU must coordinate with other SUs to share the available spectrum resources and avoid interference. Thus, medium access control (MAC) protocols and functions must work in tandem with the spectrum sensing process [11,12]. This function may include economic models for spectrum sharing and trading.

4. *Spectrum mobility*: Unlike traditional wireless communication systems, handoff does not only arise due to the mobility of the SU but also due to the behavior of the PU. Thus, the SU must seamlessly switch between available channels whenever a PU reclaims a channel that is being used by the SU. Also, the SU may move from one geographical region to another where a channel in the new region is unavailable (i.e., occupied by a PU or an SU). Thus, intelligent handoff decisions must be made.

Each component of the spectrum management process must coordinate with the other components. Therefore, cross-layer design is vital to enable effective implementation of CRNs. For instance, spectrum mobility will initiate handoff if the spectrum decision tool announces that the PU has returned. Such announcement cannot be made until the spectrum decision tool obtains the sensing results from the spectrum sensing process. This shows that the handoff process directly depends on the physical-layer sensing process.

1.4   PARADIGMS OF COGNITIVE RADIO

The CR paradigms can be classified into three main categories: interweave, underlay, and overlay paradigms [13,14]. In this section, we highlight the similarities and differences between these paradigms.

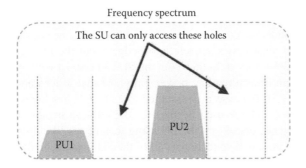

FIGURE 1.4 In the interweave paradigm, the CR user cannot coexist with the PU.

1.4.1 Interweave Paradigm

This is arguably the simplest paradigm, and it was initially what motivated the introduction of CR. In this paradigm, the SU opportunistically accesses unoccupied channels, that is, the SU merely focuses on detecting holes (i.e., unoccupied channels) in the spectrum. In other words, the CR must not coexist with the PU. This is illustrated in Figure 1.4. This paradigm is also referred to as *opportunistic spectrum access*.

1.4.2 Underlay Paradigm

Unlike the interweave paradigm, in underlay paradigm, the SU can coexist with the PU as long as the SU transmits below an *interference threshold* or the *interference temperature* [15]. For instance, the SU can spread its signal over a wide frequency band, and transmit below the noise floor of the PU as illustrated in Figure 1.5. The advantage of this paradigm is that it does not necessarily require to perform spectrum sensing. Indeed, the SU can benefit in scenarios where PUs are almost always present in some systems compared to the interweave paradigm. However, the SU must restrict its transmission power to meet the predetermined threshold. Therefore, the SU may be forced to transmit over short distances.

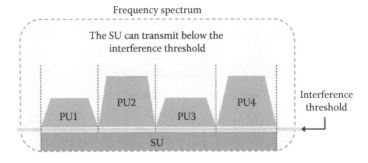

FIGURE 1.5 In underlay paradigm, the CR must transmit below the interference threshold.

1.4.3 Overlay Paradigm

Similar to the underlay paradigm, in overlay, the CR can coexist with the PU. However, the SU must have complete knowledge about the PU signal such as the *message* and *codebook* used. In this case, the SU can coexist with the PU, and more importantly, it can transmit at any power (provided that it does not cause harmful interference to PUs). Because it is assumed that the CR has knowledge about the message and the codebook used by the PU, the CR can exploit this information to cancel interference caused by the PU using techniques like dirty paper coding. Additionally, the SU can help relay part/all of the PU message to the PU's final destination, and this helps offset the interference of the PU [14]. The major drawback of this paradigm is that it requires a complete knowledge of the PU signal, and this assumption may be impractical because significant coordination between numerous PU networks and the CRN must be employed.

1.4.4 Summary

Each paradigm has its merits and poses its challenges. Table 1.1 summarizes the main differences between these three paradigms.

Intuitively, combining these paradigms together can provide further gains to CRNs such as improving the network's aggregate throughput. For instance, a hybrid scheme between interweave and underlay can be used such that the SU can transmit at high powers in the unoccupied band and transmit at lower powers in the bands that are being occupied by PUs. This scheme is commonly known as *sensing-based spectrum sharing* [16]. Figure 1.6 illustrates this concept where there are two CRs: SU1 that follows the hybrid scheme and SU2 that follows the underlay scheme. At time instant T_1, both CRs transmit below the interference threshold due to the presence of PU1 and PU2. If the PUs vacate the channel, as shown in time instant T_2, then unlike SU2, SU1 will adapt its power transmission (i.e., it will increase its transmitted power) as if it is following an interweave paradigm.

1.5 ORGANIZATION OF THIS BOOK

The book tries to address the main challenges and techniques in the spectrum cycle focusing on cooperative approaches to address these challenges. The reader needs to have a basic knowledge of wireless communications systems. It is recommended to

TABLE 1.1

Comparison between the Cognitive Radio Paradigms

Paradigm	Interweave	Underlay	Overlay
Coexistence with the PU	No	Yes	Yes
Power adaptation	Not required	Required	Not required
Required knowledge	Spectrum opportunities	Interference level	Complete knowledge of the PU message and codebook

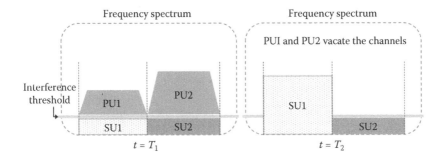

FIGURE 1.6 Illustration of the hybrid scheme: CR1 follows the hybrid scheme, and CR2 follows the underlay paradigm.

read the chapters in their numerical order; however, this is not a requirement as the chapters have been built with minimal interdependence in terms of mathematical derivations or protocol design. However, they are dependent conceptually.

Chapter 2 investigates the major spectrum sensing techniques in both single-band and multiband cases with an in-depth discussion and comparison between the different approaches.

Chapter 3 presents the basic principles of cooperative spectrum sensing. The chapter also presents some advanced combining and cooperative diversity techniques for spectrum sensing and acquisition.

Chapter 4 covers cooperative spectrum acquisition in the cases of spectrum interweave and underlay, where interference is considered as a design parameter.

Chapter 5 discusses the different trade-offs that need to be taken into account during the design and operating phases of spectrum sensing and acquisition techniques and how the performance can be affected, for example, in terms of delay, throughput, or power efficiency.

Chapter 6 addresses the spectrum mobility problem where different handoff techniques and strategies are presented along with an in-depth comparison.

Chapter 7 covers medium access protocols in the context of CRNs. In particular, it addresses cooperative and noncooperative media access control (MAC)–layer protocols that can be used in the absence of a common control channel. It tackles the challenges facing their design and the special requirements that need to be met.

Chapter 8 is devoted to cognitive radio ad hoc networks (CRAHNs). It presents the main differences between CRAHN and regular CRNs. The chapter addresses the spectrum sharing problem and the basics of scaling laws in CRAHN models. A special focus is given to local control schemes and fairness protocols for spectrum sharing.

Chapter 9 investigates MAC protocols in CRAHNs where mobility and primary exclusive regions are considered. An in-depth comparison between the different protocols is provided, and several examples are given for illustration.

Chapter 10 covers the network layer where a classification of routing protocols and their challenges are presented. In particular, the chapter investigates distributed and centralized approaches, mobility issues, as well as highly dynamic networks.

Chapter 11 presents an economic framework for CRNs mainly based on game theory. Different strategies and market models are investigated including fixed-price markets, single auctions, and double auctions, where the physical-layer parameters and interference levels are considered as part of the economic framework.

Chapter 12 gives an overview of security issues in CRNs. It covers trust as the metric used to determine if users in the network are good or bad. Several types of attacks are investigated such as route disruption, jamming, and PU emulation attacks. Furthermore, this chapter provides an in-depth discussion on how these attacks affect the network and what measures can be taken to mitigate their risks.

REFERENCES

1. C. Fortuna and M. Mohorcic, Trends in the development of communication networks: Cognitive networks, *Computer Networks*, 53 (9), 1354–1376, June 2009.
2. D. Grace, *Introduction to Cognitive Communications*, John Wiley & Sons, Ltd., 2012.
3. Q. Zhao and B. Sadler, A survey of dynamic spectrum access: Signal processing, networking, and regulatory policy, *IEEE Signal Processing Magazine*, 24 (3), 79–89, May 2007.
4. S. Haykin, Cognitive radio: Brain-empowered wireless communications, *IEEE Journal on Selected Areas in Communications*, 23 (2), 201–220, February 2005.
5. W.-Y. Lee, M. C. Vuran, S. Mohanty, and I. F. Akyildiz, A survey on spectrum management in cognitive radio networks, *IEEE Communications Magazine*, 46 (4), 40–48, April 2008.
6. B. Razavi, Cognitive radio design challenges and techniques, *IEEE Journal of Solid-State Circuits*, 45 (8), 1542–1553, August 2010.
7. G. Hattab and M. Ibnkahla, Multiband spectrum access: Great promises for future cognitive radio networks, *Proceedings of the IEEE*, 102 (3), 282–306, March 2014.
8. E. Axell, G. Lues, E. Larsson, and H. V. Poor, Spectrum sensing for cognitive radio: State-of-the-art and recent advances, *IEEE Signal Processing Magazine*, 29 (3), 101–116, May 2012.
9. B.-H. Juang, J. Ma, and G. Y. Li, Signal processing in cognitive radio, *Proceedings of the IEEE*, 97 (5), 805–823, May 2009.
10. T. Yucek and H. Arslan, A survey of spectrum sensing algorithms for cognitive radio applications, *Communications Surveys and Tutorials*, 11 (1), 116–130, 2009.
11. A. De Domenico, E. C. Strinati, and M.-G. D. Benedetto, A survey on MAC strategies for cognitive radio networks, *IEEE Communications Surveys and Tutorials*, 14 (1), 21–44, 2012.
12. C. Cormio and K. R. Chowdhury, A survey on MAC protocols for cognitive radio networks, *Ad Hoc Networks*, 7 (7), 1315–1329, September 2009.
13. J. Peha, Approaches to spectrum sharing, *IEEE Communications Magazine*, 43 (2), 10–12, February 2005.
14. A. Goldsmith, S. Jafar, I. Maric, and S. Srinivasa, Breaking spectrum gridlock with cognitive radios: An information theoretic perspective, *Proceedings of the IEEE*, 97 (5), 894–914, May 2009.
15. T. Clancy, Achievable capacity under the interference temperature model, in *Proceedings of the IEEE 26th International Conference on Computer Communications (INFOCOM'7)*, Anchorage, AK, May 2007.
16. H. Garg, L. Zang, X. Kang, and Y.-C. Liang, Sensing-based spectrum sharing in cognitive radio networks, *IEEE Transactions on Vehicular Technology*, 58 (8), 4649–4654, October 2009.

2 Spectrum Sensing

2.1 INTRODUCTION

Spectrum sensing is an essential part of a cognitive radio system [1–6]. Primary users (PUs) are licensed users that have priority or legacy rights on the usage of a specific part of the spectrum. Secondary users (SUs) have lower priority and have to exploit the spectrum without harming PUs. This requires SUs to have cognitive radio capabilities like sensing the spectrum usage known as spectrum sensing [8]. Spectrum sensing is the task of obtaining awareness about the spectrum usage and existence of PUs in a given geographical area. This awareness can be obtained, for example, by using local spectrum sensing, geolocation, or beacons. Spectrum sensing takes into account several considerations such as the spectrum usage characteristics in time, space, frequency, and code, types of signals occupying the spectrum, modulation techniques, waveforms, etc. This chapter investigates the major spectrum sensing techniques in both single-band (SB) and multiband cases with an in-depth discussion and comparison between the different approaches based on the recent study in [80].

2.2 SPECTRUM SENSING

In cognitive radio networks, SUs must reliably detect the presence of the PUs without causing any interference to them, and this is inherently a challenging task, since the detection is done independently by SUs in order not to alter the PU's network infrastructure. The spectrum sensing problem can be described by the classical binary hypothesis testing problem as

$$H_0 : y = v$$
$$H_1 : y = x + v$$

(2.1)

where
$y = [y(1)y(2) \cdots y(N)]^T$ is the received signal at the SU receiver
$x = [x(1)x(2) \cdots x(N)]^T$ is the transmitted PU signal
v is a zero-mean additive white Gaussian noise (AWGN) with variance $\sigma^2 I$
I is the identity matrix
H_0 and H_1 indicate the absence and the presence of the PU, respectively

Typically, to decide between the two hypotheses, we compare a test statistic with a predefined threshold, λ. Mathematically, this is written as

$$T(y) < \lambda \quad H_0$$
$$T(y) \geq \lambda \quad H_1$$

(2.2)

where $T(y)$ is a test statistic (such as the likelihood ratio test).

There are two design elements in spectrum sensing. First, formulating a proper test statistic that reliably gives correct information about the spectrum occupancy. Second, setting a threshold value that differentiates between the two hypotheses. In the following, we discuss coherent detection, energy detection, and feature detection, which are among the most well-known algorithms in SB spectrum sensing [8,9].

2.2.1 MATCHED FILTERING (COHERENT DETECTOR)

Matched filtering detection requires the SU to demodulate received signals, due to which it requires perfect knowledge of the PU's signaling features such as bandwidth, operating frequency, modulation type and order, pulse shaping, and frame format. It is known to be the optimum method for detection of PUs when the transmitted signal is known. The implementation complexity is very high due to matched filter operations.

When the SU has a perfect knowledge of the PU signal structure, it can correlate the received signal with a known copy of the PU signal, as shown in Figure 2.1. The test statistic is given by

$$T(y) = \Re\left[x^H y\right]$$

(2.3)

where

$\Re$ is the real part

$(.)^H$ is the Hermitian (conjugate transpose) operator.

The coherent detector is optimal in the sense it maximizes the signal-to-noise ratio (SNR). It quickly achieves high processing gain (i.e., it requires fewer samples

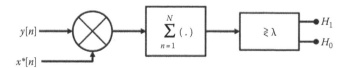

FIGURE 2.1   Coherent detector.

compared to other detectors). However, there are several drawbacks of this spectrum sensing algorithm. Since the SU must detect different bands in the spectrum, it
requires knowledge of each signal structure of these bands, which is usually infeasible for the SU to obtain. If it does obtain these information, this requires a dedicated receiver for each signal type for successful demodulation, which is difficult to
implement [12]. Also, it is susceptible to time synchronization errors especially in
the low SNR regime [13].

2.2.2 Energy Detector

The most standard way of sensing the spectrum is the energy detector–based sensing. It is preferred over other methods because of its low computational and implementation complexities. The advantage of energy detector is that the receiver does
not need any knowledge on the PU's signal. The spectrum occupancy is determined
by comparing the output of the energy detector with a threshold. The main challenge with energy detector–based sensing is the selection of the threshold for detecting the PUs.

The energy detector simply computes the energy of the received signal over a
time window, as shown in Figure 2.2. The test statistic for a typical energy detector
is expressed as

$$T(y) = \frac{1}{\sigma^2}\|y\|^2 \tag{2.4}$$

It is shown that this is optimal for detecting independent and identically distributed
(i.i.d) samples from a zero-mean constellation signal [11]. However, such simple
detector has several drawbacks. First, the threshold value inherently depends on
the noise variance, σ^2. Thus, any estimation error on σ^2 would deteriorate the performance of the energy detector, and this typically occurs at low SNR regime. In
fact, this detector is limited by the SNR wall phenomenon where the detection
becomes impossible below a certain SNR value, regardless of the number of collected samples, due to noise power uncertainties [14]. Second, this detector lacks
the capability of discriminating noise from interferences caused by other sources
(e.g., other SUs). Third, if the collected samples are highly correlated, the performance is degraded [15]. Figure 2.3 illustrates the performance of the energy detector (probability of detection P_d as a function of the probability of false alarm P_{fa})
for different SNR values [81]. The reader can consult [81] for a detailed analytical
derivation of the performance.

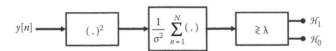

FIGURE 2.2   Energy detector.

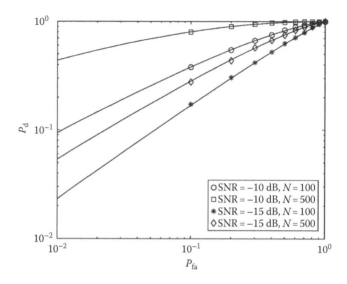

FIGURE 2.3 Performance of the energy detector for different SNR values and different number of samples, N.

2.2.3 FEATURE DETECTION

In practical wireless communication systems, the transmitted signals are deliberately embodied by some unique features to assist the receiver in detection [16]. These features are due to the redundancy added to the transmitted signal, and they provide better robustness against noise uncertainties [11]. These features could be detected, for example, by cyclostationarity, the second-order moments, or by the covariance matrix of the received signals.

2.2.3.1 Cyclostationarity-Based Sensing

This type of detection uses the cyclostationary features caused by the periodicity in the signal or in its statistics [17]. For example, Figure 2.4 illustrates a first-order cylcostationarity detector that tries to exploit the periodicity of the mean of the received signal. Higher-order cyclostationarity has also been investigated in the literature. For instance, in second-order detectors, the periodicity in the autocorrelation function (ACF) is utilized as shown in Figure 2.5.

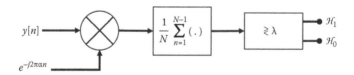

FIGURE 2.4 First-order cyclostationarity detector.

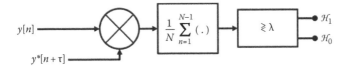

FIGURE 2.5 Second-order cyclostationarity detector.

Due to this statistical periodicity, the ACF can be written as

$$R_y^\alpha(\tau) = E[y(n)^* y(n+\tau) e^{j2\pi\alpha n}]$$ (2.5)

where
  $E[.]$ is the expectation operator
  α is known as the cyclic frequency
  n represents the discrete time

The cyclic spectral density is expressed as

$$S_y(f,\alpha) = \int_{-\infty}^{+\infty} R_y^\alpha(\tau) e^{-j2\pi f \tau}\, d\tau$$ (2.6)

The binary hypothesis testing problem in (2.1) becomes

$$H_0 : S_y(f,\alpha) = S_v(f,\alpha)$$ (2.7a)

$$H_1 : S_y(f,\alpha) = S_x(f,\alpha) + S_v(f,\alpha)$$ (2.7b)

In practice, few cyclic frequencies are sufficient for a good performance, and the more cyclic frequencies are used, the further the improvements [18]. The cyclostationarity detector outperforms the energy detector in several aspects. For instance, noise signals are usually considered uncorrelated, and they do not exhibit periodic behaviors (i.e., $S_v(f,\alpha)=0$ for $\alpha \neq 0$) [19]. This helps discriminate noise from the PU signals with periodic statistical characteristics. In fact, this may help differentiate between different PU signals since different signal structures have different cyclic frequencies. Figure 2.6 illustrates the performance of the first-order cyclostationarity detector, where the periodicity in the mean of the received signal is utilized. It should be noted that the cyclostationarity detector outperforms the energy detector especially for low SNR levels.

However, this detection scheme requires knowledge of the cyclic frequency (i.e., some knowledge of the PU signal). Also, it demands more processing complexity, sensing time, and power consumption compared to other detectors. Furthermore, an SNR wall may arise in cyclostationarity detector under fading channels (e.g., frequency selective fading channels), yet it is less severe than the SNR wall of the

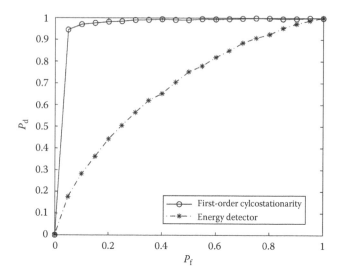

FIGURE 2.6 Probability of detection as a function of the probability of false alarm for the first-order cylcostationarity and the energy detectors, SNR = −10 dB and number of samples $N = 50$.

energy detector [20]. Concerning hardware implementation, it is shown in [21] that this detector is susceptible to sampling clock offsets.

2.2.3.2 Second-Order Moments Feature Detector

If it is assumed that the PU signal follows a Gaussian distribution, then the detector can exploit the structure of the ACF of that signal. The rationale is as follows. Gaussian signals maximize entropy and can be completely characterized by their mean and the variance [22]. It is further reasonable to assume that the PU signal is zero mean. Thus, it is sufficient to merely calculate the second-order moment of the PU-transmitted signal. The test statistic is simply expressed as

$$T(y) = E\left[yy^H \right] \qquad (2.8)$$

2.2.3.3 Covariance Matrix Detector

This detector exploits the eigenvalue structure of the covariance matrix of PU signals. It has the ability to distinguish two different PU signals. Nevertheless, some knowledge of the transmitted signal is required. However, covariance-based algorithms have relaxed this assumption as shown in [15,23]. For instance, in [23], it is demonstrated that the covariance matrices of the transmitted signal and noise are not alike. Similarly, in [15], the test statistic is simply the ratio of the maximum and minimum eigenvalues of the PU signal's sample covariance matrix where prior knowledge of the PU signal is not required. The interested reader may refer to [24] for several test statistics that are basically functions of the autocorrelation to exploit

TABLE 2.1
Comparison between Spectrum Sensing Techniques

Technique	Required Information		Distinguish PU from		Comments
	σ^2	x	Noise	Other Signals	
Coherent detector	No	Yes	Yes	Yes	Maximizes SNR but requires synchronization
Energy detector	Yes	No	SNR dependent	No	Relatively simple
Cyclostationarity	No	Yes	Yes	Yes	Complex processing and high sensing time
Second-moment detector	No	Yes	Yes	Yes	For Gaussian signals
Covariance detector	No	No	Yes	Yes	Some test statistics require knowledge of x

the second-order moments, or for test statistics that are functions of the eigenvalues and the trace of the covariance matrix of the PU signal.

2.2.4 COMPARISON

An overall comparison between these techniques is provided in Table 2.1. Obviously, there is no detector that outperforms the others in everything, and a trade-off must always be made. The choice of the detector depends on many factors such as how much information the SU has a priori about the PU signal. For example, coherent detection is preferred when the SU has full knowledge (e.g., bandwidth, carrier frequency, modulation, and packet format), whereas feature detectors could be used when partial knowledge is available (e.g., pilots, cyclic prefixes, and preambles). If the sensing duration or power consumption is not a constraint, then cyclostationarity can be implemented due to its robustness against noise uncertainties. If no prior knowledge is available, or if a simple algorithm is desired, then the energy detector is preferred. Nevertheless, some advancements are required to improve its robustness at low SNR regime such as advanced or adaptive estimation techniques of the noise power [25].

2.2.5 DESIGN TRADE-OFF AND CHALLENGES

Spectrum sensing for cognitive radios has certain trade-offs [7] as described in the following.

2.2.5.1 Hardware Requirements

Spectrum sensing requires a very high sampling rate, high-resolution analog-to-digital converters (ADCs), and high-speed digital signal processors (DSPs). The receivers discussed so far are capable of processing the narrowband baseband signals at low complexity, but the complexity and power consumption increase in wideband sensing. The radio frequency (RF) components like antennas and power amplifiers that operate at large operating bandwidths are required. Furthermore, high-speed processing units (DSPs or field programmable gate arrays [FPGAs]) are needed for computations.

2.2.5.2 Hidden Primary User Problem

The hidden PU problem occurs when the SU is unable to see the PU; that is, the PU is not in the line of sight. This may be due to severe multipath fading, shadowing, etc. This results in unwanted interference to the PU (receiver) as the primary transmitter's signal could not be detected, and hence, the SU is allowed to transmit. Cooperative sensing is proposed in the literature for handling hidden PU problem.

2.2.5.3 Primary Users with Spread Spectrum

PUs that use spread spectrum signaling are difficult to detect as the power of the PU is distributed over a wide frequency range even though the actual information bandwidth is much narrower. This can be solved by the knowledge of the spreading pattern along with perfect synchronization to the signal.

2.2.5.4 Sensing Duration and Frequency

Selection of sensing parameters results in a trade-off between the sensing time and reliability of sensing. Sensing frequency (how often cognitive radio should perform spectrum sensing) is a design parameter that needs to be chosen carefully. The optimum value depends on the capabilities of cognitive radio itself and the temporal characteristics of PUs in the environment. PUs are the licensed users and can claim their frequency bands being used by an SU anytime.

2.2.5.5 Decision and Fusion Center

Cooperative spectrum sensing [10,11] enhances the performance of cognitive radio networks. However, sharing information among cognitive radios and combining their results, that is, the decision fusion, is a challenging task. The shared information can be in the form of soft or hard decisions made by each cognitive device. Soft information–combining method outperforms hard information–combining method in terms of the probability of misdetection. On the other hand, hard decisions are found to perform as good as soft decisions when the number of cooperating users is high.

2.2.5.6 Security

A selfish or malicious user within a cognitive radio may act as a PU by modifying its characteristics. This results in interference for PUs as well as SUs. This behavior is known as primary user emulation (PUE) attack.

2.3 MULTIBAND SPECTRUM SENSING

Here, we describe the multiband detection problem and its typical applications. We briefly describe the best physical-layer candidate for multiband cognitive radio networks (MB-CRNs). Then, we analyze the recent advancements of multiband spectrum sensing algorithms. We classify these algorithms into three broad categories, namely, serial spectrum sensing, parallel spectrum sensing, and novel multiband spectrum sensing.

2.3.1 Introduction

MB-CRNs have recently caught the attention of several research organizations, since they can significantly enhance the SUs' throughput. There are several scenarios where MB-CRN can be encountered such as

- Many modern communication systems and applications require a wideband access (e.g., ultra-wideband communications). The wideband spectrum can be divided into multiple subbands or subchannels. Thus, the problem becomes a multiband detection problem.
- When an SU wants to minimize the data interruptions due to the return of PUs to their bands, seamless handoff from one band to another becomes vital. Therefore, the SU must have backup channels besides those channels it has already accessed. With MB-CRNs, the SU does not only have a set of candidate channels, but it can also reduce handoff frequency.
- When an SU wants to achieve higher throughput or maintain a certain Quality of Service (QoS), then it may transmit over a larger bandwidth, and this is primarily enabled by accessing multiple bands.
- In cooperative communications, multiple SUs may share their detection results among each other. However, if each SU monitors a subset of subchannels, and then shares its results with others, then the entire spectrum can be sensed, and consequently, more opportunities are explored for spectrum access.

Consider Figure 2.7 where it is assumed that the wideband spectrum can be divided into M nonoverlapping subchannels (or subbands). For simplicity, we assume that each subband has the same bandwidth. Clearly, the SU's primary task is to determine which subchannels are available for spectrum access. This is, in general, a challenging task, since the available bands are not necessary contiguous, and the activity of the PUs might be correlated across these bands (e.g., the PUs in wireless local area networks (WLAN) and broadcast television [26]). In addition, each particular band is considered occupied even if a small portion of it is only being used. For example,

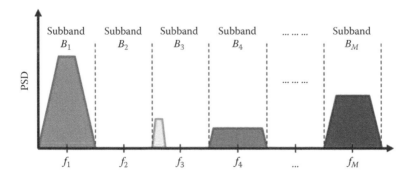

FIGURE 2.7 Illustration of multiband spectrum with M channels.

in IEEE 802.22, the 6 MHz channel must not be accessed, under the interweave paradigm, when it is being used by a wireless microphone that consumes only 200 KHz [27]. In Figure 2.7, a PU occupies a small portion of subband B3, and hence, it must not be used by the SU when it follows the interweave paradigm.

Orthogonal frequency division multiplexing (OFDM) is believed to be a strong physical-layer candidate for MB-CRNs due to its flexibility and ability to generate subcarriers that can fit into noncontiguous subchannels [28,29]. It is basically a multicarrier modulation technique that deploys multiple orthogonal subcarriers to carry the information over the wideband spectrum. By taking care of the amplitude of these subcarriers, SUs' signals can be shaped such that they do not interfere with PUs. For example, if we look back at Figure 2.7, using OFDM, the SU can nullify (or suppress) the subcarriers at f_1, f_3, and f_4 to avoid interference with users that occupy them. Other advantages of OFDM signals include achieving the Shannon capacity in a fragmented spectrum (i.e., noncontiguous bands), robustness against multipath, and scalability to multiantenna systems [28]. This modulation scheme also brings few challenges such as the high peak-to-average power ratio (PAPR), which is a cumbersome for power amplifiers. Other OFDM CR-specific challenges are discussed in [29] such as the synchronization requirements that maintain the orthogonality of subcarriers among the SUs and power leakage of some of the subcarriers to adjacent nullified subcarriers. If we assume that the subbands are independent, then the MB sensing problem reduces to a binary hypothesis for each one. Mathematically, this is expressed as

$$H_{0,m} : y_m = v_m, \quad m = 1, 2, \ldots, M \tag{2.9a}$$

$$H_{1,m} : y_m = x_m + v_m, \quad m = 1, 2, \ldots, M \tag{2.9b}$$

where individual subband m is indicated by subscript m. The decision rule for each band is

$$T(y_m) < \lambda_m \quad H_{0,m} \tag{2.10a}$$

$$T(y_m) \geq \lambda_m \quad H_{1,m} \tag{2.10b}$$

While SB sensing constitutes the building block of MB spectrum sensing, many modifications and advancements are required to put SB sensing into feasible implementation for MB sensing. In the following section, we present three main sensing techniques for MB access.

2.3.2   SERIAL SPECTRUM SENSING TECHNIQUES

In serial spectrum sensing, any of the aforementioned SB detectors can be used to sense multiple bands one at a time using reconfigurable bandpass filters (BPFs) or tunable oscillators.

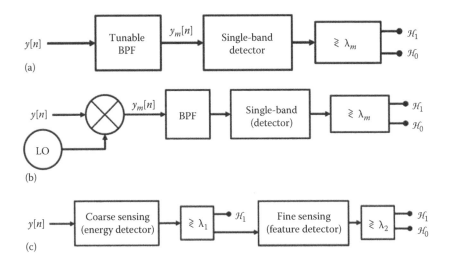

FIGURE 2.8 (a) A reconfigurable BPF; (b) a local oscillator; (c) a two-stage serial spectrum sensing scheme.

2.3.2.1 Reconfigurable Bandpass Filter

A reconfigurable bandpass filter (BPF) can be implemented at the receiver front end to pass one band at a time, and then, an SB detector is used to determine whether that particular band is occupied or not, as shown in Figure 2.8a. Clearly, this requires a wideband receiver front end, which sets several challenges for hardware implementation due to the high sampling rates. In addition, controlling the cutoff frequency and the filter's bandwidth is a challenging design issue [30].

2.3.2.2 Tunable Oscillator

Another approach is based on tunable local oscillator (LO) that down-converts the center frequency of a band to a fixed intermediate frequency, as shown in Figure 2.8b. This will significantly reduce the sampling rate requirement. Nevertheless, tuning and sweeping the reconfigurable filters or oscillators would slow the detection process, and this makes them undesirable.

2.3.2.3 Two-Stage Sensing

Spectrum sensing can be done over two stages. A coarse sensing is first performed and followed by a fine sensing stage if necessary [31,32]. The block diagram of this scheme is presented in Figure 2.8c. For instance, in [31], an energy detector is used in both stages. In the coarse stage, a quick search is done over a wide bandwidth, and in the fine stage, the sensing is done over the individual candidate subbands in that bandwidth. Simulation results show that two-stage spectrum sensing provides faster searching time compared to one-stage-based searching algorithms when the PU activity is high. In [32], the coarse stage is based on energy detection due to its fast processing. If the test statistic is larger than a predefined threshold, then the band is considered occupied. Otherwise, a fine stage is performed where a cyclostationarity detector is implemented due to its robustness at low SNR. Simulation results show tangible improvements on both detection reliability and sensing time compared to single-band detectors, especially if the vacancy levels are high.

2.3.2.4 Other Algorithms

There are several other techniques that are used to serially sense multiple bands such as sequential probability ratio tests (SPRTs). SPRTs have been extensively used to provide efficient, yet fast channel search algorithms. Unlike conventional test statistics that use a fixed number of samples, SPRTs tend to reduce the average number of samples required to be collected to achieve a certain performance. The basic principle is to collect samples as long as $a < T(y) < b$, where a and b are predetermined bounds. A decision will be made once the test statistic is outside these bounds (particularly, we decide H_0 if $T(y) < a$ and H_1 if $T(y) > b$) [10, ch. III].

For instance, Dragalin proposed in [33] an SPRT searching algorithm where it is assumed that there is only one available channel for spectrum access. This assumption implies that the channel occupancy is correlated as well as the SU spectrum access is eventually limited to one channel. In [34], a Bayesian-based SPRT is adopted, yet it is usually infeasible to implement since the Bayesian framework requires both prior probabilities of the PU signals and some cost structures [10, ch. II]. Also, it is not possible to retest the channels again because the authors assume that the number of channels is infinite, which is not practical for CRNs. The limitations of these works have motivated the work in [35] where it is assumed that the number of channels is finite and no specific cost structures are required. Two efficient algorithms are analyzed based on SPRT and energy detection, and both algorithms reduce the sensing time compared to Dragalin's algorithm. An agile multidetector, called iDetector, is proposed in [36]. This detector intelligently sets a detection algorithm based on the availability of PU signal information at the SU side. For example, if no information can be obtained for some bands, then a combination of energy and cyclostationarity detection is used, and for bands where such information is easier to obtain, the coherent detector can be used. The detection reliability is comparable with the cyclostationarity detector, yet the detection time is significantly shorter. Obviously, integrating all these detectors in one receiver imposes higher costs and complexity. In general, serial spectrum sensing has high average searching time when the presence probability of the PU is high [35]. This has motivated the researchers to come up with more advanced receivers for MB spectrum sensing.

2.3.3 Parallel Spectrum Sensing (Multiple Single-Band Detectors)

Here, the SU is equipped with multiple SB detectors such that each one senses a particular band. This can be done using a filter bank as shown in Figure 2.9a. It consists of multiple BPFs, each with a certain center frequency followed by SB detectors [37]. Even though this filter bank merely considers one type of multiple SB detectors (i.e., *homogeneous* structure), we can extend this principle to have *heterogeneous* structure with different multiple SB detectors. For instance, since pilots are being used in TV bands, we can use multiple SB feature detectors over these bands. For bands with unknown signal structures, energy detectors can be implemented. However, filter

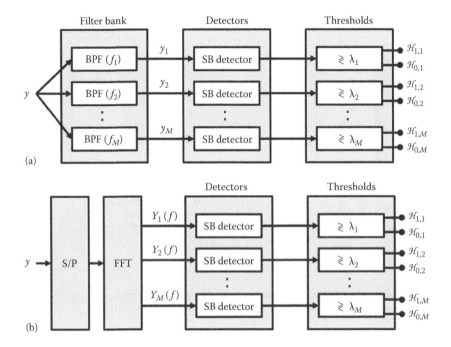

FIGURE 2.9 (a) Filter bank structure; (b) frequency-based parallel SB detectors.

banks demand many RF components, which increase the cost as well as the size of the receiver. The complexity will also be further increased if the detectors are not of the same type. The wideband spectrum can be decomposed in the frequency domain, as shown in Figure 2.9b. This is done using a serial-to-parallel (S/P) converter and fast Fourier transform (FFT) before feeding the signal to SB detectors. The energy detector is the most commonly used here, because it is easy to compute the energy in the frequency domain by analyzing the power spectral density (PSD) [26,38–42]. This can be expressed as

$$T(y_m) = \sum_{n=1}^{N} |Y_m(n)|^2, \quad m = 1, 2, \ldots, M \tag{2.11}$$

where $Y_m(n)$ is the frequency domain representation of the received signal $\mathbf{y}$ at the mth subchannel, and N here is interpreted as the FFT size. Clearly, each subchannel has its own threshold (in a vector form we have $\lambda = [\lambda_1, \lambda_2, \ldots, \lambda_M]$). In [38], Quan et al. propose a *multiband joint detector* (MJD) to maximize the CRN's throughput. It is demonstrated that when the thresholds are jointly optimized, significant throughput gains can be attained compared to a uniform threshold approach (i.e., $\lambda = \lambda_1$). The proposed algorithm intelligently assigns higher thresholds for the bands with higher opportunistic rate and lower thresholds for the bands that require higher PU protection. The former helps minimize transmission interruptions, and the latter helps reduce interference with PUs.

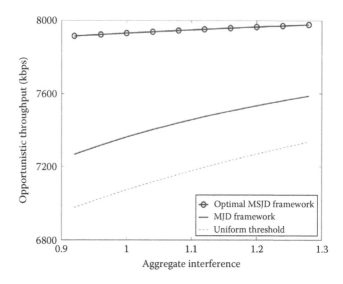

FIGURE 2.10   Throughput performance of MSJD and MJD. (From Paysarvi-Hoseini, P. and Beaulieu, N., *IEEE Trans. Signal Process.*, 59(3), 1170, March 2011.)

The MJD has become a benchmark in wideband spectrum sensing, and many works have been recently proposed to improve it. In particular, the MJD lacks the periodic sensing, an important requirement in spectrum sensing, and hence the authors in [39] have proposed a *multiband sensing-time-adaptive joint detector* (MSJD) where a dynamic sensing time is proposed. This remarkably improves the throughput of the network compared to the MJD. Figure 2.10 shows the throughput versus aggregate interference in the PU network when we use uniform threshold, MJD, and MSJD. Clearly, joint optimization of the thresholds provides significant throughput enhancements in comparison to the uniform threshold. The throughput can be further improved by allocating dynamic sensing time as shown by the MSJD performance. This shows that sensing time is very critical in multiband detection.

The MSJD is also investigated in [40] under different channel models. In [41], the authors present a cost-effective and time-effective MJD to reduce the system's complexity. A resource allocation strategy for MJD is investigated in [42], whereas hardware implementations are presented in [43,44]. An OFDM-based system with MJD is studied in [45]. In addition, a two-stage spectrum sensing is proposed in [46] for an OFDM-based system such that the maximum likelihood estimate is used to estimate the number of PUs in the first stage, and in the second stage, a wideband energy detector is used to detect if they are actually present in the suspected bands. Finally, a modified MJD is proposed in [47] where the energy detectors are replaced by coherent detectors.

The previous techniques assume that the subchannels are independent, which is not generally the case in practice, since the subchannels may be correlated. This may arise when the PU transmits over multiple channels, so that the occupancy of one channel is correlated with the neighboring channels. Another scenario arises when the PU uses high power in one of the channels such that the neighboring channels experience adjacent channel interference (ACI), and this leads to some correlation among this set of channels. In [48], the authors investigate the impact of noise power uncertainties when it is correlated between the subchannels. It is demonstrated that losing the independency between the subchannels makes the complexity of the detection problem to exponentially increase as the number of subchannels increases. To mitigate such high complexity, the authors in [26] propose a linear energy combiner where Equation 2.11 becomes

$$T(y_m) = \sum_{n=1}^{N} w_n |Y_m(n)|^2, \quad m = 1, 2, \ldots, M \qquad (2.12)$$

where $\{w_n\}$ are weighting coefficients that must be optimized. It is demonstrated that this energy combiner outperforms the MJD in terms of detection reliability when the occupancies of the bands are correlated. Nevertheless, the authors assume that the correlation model is known *a priori*, and it follows a homogeneous Markov chain. Therefore, for practical implementations, further work is required to come up with efficient algorithms that are robust against different subchannel correlation models.

2.3.4 WAVELET SENSING

One of the assumptions that have been made in the aforementioned techniques is that the SU knows the number of subbands, M, and their corresponding locations at $f_1, f_2, \ldots, f_M$. However, in practice, this assumption is not very practical, since the CRN must be able to support heterogeneous technologies that have different requirements (e.g., transmission schemes and bandwidth). To overcome this problem, wavelet-based detectors have become a good candidate due to their ability to detect and analyze the singularities in the spectrum [49]. These singularities have important interpretations since they occur at the edges of the subbands (i.e., when we transit from one band to the neighboring bands). Tian and Giannakis have used the wavelet transform for MB spectrum sensing in [50], where they propose a continuous wavelet transform (CWT) that is carried out in the frequency domain to detect the singularities of the wideband spectrum. In other words, using the CWT, the authors have successfully determined the boundaries of the subbands without prior knowledge of the number of subbands and their corresponding center frequencies. Once these edges are determined, the PSD is estimated to determine which subchannels are vacant for opportunistic access. This type of spectrum sensing is referred to as *edge detection*.

Mathematically, the CWT is expressed as

$$W_s(f) = S(f) * \psi_s(f) \qquad (2.13)$$

where
* $*$ is the convolution operator
* $S(f)$ is the PSD as a function of frequency

$$\psi_s(f) = \frac{1}{s}\psi\left(\frac{f}{s}\right) \tag{2.14}$$

is called the wavelet smoothing function and $s = 2^j$, $j = 1, 2, \ldots, J$ is called the dilation factor.

The authors in [50] have used the derivatives of the CWT since they sharpen the edges to help better characterize them. This method is called *wavelet modulus maxima* (WMM). To further enhance the peaks caused by the edges and to suppress noise, the *wavelet multiscale product* (WMP) can be used. It is simply the product of the J first-derivative CWTs, and it is expressed as

$$U_J = \prod_{j=1}^{J} W'_{s_j}(f) \tag{2.15}$$

where $W'_{s_j}(f)$ is the first derivative of $W_{s_j}(f)$.

It should be noted that increasing J further improves the reliability of edge detection, and this is at the expense of additional complexity. One of the challenges of this technique, however, is that these sharp edges do not merely arise at the boundaries of the subbands but also arise due to other sources (e.g., impulsive noise and spectral leakage). These undesired edges may degrade the boundary estimation. For example, consider the spectrum shown in Figure 2.11. There are three edges at the boundaries of f_1, f_2, and f_3 bands, and one another edge due to the impulsive noise. The three true edges are correctly detected using WMP, and if a larger number of products is taken (i.e., larger J), then the detection is further improved. However, the edge estimation is not completely correct because the impulsive noise provides a false edge, and thus, the edge detection estimation is incorrect. To alleviate false edges, Zeng et al. propose a robust algorithm in [51] where the local maxima of (2.15) are compared to a threshold, δ to limit the number of edges. That is, a local maximum will only be counted as an edge if it is larger than δ. Otherwise, it will be neglected. Since the local maximum depends on the shape of the wavelet and the PSD at that point, then δ is not fixed. To mitigate this variability, the WMP in (2.15) is normalized by the mean of the PSD. However, since a threshold is introduced, the SU may miss an actual subchannel boundary when this boundary is heavily corrupted by noise (i.e., the local maximum at this boundary would be less than δ). So back to Figure 2.11, using the threshold can successfully ignore the impulsive noise, but if an actual edge has a PSD below δ such as the edge at f_3 band, then it will be missed too, and thus, the estimation will be degraded.

Another algorithm is proposed in [52] where the *wavelet multiscale sum* (WMS) is used instead of WMP. That is, (2.15) becomes

$$U_J = \sum_{j=1}^{J} W'_{s_j}(f) \tag{2.16}$$

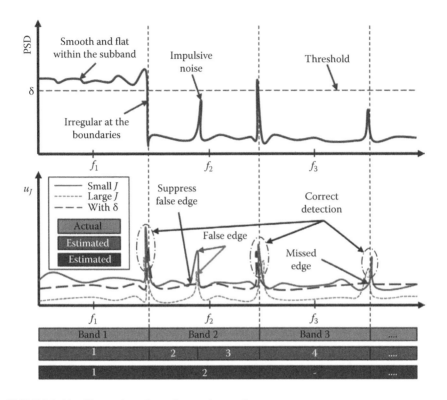

FIGURE 2.11 Illustration of wavelet sensing performance.

The reason of using the summation over the product is that narrowband signals with slow variations of the PSD are not detected by the multiscale product, because they are attenuated when the multiplication operation is used. In such conditions, the multiscale sum is shown to provide a better performance [52]. While it is believed that increasing J in WMS smoothes the edges [52], the authors in [53] demonstrate that the reason of this degradation is attributed to the orthogonal wavelet family used in [52], and to alleviate this degradation (or smoothing), nonorthogonal smoothing functions must be implemented instead. This is at the expense of higher miss detection in low SNR regions. Finally, in [54], a two-stage sensing is proposed: A coarse sensing, where the wavelet transform is implemented to identify the set of candidate subchannels. This is followed by a fine sensing stage that exploits signal features to determine which subchannels are unoccupied.

Figure 2.12 illustrates the performance of different wavelet-based techniques. Here, the spectrum consists of a wideband signal, a narrowband signal, and an impulsive noise. The figure shows that when a higher scale is used (e.g., $J = 4$), the noise is reduced as shown in WMP and WMS compared to CWT. It can be observed that WMP has good rejection to false edges as we observe that the impulsive noise is eliminated. However, if the signal is narrowband, the WMP may reject it as well. This demonstrates the drawback of WMP, which has a poor performance in detecting

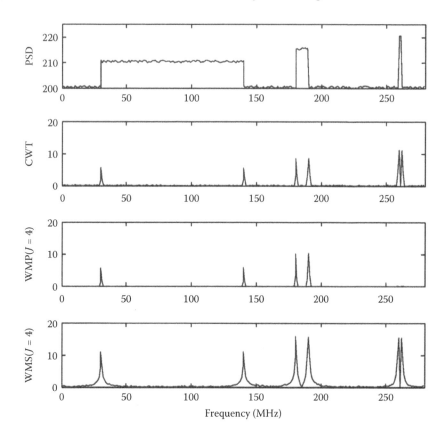

FIGURE 2.12   Wavelet-based technique using Gaussian smoothing function.

narrowband signals. To overcome this, WMS can be used. Unfortunately, WMS has very poor impulsive noise rejection, which is illustrated in the figure.

To summarize, WMM, WMP, and WMS each have their own advantages and disadvantages. Further advancements are required to provide a robust algorithm that successfully detects subband edges and neglects false edges at low complexity. Also, different smoothing functions should be studied to analyze their impact on the quality of edge detection.

2.3.5   COMPRESSED SENSING

Conventionally, to successfully reconstruct the received signal, the sampling rate must be at least as twice the maximum frequency component in the signal (also known as the Nyquist rate) [55]. For instance, if the wideband spectrum of interest has a 3 GHz bandwidth, then the sampling rate must be at least 6 GHz, which is very challenging in terms of feasible implementation and signal processing. Thus, one can ask, can we sample below this rate, and yet successfully recover the signal? The answer is yes. This is enabled by compressive sampling (CS) techniques (also known as compressive sensing).

Compressive sensing has become an active area of research due to its capability to tangibly reduce the sampling rate when the signal is *sparse* in a certain domain [56–58]. For example, signal sparsity in frequency domain indicates that the signal has relatively less significant frequency components compared to its bandwidth. In other words, it has a lower information rate (how much information the signal has) compared to its Nyquist rate. Since the wideband spectrum is underutilized (recall that this is the main motivation of introducing CR), CS appears to be a good candidate for MB spectrum sensing [59–63]. Mathematically, assume that the received $N \times 1$ discrete-time signal $\mathbf{y}$ can be written as

$$\mathbf{y} = \Psi \mathbf{s}, \qquad (2.17)$$

where
  Ψ is an $N \times N$ sparsity basis matrix
  $\mathbf{s}$ is an $N \times 1$ weighting vector

The signal $\mathbf{y}$ is said to be L-sparse if it can be represented by linear combination of only L basis vectors (i.e., only L elements in $\mathbf{s}$ are nonzero) [57]. For $L \ll N$, $\mathbf{y}$ is said to be *compressible* if it has few large coefficients and the rest are small or zero coefficients. The compressive sensing problem can be described by [57].

$$\mathbf{z} = \Phi \mathbf{y} = \Phi \Psi \mathbf{s}, \qquad (2.18)$$

where
  $\mathbf{z}$ is an $O \times 1$ measurement vector
  Φ is an $O \times N$ *measurement matrix* and is nonadaptive (i.e., its columns are fixed and independent of $\mathbf{y}$)

The CS problem is to design a stable Φ such that we reduce the dimension of $\mathbf{y} \in R^N$ to $\mathbf{z} \in R^O$ and we do not lose the signal information. Also, we need a reconstruction algorithm to recover $\mathbf{y}$ from only $O \approx L$ measurements of $\mathbf{z}$. Note that since $O \ll N$, we have infinitely many solutions to (2.18), and hence, there are several existing sparse reconstruction algorithms to obtain the optimum solution (e.g., *basis pursuit* and *orthogonal matching pursuit* (see [58,64] and references therein)).

In [59], frequency sparsity is exploited for MB spectrum sensing. Nevertheless, the spectrum occupancy is measured based on the PSD, which is still prone to estimation errors due to noise uncertainty. To overcome this, the authors in [60] exploit the cyclic sparsity where a sub-Nyquist cyclostationarity detector is used. It is shown that not only the frequency spectrum is sparse but also the 2D cyclic spectrum is sparse too, since some cyclic frequencies may not be occupied by any PUs. Thus, we can solely reconstruct the cyclic spectrum using the sub-Nyquist compressive samples without the need to recover the original signal or its frequency response. In addition, the proposed cyclic-CS (C-CS) can be used to estimate the PSD at lower complexity if the PU signals are assumed to be stationary. The advantage of C-CS is that it is robust to noise uncertainty, since noise is noncyclic (i.e., it does not appear when the cyclic frequency $\alpha \neq 0$).

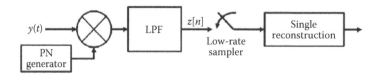

FIGURE 2.13 Compressed sensing: The AIC (*t* denotes continues time domain).

The previous techniques assume that the signals are discrete. For analog signals, an analog-to-information converter (AIC) is used (also known as the *random demodulator*) [64,65]. The AIC basically extracts the information of the signal, and because the information rate of a sparse signal is less than its Nyquist rate, AIC promises to reduce the sampling burden. The structure of the AIC is shown in Figure 2.13. The received signal is modulated by a pseudorandom number (PN) generator to spread the frequency content of the signal so that it is not destroyed by the low-pass filter (LPF). The signal is then sampled at a lower rate using any conventional analog-to-digital convertor (ADC). Then, using the appropriate CS algorithms, the signal can be successfully reconstructed from these partial measurements. An extension to this is presented in [66] where multiple SUs cooperate to improve the detection reliability. The drawback of this converter is that the PN generator is required for compression, and thus, in order to exploit the spatial diversity in CRNs, each SU should have a separate compression device. In other words, synchronization among the SUs is required because asynchronized Φ may degrade the spectrum reconstruction. To alleviate this deficiency, parallel sampling channels could be used. For instance, in [67], the authors propose a multirate asynchronous sub-Nyquist sampling (MASS) system, as shown in Figure 2.14. That is, the system consists of M sub-Nyquist sampling branches, with each having a different low sampling rate, f_s. This structure does not require synchronization for generating Φ and has higher energy efficiency and better data compression capability compared to AIC [67].

To summarize, CS is expected to significantly reduce the sampling rate, and hence, the stringent requirements of ADC and receiver front end might be relaxed. In fact, it is shown in [65] that the number of samples is proportional to the information rate of the signal instead of its bandwidth (or its Nyquist rate). Nevertheless, there are some challenges to be further studied. For instance, careful analysis must be done regarding the SNR because it is expected to be degraded when the signal's

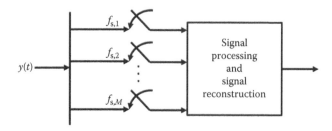

FIGURE 2.14 Compressed sensing: The multirate sub-Nyquist sampling system.

information is compressed into smaller number of samples. Also, nonidealities in hardware will introduce noise to these measurements (e.g., the jitter in the PN generator and mixer's nonlinearity) [68]. In addition, the reconstruction process is highly nonlinear unlike the Shannon sampling procedure where the signal can be linearly reconstructed from its samples [64]. This means that to reduce the burden on the ADC, we need to pay it off with more complex signal processing (a trade-off between software and hardware). In addition, all the aforementioned techniques assume that the sparsity basis is known. This opens a future research direction for robust CS with unknown basis. Finally, we have mentioned earlier that CS is a good candidate due to the spectrum sparsity. However, CRNs are expected to utilize these unoccupied bands, and hence, upon their successful implementation, it is expected that the occupancy rate becomes higher (i.e., the wideband spectrum becomes less sparse), and thus, CS loses its capability to compress the signals. In other words, CS may provide a solution to the spectrum scarcity problem, and ironically, this solution would hinder implementing CS afterward!

2.3.6 ANGLE-BASED SENSING

Most common spectrum sensing techniques are used to exploit the available opportunities in time, frequency, or space. That is, not all subchannels are occupied at the same time (opportunity in frequency domain), not all of them are permanently occupied (opportunity in time domain), and due to the propagation losses in the wireless channels, the same channels may be reused in different geographical regions (opportunity in space). In [25], a *multidimensional* opportunistic access is outlined where new dimensions could be exploited such as the code domain (due to the advancements in coding and spread spectrum techniques), or the direction of arrival (DOA) domain (due to advancements in multiantenna technologies such as multiple-input multiple-output [MIMO] systems, beam forming, and array processing).

The basic principle is that if the SU has a knowledge of the azimuth angle of the PUs, then when the PU transmits in a certain direction, the SU can simultaneously transmit in another direction, over the same band and geographical area, as illustrated in Figure 2.15. DOA-based spectrum sensing is investigated for SB-CR

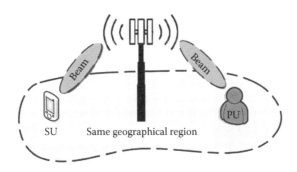

FIGURE 2.15 Angle-based sensing.

in [69], and it is extended for MB-CR in [70]. DOA-based sensing can estimate the location of the occupied subbands by estimating the PU signals' DOA. The drawback is that the SU must be equipped with a multiantenna receiver where array processing becomes essential.

2.3.7 BLIND SENSING

Blind sensing (BS) refers to the sensing problem when the structure of the received signal is unknown at the SU side. The energy detector can be considered, to a certain extent, as a blind detector since it does not require any prior knowledge of the PU signal, yet it still requires a good estimation of the noise variance. A more robust blind detector is presented in [71] where a blind MJD exploits the eigenvalues of the covariance matrix of the received signal without estimating the noise variance. For compressive sensing, the authors in [72] propose a blind compressive sensing (BCS) algorithm where the sparsity basis, Ψ, is unknown. They have shown that if the bases are orthogonal, the BCS algorithm performs very well. A multiband blind reconstruction of analog compressed signals is also investigated in [73].

2.3.8 OTHER ALGORITHMS

Other algorithms are based on the information theoretic criteria (ITC) such as the Akaike information criterion and the minimum description length criterion, which are both conventionally used for model selection problems (e.g., selection of the best model that fits the observed data among a family of models) [74]. These algorithms have been used to calculate the number of independent eigenvalues in the covariance matrix of the observed data, which is shown to be proportional to the number of source signals in array signal processing. Recently, this work has been extended to spectrum sensing [75–77]. With these techniques, the number of eigenvalues is shown to be proportional to the occupancy of the subchannel as well as the number of PU-transmitted signals. Other ITC-based algorithms are presented for multiband detection in [78,79].

2.3.9 COMPARISON

Each multiband sensing technique poses some advantages and disadvantages as shown in Table 2.2. Typically, serial spectrum sensing is relatively simple to implement. However, it is relatively slow, and this is undesirable especially when we have many subchannels. Some techniques have been proposed to make such algorithms faster, such as the two-stage sensing, which requires additional components, and the SPRT-based sensing, which has some practical disadvantages [10, ch. III]. For example, SPRTs' good performance is limited for i.i.d samples. Also, the number of required samples is a random variable, and it is usually unbounded. Thus, some truncation techniques to limit the number of collected samples are necessary. In addition, SPRTs usually require prior knowledge of the PU signals, which is not easy to obtain.

TABLE 2.2

Comparison between Multiband Spectrum Sensing Algorithms

Category	Sensing Algorithm	Advantages	Drawbacks
Serial-based detectors	Reconfigurable BPF	Simple	High sampling rates and slow
	Tunable oscillator	Reduce the sampling rate	Slow sweeping and tuning
	Two-stage sensing	Faster sensing and better detection	Complex and more expensive
	SPRTs	Faster sensing	They have practical challenges
	i-Detector	Faster sensing and good detection	Higher complexity and cost
Parallel-based detectors	Filter bank	Relatively simple	High cost
	Frequency-based sensing	Enhanced performance	High cost and processing complexity
Wideband detectors	Wavelet sensing	Used when boundaries are unknown	May detect false edges
	Compressive sensing	Tangibly reduce the sampling rate	Requires knowledge of Ψ
	Angle-based sensing	Exploits new dimension for spectrum access	Requires multiantenna system
	Blind sensing	Good in the absence of prior knowledge	Requires good estimation techniques

2.4 CONCLUSION

Spectrum sensing is a key component of the spectrum cycle in cognitive radio. This chapter covered different spectrum sensing techniques used in cognitive radio networks. These techniques vary from simple and computationally efficient algorithms, such as the energy detector, to more sophisticated and computationally complex algorithms such as two-stage serial sensing techniques. The network designer has to consider trade-offs between spectrum sensing efficiency, computational complexity, and cost. Chapter 3 will be devoted to cooperative spectrum sensing, which offers new horizons for further enhancements of sensing performance. Chapter 4 will cover the performance measures and design trade-offs in spectrum sensing.

REFERENCES

1. I. Mitola, J., Cognitive radio for flexible mobile multimedia communications, in *Proceedings Of the IEEE International Workshop on Mobile Multimedia Communications*, San Diego, CA, November 1999, pp. 3–10.
2. FCC, Notice of proposed rule-making and order: Facilitating opportunities for flexible, efficient, and reliable spectrum use employing cognitive radio technologies, February 2005.
3. S. Haykin, Cognitive radio: Brain-empowered wireless communications, *IEEE Journal on Selected Areas in Communications*, 23 (2), 201–220, December 2005.

4. E. Biglieri, A. Goldsmith, L. Greenstein, N. Mandayam, and H. Poor, *Principles of Cognitive Radio*, Cambridge University Press, New York, 2012.

5. A. El-Mougy, M. Ibnkahla, G. Hattab, and W. Ejaz, Reconfigurable dynamic networks, *The Proceedings of the IEEE* (submitted for publication), 2014.

6. I. F. Akyildiz, B. F. Lo, and R. Balakrishnan, Cooperative spectrum sensing in cognitive radio networks: A survey, *Physics Communications*, 4 (1), 40–62, March 2011.

7. K. Letaief and W. Zhang, Cooperative communications for cognitive radio networks, *Proceedings of the IEEE*, 97 (5), 878–893, May 2009.

8. A. Hulbert, Spectrum sharing through beacons, in *Proceedings of the 16th IEEE International Symposium on Personal, Indoor and Mobile Radio Communications (PIMRC'05)*, Berlin, Germany, vol. 2, September 2005, pp. 989–993.

9. S. Mangold, Z. Zhong, K. Challapali, and C.-T. Chou, Spectrum agile radio: Radio resource measurements for opportunistic spectrum usage, in *Proceedings of the IEEE Global Telecommunications Conference (GLOBECOM'04)*, Dallas, TX, vol. 6, November 2004, pp. 3467–3471.

10. H. Poor, *An Introduction to Signal Detection and Estimation*, Springer, New York, 1994.

11. A. Sahai, N. Hoven, and R. Tandra, Some fundamental limits on cognitive radio, in *Proceedings of the 42nd Allerton Conference on Communications, Control, and Computing*, Monticello, IL, October 2004.

12. D. Cabric, S. Mishra, and R. Brodersen, Implementation issues in spectrum sensing for cognitive radios, in *Proceedings of the 38th Asilomar Conference on Signals, Systems and Computers (ASILOMAR'04)*, Pacific Grove, CA, vol. 1, November 2004, pp. 772–776.

13. D. Cabric, A. Tkachenko, and R. Brodersen, Spectrum sensing measurements of pilot, energy, and collaborative detection, in *Proceedings of the IEEE Military Communications Conference (MILCOM'06)*, Washington, DC, October 2006, pp. 1–7.

14. R. Tandra and A. Sahai, SNR walls for signal detection, *IEEE Journal of Selected Topics in Signal Processing*, 2 (1), 4–17, February 2008.

15. Y. Zeng and Y.-C. Liang, Eigenvalue-based spectrum sensing algorithms for cognitive radio, *IEEE Transactions on Communications*, 57 (6), 1784–1793, June 2009.

16. M. Ibnkahla, *Wireless Sensor Networks: A Cognitive Perspective*, Taylor & Francis, Boca Raton, FL, 2012.

17. W. Gardner, Exploitation of spectral redundancy in cyclostationary signals, *IEEE Signal Processing Magazine*, 8 (2), 14–36, April 1991.

18. J. Lunde´n, V. Koivunen, A. Huttunen, and H. Poor, Collaborative cyclostationary spectrum sensing for cognitive radio systems, *IEEE Transactions on Signal Processing*, 57 (11), 4182–4195, November 2009.

19. P. Peebles, *Probability, Random Variables and Random Signal Principles*, McGraw-Hill Education, New York, 2002.

20. R. Tandra and A. Sahai, SNR walls for feature detectors, in *Proceedings of the IEEE International Symposium on New Frontiers in Dynamic Spectrum Access Networks (DySPAN'07)*, Dublin, Ireland, April 2007, pp. 559–570.

21. A. Tkachenko, D. Cabric, and R. Brodersen, Cyclostationary feature detector experiments using reconfigurable BEE2, in *Proceedings of the IEEE International Symposium on New Frontiers in Dynamic Spectrum Access Networks (DySPAN'07)*, Dublin, Ireland, April 2007, pp. 216–219.

22. T. Cover and J. Thomas, *Elements of Information Theory*, John Wiley & Sons, Hoboken, NJ, 2006.

23. Y. Zeng and Y.-C. Liang, Covariance based signal detections for cognitive radio, in *Proceedings of the IEEE International Symposium on New Frontiers in Dynamic Spectrum Access Networks (DySPAN'07)*, Dublin, Ireland, April 2007, pp. 202–207.

24. E. Axell, G. Leus, E. Larsson, and H. Poor, Spectrum sensing for cognitive radio: State-of-the-art and recent advances, *IEEE Signal Processing Magazine*, 29 (3), 101–116, May 2012.

25. T. Yucek and H. Arslan, A survey of spectrum sensing algorithms for cognitive radio applications, *IEEE Communications Surveys and Tutorials*, 11 (1), 116–130, January 2009.

26. K. Hossain and B. Champagne, Wideband spectrum sensing for cognitive radios with correlated subband occupancy, *IEEE Signal Processing Letters*, 18 (1), 35–38, January 2011.

27. G. Ko, A. Franklin, S.-J. You, J.-S. Pak, M.-S. Song, and C.-J. Kim, Channel management in IEEE 802.22 WRAN systems, *IEEE Communications Magazine*, 48 (9), 88–94, September 2010.

28. H. Tang, Some physical layer issues of wide-band cognitive radio systems, in *Proceedings of IEEE International Symposium on New Frontiers in Dynamic Spectrum Access Networks (DySPAN'05)*, Baltimore, MD, November 2005, pp. 151–159.

29. H. Mahmoud, T. Yucek, and H. Arslan, OFDM for cognitive radio: Merits and challenges, *IEEE Wireless Communications Magazine*, 16 (2), 6–15, April 2009.

30. H. Joshi, H. H. Sigmarsson, S. Moon, D. Peroulis, and W. Chappell, High-Q fully reconfigurable tunable bandpass filters, *IEEE Transactions on Microwave Theory and Techniques*, 57 (12), 3525–3533, December 2009.

31. L. Luo, N. Neihart, S. Roy, and D. Allstot, A two-stage sensing technique for dynamic spectrum access, *IEEE Transactions on Wireless Communications*, 8 (6), 3028–3037, June 2009.

32. S. Maleki, A. Pandharipande, and G. Leus, Two-stage spectrum sensing for cognitive radios, in *Proceedings of the IEEE International Conference on Acoustic Speech and Signal Processing (ICASSP'10)*, Dallas, TX, March 2010, pp. 2946–2949.

33. V. Dragalin, A simple and effective scanning rule for a multi-channel system, *Metrika*, 43 (1), 165–182, 1996.

34. L. Lai, H. Poor, Y. Xin, and G. Georgiadis, Quickest search over multiple sequences, *IEEE Transactions on Information Theory*, 57 (8), 5375–5386, August 2011.

35. Y. Xin, G. Yue, and L. Lai, Efficient channel search algorithms for cognitive radio in a multichannel system, in *Proceedings of the IEEE Global Telecommunications Conference (GLOBECOM'10)*, Miami, FL, December 2010, pp. 1–5.

36. W. Ejaz, N. Ul Hasan, and H. S. Kim, iDetection: Intelligent primary user detection for cognitive radio networks, in *Proceedings of the Sixth International Conference on Next Generation Mobile Applications, Services and Technologies (NG-MAST'12)*, Paris, France, September 2012, pp. 153–157.

37. B. Farhang-Boroujeny, Filter bank spectrum sensing for cognitive radios, *IEEE Transactions on Signal Processing*, 56 (5), 1801–1811, May 2008.

38. Z. Quan, S. Cui, A. Sayed, and H. Poor, Optimal multiband joint detection for spectrum sensing in cognitive radio networks, *IEEE Transactions on Signal Processing*, 57 (3), 1128–1140, March 2009.

39. P. Paysarvi-Hoseini and N. Beaulieu, Optimal wideband spectrum sensing framework for cognitive radio systems, *IEEE Transactions on Signal Processing*, 59 (3), 1170–1182, March 2011.

40. S. Farooq and A. Ghafoor, Multiband sensing-time-adaptive joint detection cognitive radios framework for Gaussian channels, in *Proceedings of the International Bhurban Conference on Applied Sciences and Technology (IBCAST'13)*, Islamabad, Pakistan, January 2013, pp. 406–411.

41. P. Paysarvi-Hoseini and N. Beaulieu, On the efficient implementation of the multiband joint detection framework for wideband spectrum sensing in cognitive radio networks, in *Proceedings of the IEEE Vehicular Technology Conference (VTC'11)*, Budapest, Hungary, September 2011, pp. 1–6.

42. C. Shi, Y. Wang, T. Wang, and P. Zhang, Joint optimization of detection threshold and throughput in multiband cognitive radio systems, in *Proceedings of the IEEE Consumer Communications and Networking Conference (CCNC'12)*, Las Vegas, NV, January 2012, pp. 849–853.

43. S. Srinu, S. Sabat, and S. Udgata, Wideband spectrum sensing based on energy detection for cognitive radio network, in *Proceedings of the World Congress on Information and Communication Technologies (WICT'11)*, Mumbai, India, December 2011, pp. 651–656.

44. M. Kitsunezuka and K. Kunihiro, Efficient spectrum utilization: Cognitive radio approach, in *Proceedings of IEEE Radio and Wireless Symposium (RWS'13)*, Austin, TX, January 2013, pp. 25–27.

45. L. Khalid, K. Raahemifar, and A. Anpalagan, Cooperative spectrum sensing for wideband cognitive OFDM radio networks, in *Proceedings of the 70th IEEE Vehicular Technology Conference (VTC'09)*, Anchorage, AK, September 2009, pp. 1–5.

46. C.-H. Hwang, G.-L. Lai, and S.-C. Chen, Spectrum sensing in wide-band OFDM cognitive radios, *IEEE Transactions on Signal Processing*, 58 (2), 709–719, February 2010.

47. M. Iqbal and A. Ghafoor, Analysis of multiband joint detection framework for waveform-based sensing in cognitive radios, in *Proceedings of the IEEE Vehicular Technology Conference (VTC'12)*, Yokohama, Japan, September 2012, pp. 1–5.

48. E. Axell and E. Larsson, A Bayesian approach to spectrum sensing, denoising and anomaly detection, in *Proceedings of the IEEE International Conference on Acoustics, Speech and Signal Processing (ICASSP'09)*, Taipei, Taiwan, April 2009, pp. 2333–2336.

49. S. Mallat and W.-L. Hwang, Singularity detection and processing with wavelets, *IEEE Transactions on Information Theory*, 38 (2), 617–643, March 1992.

50. Z. Tian and G. B. Giannakis, A wavelet approach to wideband spectrum sensing for cognitive radios, in *Proceedings of the First International Conference on Cognitive Radio Oriented Wireless Networks and Communications (CrownCom'06)*, Mykonos, Greece, June 2006, pp. 1–5.

51. Y. Zeng, Y.-C. Liang, and M. W. Chia, Edge based wideband sensing for cognitive radio: Algorithm and performance evaluation, in *Proceedings of the IEEE International Symposium on New Frontiers in Dynamic Spectrum Access Networks (DySPAN'11)*, Aachen, Germany, May 2011, pp. 538–544.

52. Y.-L. Xu, H.-S. Zhang, and Z.-H. Han, The performance analysis of spectrum sensing algorithms based on wavelet edge detection, in *Proceedings of the Fifth International Conference on Wireless Communications, Networking and Mobile Computing (WiCom'09)*, Beijing, China, September 2009, pp. 1–4.

53. S. El-Khamy, M. El-Mahallawy, and E. Youssef, Improved wideband spectrum sensing techniques using wavelet-based edge detection for cognitive radio, in *Proceedings of the International Conference on Computing, Networking and Communications (ICNC'13)*, San Diego, CA, January 2013, pp. 418–423.

54. Y. Hur, J. Park, W. Woo, K. Lim, C.-H. Lee, H. Kim, and J. Laskar, A wideband analog multi-resolution spectrum sensing (MRSS) technique for cognitive radio (CR) systems, in *Proceedings of the IEEE International Symposium on Circuits and Systems (ISCAS'06)*, Kos, Greece, May 2006, pp. 4090–4093.

55. C. Shannon, Communication in the presence of noise, *Proceedings of the IEEE*, 72 (9), 1192–1201, September 1984.

56. D. Donoho, Compressed sensing, *IEEE Transactions on Information Theory*, 52 (4), 1289–1306, April 2006.

57. R. Baraniuk, Compressive sensing [lecture notes], *IEEE Signal Processing Magazine*, 24 (4), 118–121, July 2007.

58. E. Candes and M. Wakin, An introduction to compressive sampling, *IEEE Signal Processing Magazine*, 25 (2), 21–30, March 2008.

59. Z. Tian and G. Giannakis, Compressed sensing for wideband cognitive radios, in *Proceedings of the IEEE International Conference on Acoustics, Speech and Signal Processing (ICASSP'07)*, Honolulu, HI, vol. 4, April 2007, pp. 1357–1360.

60. Z. Tian, Y. Tafesse, and B. Sadler, Cyclic feature detection with sub-Nyquist sampling for wideband spectrum sensing, *IEEE Journal of Selected Topics in Signal Processing*, 6 (1), 58–69, February 2012.

61. Z. Tian, Compressed wideband sensing in cooperative cognitive radio networks, in *Proceedings of the IEEE Global Telecommunication Conference (GLOBE-COM'08)*, Washington, DC, December 2008, pp. 1–5.

62. Z. Fanzi, C. Li, and Z. Tian, Distributed compressive spectrum sensing in cooperative multihop cognitive networks, *IEEE Journal of Selected Topics in Signal Processing*, 5 (1), 37–48, February 2011.

63. Y. Polo, Y. Wang, A. Pandharipande, and G. Leus, Compressive wide- band spectrum sensing, in *Proceedings of the IEEE International Conference on Acoustics, Speech and Signal Processing (ICASSP'09)*, Taipei, Taiwan, April 2009, pp. 2337–2340.

64. J. Tropp, J. Laska, M. Duarte, J. Romberg, and R. Baraniuk, Beyond Nyquist: Efficient sampling of sparse bandlimited signals, *IEEE Transactions on Information Theory*, 56 (1), 520–544, January 2010.

65. S. Kirolos, J. Laska, M. Wakin, M. Duarte, D. Baron, T. Ragheb, Y. Massoud, and R. Baraniuk, Analog-to-information conversion via random demodulation, in *Proceedings of the IEEE Dallas/CAS Workshop on Design, Applications, Integration and Software*, Richardson, TX, October 2006, pp. 71–74.

66. Y. Wang, A. Pandharipande, Y. L. Polo, and G. Leus, Distributed compressive wideband spectrum sensing, in *Information Theory and Applications Workshop*, La Jolla, CA, February 2009, pp. 2337–2340.

67. H. Sun, W.-Y. Chiu, J. Jiang, A. Nallanathan, and H. Poor, Wideband spectrum sensing with sub-Nyquist sampling in cognitive radios, *IEEE Transactions on Signal Processing*, 60 (11), 6068–6073, November 2012.

68. S. Kirolos, T. Ragheb, J. Laska, M. Duarte, Y. Massoud, and R. Baraniuk, Practical issues in implementing analog-to-information converters, in *Proceedings of the Sixth International Workshop on System-on-Chip for Real-Time Applications*, Le Caire, Egypt, December 2006, pp. 141–146.

69. J. Xie, Z. Fu, and H. Xian, Spectrum sensing based on estimation of direction of arrival, in *Proceedings of the International Conference on Computational Problem-Solving (ICCP'10)*, Cambridge, MA, December 2010, pp. 39–42.

70. A. Mahram, M. Shayesteh, and S. Kordan, A novel wideband spectrum sensing algorithm for cognitive radio networks based on DOA estimation model, in *Proceedings of the Sixth International Symposium on Telecommunications (IST'12)*, Copenhagen, Denmark, November 2012, pp. 359–362.

71. T.-X. Luan, L. Dong, K. Xiao, and X.-D. Zhang, Blind multiband joint detection in cognitive radio networks based on model selection, in *Proceedings of the Sixth International Conference on Wireless Communications Networking and Mobile Computing (WiCOM'10)*, Chengdu, China, September 2010, pp. 1–4.

72. S. Gleichman and Y. Eldar, Multichannel blind compressed sensing, in *Proceedings of the IEEE Sensor Array and Multichannel Signal Processing Workshop (SAM'10)*, Jerusalem, Israel, October 2010, pp. 129–132.

73. M. Mishali and Y. Eldar, Blind multiband signal reconstruction: Compressed sensing for analog signals, *IEEE Transactions on Signal Processing*, 57 (3), 993–1009, March 2009.

74. R. Wang and M. Tao, Blind spectrum sensing by information theoretic criteria for cognitive radios, *IEEE Transactions on Vehicular Technology*, 59 (8), 3806–3817, October 2010.

75. M. Haddad, A. Hayar, M. H. Fetoui, and M. Debbah, Cognitive radio sensing information-theoretic criteria based, in *Proceedings of the Second International Conference on Cognitive Radio Oriented Wireless Networks and Communications (CrownCom'07)*, Orlando, FL, September 2007, pp. 241–244.

76. B. Zayen, A. Hayar, and D. Nussbaum, Blind spectrum sensing for cognitive radio based on model selection, in *Proceedings of the Third International Conference on Cognitive Radio Oriented Wireless Networks and Communications (CrownCom'08)*, Singapore, May 2008, pp. 1–4.

77. B. Zayen, A. Hayar, and K. Kansanen, Blind spectrum sensing for cognitive radio based on signal space dimension estimation, in *Proceedings of the IEEE International Conference on Communications (ICC'09)*, Alabama, FL, June 2009, pp. 1–5.

78. S. Liu, J. Shen, R. Zhang, Z. Zhang, and Y. Liu, Information theoretic criterion-based spectrum sensing for cognitive radio, *IET Communications*, 2 (6), 753–762, July 2008.

79. Y. Jing, X. Yang, L. Ma, J. Ma, and B. Niu, Blind multiband spectrum sensing in cognitive radio network, in *Proceedings of the Second International Conference on Consumer Electronics, Communications and Networks (CECNet'12)*, Hubei, China, April 2012, pp. 2442–2445.

80. G. Hattab and M. Ibnkahla, Multiband cognitive radio: Great promises for future radio access, *Proceedings of the IEEE*, 102 (3), 282–306, March 2014.

81. A. Abu Alkhair, Cooperative cognitive radio networks: Spectrum acquisition and co-channel interference effect, PhD dissertation, Department of Electrical and Computer Engineering, Queen's University, Kingston, Ontario, Canada, February 2013.

3 Cooperative Spectrum Acquisition

3.1 INTRODUCTION

Cooperative approaches in cognitive radio (CR) have emerged in the past few years as very promising techniques to enhance and improve the performance of cognitive radio networks (CRNs).

To show the need for cooperative spectrum sensing (CSS), consider, for example, the hidden terminal problem, which is one of the common challenges in wireless communication systems. This phenomenon arises from the random nature of the wireless channel where a secondary user (SU) could be shadowed by an object or in a deep fade as illustrated in Figure 3.1a [3]. As a result, the SU would decide that there is a spectrum opportunity even though the primary user (PU) is present. To mitigate this issue, a collaborative work between several SUs is required, and this is referred to as CSS [1,2]. Consider the situation illustrated in Figure 3.1b. The secondary user (SU2) has sensed the presence of the PU while the other one (SU1) could not because it is shadowed by a building. If the two secondary users share their spectrum sensing results with one another, then SU1 would not access the spectrum. This is the basic principle of cooperative sensing where the SUs in a certain geographical region would cooperate together to effectively improve the sensing reliability.

Cooperative sensing can be done *externally* or *internally*. In the former, external entities, such as sensing nodes, perform spectrum sensing and collaboratively share the sensing results with a central entity which shares the final decision with the SUs in vicinity [9]. Clearly, this saves power consumption since the SUs do not sense the spectrum, yet a new network infrastructure must be established for these sensor nodes. On the other hand, the SUs independently cooperate in internal cooperation. This cooperation can be either *distributed* or *centralized* [4]. In the former, even though the SU shares the sensing information with others, it solely makes its own decision whether to access a subchannel or not, whereas in the latter, the SUs share the information with a fusion center (FC), which can also act as a particular CR station (which is assumed in this chapter). This unit makes a final decision and feeds it back to the SUs in vicinity. The centralized scheme is prone to a single point of failure since if something goes wrong with the FC (e.g., its energy is depleted or it is in deep fade and so on), then the entire cooperating network fails, and the other SUs would be inaccessible to the spectrum information. Thus, the SUs must efficiently rotate the cluster-head role among each other. This is not the case in distributed method where a backbone is not required. Intuitively, dealing with distributed networks is more difficult since they lack a control center. Another key issue is how

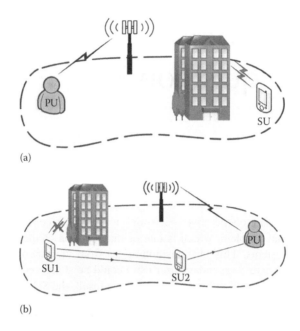

(a)

(b)

FIGURE 3.1   (a) An illustration of the hidden terminal problem. (b) The cooperative sensing could mitigate the hidden terminal problem.

to combine the collected information from the cooperating SUs who participate in sensing the spectrum.

This chapter is organized as follows. Section 3.2 presents the basic principles of CSS (based on the results of [20] and [77]). This includes basic combining techniques as well as the multiband case. Sections 3.3 and 3.4 present some advanced combining and cooperative diversity techniques based on the results of [74–76].

3.2   BASICS OF COOPERATIVE SPECTRUM SENSING

This section presents the basics of CSS. We discuss the major techniques, namely, hard combining where the SU shares a one-bit decision, soft combining where the SU shares its actual test statistics, and the hybrid combining which is a compromise between the two. Then we explore the case of multiband cognitive radio [20,77].

Figure 3.2 shows a PU, K secondary users (SU_k, $k = 1, 2, \ldots K$), and an FC, which acts also as a secondary user (SU_0).

3.2.1   HARD COMBINING

In this technique, the SU merely sends its final one-bit decision to the other SUs (distributed cooperation) or to the FC (centralized cooperation). If we have K SUs, the final decision metric is expressed as [10]

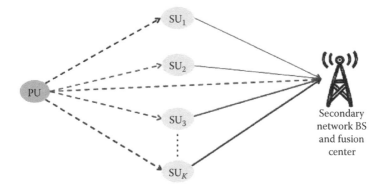

FIGURE 3.2   Network of CR transmitters and FC.

$$D = \sum_{k=0}^{K} d_k \qquad (3.1)$$

$$\begin{aligned} H_1 &: \quad D \geq \lambda \\ H_0 &: \quad D < \lambda \end{aligned} \qquad (3.2)$$

where $d_k \in \{0,1\}$ is the final decision made by the kth SU such that 0 and 1 indicate the absence and the presence of the PU, respectively. This is basically a logical decision metric such that

- If $\lambda = 1$, then (3.2) is an OR-logic rule (i.e., the PU is considered present if only one SU sends *1*).
- If $\lambda = K + 1$, then (3.2) is an AND-logic rule (i.e., the PU is considered present if *all* the SUs send *1*).
- If $\lambda = \lceil (K + 1)/2 \rceil$ ($\lceil x \rceil$ denotes the smallest integer not less than x), (3.2) becomes a majority rule (i.e., the PU is considered present if the majority of the SUs send *1*).

Note that the OR-logic rule guarantees minimum interference to the PU since only a single *1* is enough to declare the band occupied, whereas the AND-logic rule guarantees higher throughput, since the band is considered reoccupied by the PU when all SUs send *1*. Figure 3.3 illustrates the receiver operating characteristics (ROC) curves of OR-rule and AND-rule under different number of users using the energy detector (*SNR* = −5 dB). We observe that cooperation significantly improves the detection performance in comparison with no cooperation. Also, increasing the number of users further improves the performance, with diminishing gains. Finally, we observe that the OR rule outperforms the AND rule.

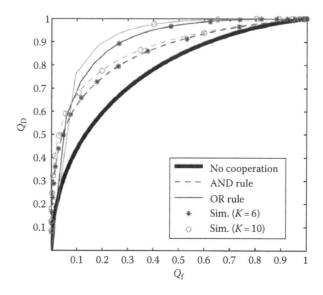

FIGURE 3.3 Comparison between OR-logic and AND-logic rules for different number of users, K.

3.2.2 SOFT COMBINING

In this technique, the SU shares its original sensing information (or original statistics) with the other SUs without locally processing them. An optimal soft combination scheme based on Neyman–Pearson (NP) criterion is proposed in [11]. It is shown that the optimal combination (OC) is actually based on the weighted summation of the observed energies from the collaborative SUs [11]. Mathematically, this is written as

$$\tau = \sum_{k=0}^{K} c_k T_k(y_k) \tag{3.3}$$

where
  $T_k(y_k)$ is the kth user test statistics
  y_k is the received signal by user k
  c_k is the weight coefficient

These weights could be based on equal gain combination (EGC) or maximal ratio combination (MRC). In the former, $c_k = 1$, and in the latter, c_k is proportional to the SNR at the kth user. It is demonstrated that the OC converges to EGC scheme in high SNR, whereas it converges to MRC in low SNR region [11].

Hard combining is simple to deploy, and more importantly, it requires lower overhead (one bit). However, since the statistics at each SU are reduced to one bit, there is an information loss that propagates to the other SUs. Therefore, the final decision is less reliable compared to soft combining. It is actually shown that soft-combining schemes outperform hard combining in terms of detection reliability [11,12]. Nevertheless, the gain of soft combining compared to hard combining diminishes as the number of cooperating SUs increases [13].

3.2.3 HYBRID COMBINING

In [11], a combination scheme that comprises both the soft and hard combining is presented, and it is referred to as *softened hard* combining scheme where the SU sends two-bit information instead of one bit. Using two bits allows having three different thresholds that decompose the energy region into four sections (i.e., 00, 01, 10, and 11). The weights are allocated proportionally to the level of the energy region (i.e., an SU will be allocated the highest weight if it is above the highest threshold). In other words, this scheme accounts for the credibility of the decisions made by the individual SUs. The most credible SU will have the highest influence in the final decision and vice versa. This scheme is less complex, has comparable performance to soft combining, and more importantly, it significantly reduces the overhead compared to soft-combining scheme. In general, increasing the number of bits will improve the performance at the expense of larger overhead. Further studies are required to determine the optimal number of bits (or thresholds) to meet a certain detection performance. Figure 3.4 illustrates the detection performance of hard combining (one bit) using OR-logic rule and hybrid combining (two bits) for $K = 4$ users. It is observed that the performance improves as we increase the number of bits. As the number of bits increases, the performance is expected to be bounded by the performance of the soft-combining scheme (where the actual statistics are sent to the FC).

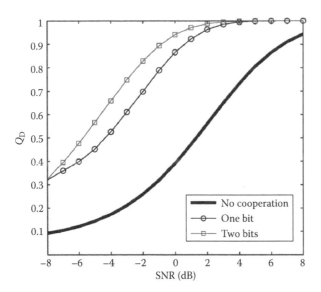

FIGURE 3.4 Comparison between hard combining and hybrid combining.

3.2.4 COOPERATIVE SPECTRUM ACCESS IN MULTIBAND COGNITIVE RADIO NETWORKS

Significant breakthroughs have been accomplished in CSS [2], yet most of them assume a single channel, and the work on cooperating SUs over multiple bands is still limited.

A cooperative compressive sensing (CS) based on hard combining is investigated in [6], where SUs individually perform CS, and then they share binary decision with one another. A more practical scenario is studied in [7] where the channel-state information (CSI) is assumed to be unknown at the SU side. A soft-combining cooperative network is proposed for CS of analog signals in [8]. Cooperation in CS provides two advantages. First, the detection performance is enhanced due to the spatial diversity, and second, as the number of cooperating SUs increases, we can increase the compression ratio (i.e., reduce the sampling rate further) without performance degradation. Cooperation is also investigated with respect to the network's throughput [4–16]. The multiband joint detector (MJD) framework is extended for multiple SUs where a *spatial-spectral joint detector* is proposed in [5,14]. In this detector, the idea is to linearly combine the soft decisions made by the individual SUs such that a joint optimization of the detection thresholds and the weighting coefficients, as expressed in (3.3), is implemented. It is demonstrated that cooperation significantly improves the throughput compared to single MJD.

Other works include [17–19]. In [17], several multiband detectors based on the generalized likelihood ratio test (GLRT) are investigated under different fading channels. In [18], the impact of noise uncertainty on cooperative sensing is studied. The authors propose an algorithm where the SU first performs multiband spectrum sensing. For subchannels that have uncertain noise power estimation, the SU cooperates with the neighboring SUs to check their decision of these uncertain channels. In [19], sequential cooperation is investigated. That is, for a certain channel, the FC will sequentially collect one-bit decisions from the SUs until a decision is made. It is shown that subchannels with uncertain SU decisions require more cooperating SUs to mitigate the uncertainties.

All of the previous works assume that each SU senses the entire spectrum before cooperation. This hefty load and the load of cooperating over the multiple bands make the implementation implausible (even if hard combining is used) since each SU must represent the M subchannels by an $M \times 1$ binary vector. Thus, instead of making the maximum gains of both paradigms, one can resort to a trade-off that makes network implementation more feasible where each SU senses a subset of M subchannels. We shall refer to this paradigm as *cooperative multiband cognitive radio networks*. Consider Figure 3.5 where each SU senses a subset of these bands such that the entire spectrum is sensed by all of them. Since we have six SUs, the maximum possible spatial diversity is six, which is attained when each SU senses all the M subchannels. This is very demanding in terms of sampling requirements, and to reduce these requirements, compressive sampling may be implemented. Alternatively, each SU can monitor a subset of these channels to reduce the sampling requirements. In Figure 3.5a, uniform diversity is achieved, with a value of 2. That is, each subchannel is being sensed by two SUs. Another approach is to use

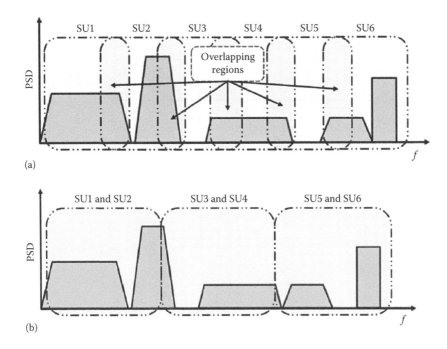

(a)

(b)

FIGURE 3.5 (a) Uniform diversity of 2. (b) Nonuniform diversity.

nonuniform diversity where the number of SUs monitoring a certain band depends on several factors. For example, if the band has higher priority (e.g., bands for public safety), or it has intense PUs activities (e.g., cellular bands), then more SUs are allocated to sense such band to improve the reliability of detection. This is demonstrated by the *overlapping regions* in Figure 3.5b. Such paradigm resembles a fundamental trade-off between spatial diversity and sampling requirements. On the other hand, this paradigm opens future directions on how to allocate channel subsets to the cooperating SUs using more advanced algorithm (e.g., adaptive algorithms) or under some constraints (e.g., SU's power budget and location).

3.3 EXAMPLES OF COOPERATIVE SPECTRUM ACQUISITION TECHNIQUES

The CSS process consists of three stages as illustrated in Figure 3.6. These are sensing period, reporting period, and broadcast period. During the sensing period, every network terminal, including the FC itself, senses the examined channel using some sensing method, like the energy detector. At the end of this period, every terminal sends a local report to the FC in a round-robin manner, that is, every terminal has a preset time slot to report. The content of the local reports can be either the local decision metric, which is the energy estimate in the energy detector case, or a binary local decision. Choosing between these two options is subject to the performance/overhead trade-off [23,24]. While sending the decision metrics requires significant

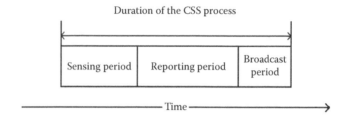

FIGURE 3.6 The three parts of the CSS process.

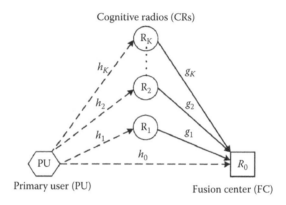

FIGURE 3.7 A block diagram of CSS.

reporting overhead, it can help the FC achieve superior detection reliability compared to the low overhead decision-based reporting [11]. In practice, however, the reporting overhead shares the data transmission due to the absence of dedicated control channels [26]. Consequently, decision-based CSS is more appealing especially for highly populated networks.

Figure 3.7 represents a typical cooperative CR network where secondary users (R_k, $k = 1,2,..., K$) try to detect the presence of a PU. Assuming energy detection, we have

$$\begin{cases} H_0^{CR} : & \phi_k = \sigma_0^2 \\ H_1^{CR} : & \phi_k = E_p \left| h_k \right|^2 + \sigma_0^2 \end{cases} \tag{3.4}$$

3.3.1 DECISION-BASED CSS USING ENERGY DETECTION

At the end of the sensing period, every terminal makes a local decision. Let $d_k \in \{0,1\}$ denote the local decision of terminal R_k, $k = 0,1,..., K$, then we have

$$d_k \begin{cases} 1 \equiv H_1^{CR}, & \text{if } \phi_k \geq \lambda \\ 0 \equiv H_1^{CR}, & \text{if } \phi_k < \lambda \end{cases} \tag{3.5}$$

where we have assumed that all terminals operate at the same individual probability of false alarm, P_{fa} and ϕ_k is the received signal to noise ratio by CR station R_k.

This assumption helps the FC treat the decisions of the various terminals equally. These local decisions are sent to the FC to make the final decision about the channel. When R_k sends d_k at time slot t_k, the FC receives

$$y_k = \begin{cases} g_k x_k + w_0, & H_0^{PU} \\ g_k x_k + h_0 x_p w_0, & H_1^{PU} \end{cases} \tag{3.6}$$

where

g_k is the coefficient of the R_k-FC reporting channel
x_k is the signal of R_k with energy E_k
x_p is the PU signal (we assume here that SU stations use the PU channel for reporting, which causes interference)
h_0 is the channel coefficient between the PU and the FC

Assuming that the FC perfectly knows g_k, and hence uses coherent detection, the received SNR and signal to interference plus noise ratio (SINR) from R_k under H_0^{PU} and H_1^{PU} are given by

$$H_0^{PU} : \quad \psi_k = \frac{E_k}{\sigma_0^2} |g_k|^2$$

$$H_1^{PU} : \quad \psi_k = \frac{E_k |g_k|^2}{E_p |h_0|^2 \sigma_0^2} \tag{3.7}$$

All reporting channels are assumed here to experience flat Rayleigh fading and binary phase shift keying (BPSK) modulation is assumed.

Due to the imperfectness of the reporting channels, the decoded versions at the FC are subject to nonnegligible decoding errors. In particular, if $\hat{d}_k$ denotes the decoded version of d_k, then we can write

$$\Pr\left[\hat{d}_k = d_k \,\middle|\, H_1^{PU}\right] = 1 - BER_{1,k} \quad \text{and}$$

$$\Pr\left[\hat{d}_k = d_k \,\middle|\, H_0^{PU}\right] = 1 - BER_{0,k} \tag{3.8}$$

where $BER_{1,k}$ and $BER_{0,k}$ are the decoding error probabilities experienced by d_k under H_1^k and H_0^k, respectively.

At the end of the reporting phase, the FC uses the set of decoded reports, $\{\hat{d}_k\}, k = 1, \ldots K$ and its own decision, d_0, to calculate the decision metric Θ as

$$\Theta = d_0 + \sum_{l=1}^{K} \hat{d}_l = \sum_{l=0}^{K} \hat{d}_l \tag{3.9}$$

where, for mathematical convenience, we have defined $\hat{d}_0 = d_0$.

The FC chooses either H_0^{FC} or H_1^{FC} by comparing Θ to Θ_{th}:

$$
\begin{aligned}
H_1^{\text{FC}} &: \quad \Theta \geq \Theta_{\text{th}} \\
H_0^{\text{FC}} &: \quad \Theta < \Theta_{\text{th}}
\end{aligned}
\tag{3.10}
$$

3.3.2 Performance Analysis

If we denote the FC decision by $D \in \{0,1\}$, then the cooperative detection and false alarm probabilities at the FC side are defined, respectively, as

$$
P_{\text{d,CSS}} = \Pr\left[D = 1 \mid H_1^{\text{PU}}\right] \quad \text{and} \quad P_{\text{fa,CSS}} = \Pr\left[D = 1 \mid H_0^{\text{PU}}\right].
$$

These two probabilities can be calculated using the total probability theorem as

$$
P_{\text{d,CSS}} = \sum_{l=\Theta_{\text{th}}}^{K+1} \Pr\left[\Theta = l \middle| H_1^{\text{PU}}\right]
\tag{3.11}
$$

$$
P_{\text{fa,CSS}} = \sum_{l=\Theta_{\text{th}}}^{K+1} \Pr\left[\Theta = l \mid H_0^{\text{PU}}\right]
\tag{3.12}
$$

where $\Pr\left[\Theta = l \mid H_1^{\text{PU}}\right]$ and $\Pr\left[\Theta = l \mid H_0^{\text{PU}}\right]$ are the probabilities that a total l out of the $K + 1$ decoded decisions, d_k, equal one.

The performance of this CSS strategy was thoroughly studied in the literature [10,24,27]. It was shown that the best detection performance is achieved when $\Theta_{\text{th}} = 1$ while the worst occurs at $\Theta_{\text{th}} = K + 1$, as Figure 3.8 shows. In this figure, we consider a $K = 6$ network sensing a particular channel where an active PU transmits at 20 dB, while reports are sent using 5 dB. The average gains of the sensing and reporting channels are considered to have a path loss exponent of 4. The network terminals are randomly distributed around the PU and FC.

The value of Θ_{th} gives the FC the ability to control the level of protection given to the PU. By increasing Θ_{th}, the FC requires more evidence to claim the presence of a PU and, hence, increases the chances of missing to detect an active PU. On the other hand, having a low Θ_{th} means that the FC claims the presence of a PU even if the evidence is mild or even low [10].

Despite the simplicity of this CSS strategy and the low overhead requirement, it suffers from a few downsides. First, the duration of the reporting phase increases linearly with the number of SUs. While this may not be a problem for low populated networks, it becomes a performance bottleneck for highly populated networks. Second, this strategy is prone to the faulty reporters problem, which results due to decoding errors. Third, the amount of CCI affecting an active

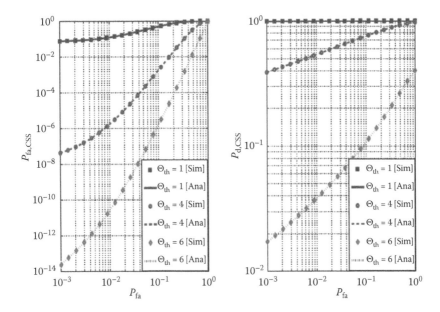

FIGURE 3.8 False alarm of the CSS scheme ($P_{fa,CSS}$) and correct decision ($P_{d,CSS}$) probabilities using different decision thresholds versus probability of false alarm of the individual CR stations (P_{fa}).

PU during the reporting phase can be nonnegligible (because the reporting stations use the PU channel).

Over the past few years, these downsides have been addressed in a number of publications. In particular, the faulty reporters' problem has been mitigated in [28–31] using clustering techniques, and in [27] using space-time coding and relaying. However, these techniques resulted in longer reporting time, which only worsens the other downsides. An attempt to reduce the CCI effect on an active PU was made in [32] where only terminals that claim the absence of the PU are allowed to report to the FC. However, since the identity of these terminals cannot be known beforehand, the duration of the reporting period cannot be reduced. On the opposite side, [33] observed that when the FC sets $\Theta_{th} = 1$, only a single transmission is needed to claim the presence of the PU. Hence, the authors proposed that all SUs, which decide the presence of the PU, transmit at the same time such that the FC will enjoy spatial diversity. While this limits the duration of the reporting phase, it causes more CCI to a potentially active PU.

3.4 COOPERATIVE TRANSMISSION TECHNIQUES

In the absence of PUs, the CRN uses the vacant channels as if they were the PUs. This means that any communication technique can be used, including multiple antenna techniques, relaying techniques, coordinated multipoint techniques, etc. However, as some of these techniques cannot be used in certain situations, the network terminals need to make intelligent decisions about the deployed transmission techniques. For instance, TV bands are known to have long wavelengths, and thus render using multiple antenna techniques infeasible [34]. Also, due to the diversity of PUs in these bands, CRNs operating over these bands have to minimize their transmission power as much as possible and have to keep it below a preset threshold [21]. For these two reasons, it is envisioned that CRNs operating over TV bands will rely on cooperative diversity techniques to extend their transmission range and achieve spatial diversity. They will also use hybrid automatic repeat request (HARQ) to achieve time diversity [35]. In the sequel, these transmission techniques shall be revised.

3.4.1   CHASE-COMBINING HYBRID AUTOMATIC REPEAT REQUEST

Automatic repeat request (ARQ) is a link-layer protocol that enhances the communication reliability by using acknowledged transmissions. According to this protocol, the transmitting terminal expects a reception acknowledgment (ACK) for every transmitted packet. If this ACK is not received during a preset time-out period, the packet is retransmitted. This process is repeated until an ACK is received or until a maximum number of retransmissions are reached.

A combination of an error correction code and ARQ is referred to as a hybrid ARQ (HARQ) protocol. Unlike ARQ, this protocol gives the receiver the ability to correct some of the detected errors. In HARQ, the receiver sends an ACK if the channel is good and a negative ACK (NACK) if the channel is bad. The channel quality is measured by comparing the instantaneously received SNR (in a CCI-free environment) or the SINR (in a CCI environment) to a preset quality threshold. If this threshold is exceeded, the channel is deemed good; otherwise, it is deemed bad. When the retransmitted signals contain the same amount of information, the receiver can use maximum ratio combining (MRC) to combine the replicas. This allows the receiver to maximize the SNR (or the SINR). This type of HARQ is referred to as chase-combining HARQ [36]. On the other hand, incremental redundancy is another HARQ protocol wherein every retransmission contains a different set of coding bits, hence giving the receiver additional evidence about the transmitted packet. Intuitively, a Chase combiner, which resembles a repetition encoder, is easier to implement [37, Ch. 6].

3.4.2   COOPERATIVE DIVERSITY

Cooperative diversity has been proposed in the pioneering works of Sendonaris and Laneman [22,38,39] as means to enhance the performance of cellular networks through user cooperation. In particular, this family of protocols allows single antenna terminals to experience space-time diversity through relaying. This notion has stimulated a tremendous number of researchers to investigate almost all aspects of these protocols and to propose a countless number of ideas suiting a wide range of wireless applications.

In general, cooperative diversity protocols can be classified based on the underlying relaying protocol into transparent protocols, for example, amplify and forward (AF), and regenerative protocols, for example, decode and forward (DF). They can also be classified based on the number of hops into dual-hop protocols and multihop protocols [40]. Recently, the impact of co-channel interference (CCI) on a number of cooperative diversity protocols has been investigated. In particular, the performance of a dual-hop relay network was studied in [41–48]. The extension to the multihop case (where CCI affects all relays and the destination) was studied in [49–52]. The DF dual-hop case in the presence of multiple relays was considered in [53–60]. Another group of researchers have focused on proposing protocols for CCI environments. These works have exploited the abundance of relay terminals in certain environments to propose relay selection strategies that account for the CCI effect. For instance, [61] proposed a modified version of the max–min relay selection criterion (see [62]) where a single relay is chosen based on the SINR of the source-relay link and the SNR of the relay-destination link. On the other hand, [63] proposed a selection strategy that allows the network to use relaying only when the relay path is better than direct transmission, while [64] proposed a relay selection criterion when AF is used in two adjacent cells.

3.4.2.1 Fixed Relaying

This primitive protocol allows the destination to enjoy spatial diversity at the cost of halving the throughput. A source terminal, S, communicates with a destination terminal, D, through the assistance of an intermediate relay terminal, R. Every transmission consumes two consecutive time slots. In the first slot, S broadcasts the signal to R and to D, while in the second slot, R forwards a regenerated replica of this signal. Consequently, D combines the two replicas and achieves a better performance. Figure 3.9 shows a flow diagram of fixed relaying (FR).

This relaying protocol suffers from error propagation and throughput reduction. Error propagation results from the fact that R has to regenerate the signal before forwarding it. While this may not be a problem when the S–R channel is good, it becomes a performance bottleneck when this channel is poor. Mathematically, analyzing error propagation probability is an involved process especially for high-constellation modulation techniques, for example, 8 PSK. For the BPSK case, however, [65] were able to obtain an accurate closed-form expression for this probability by comparing the SNRs of the R–D and S–D channels.

A widely used remedy for error propagation problem is threshold-based relaying where R uses DF only if the S–R channel is in a good condition [22,65,66]. While this does not totally eliminate error propagation, it significantly reduces it. When the

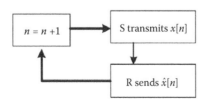

FIGURE 3.9   Flow diagram of FR.

S–R channel is poor, R can still use AF [67], or it can remain silent while S retransmits the signal. A protocol that uses this latter option is studied next.

3.4.2.2 Selection Relaying

Selection relaying (SR) measures the quality of the S–R channel by comparing the
instantaneously received SNR/SINR to a preset quality threshold, λ_r. The S–R channel is considered good only if this condition is met or is exceeded. When this is not
the case, R remains silent and S intervenes by retransmitting the signal in the following time slot, assuming the channel condition changes [22]. Figure 3.10 shows a flow
diagram of this protocol.

This protocol suffers from a minimal error propagation effect while enjoying
diversity gain at all times. As a result, it still suffers from a 50% throughput loss.
This loss can be mitigated by extending the decision-based relaying to the destination side.

3.4.2.3 Incremental Relaying

Incremental relaying (IR) waives diversity gain when the S–D channel is in a good
condition. In this case, D is likely to successfully decode the signal, and hence, assistance is not needed. Similar to SR, IR measures the quality of the S–D channel by
comparing the instantaneously received SNR/SINR to a preset threshold λ_d. When
this threshold is met or exceeded, D sends an ACK asking for a new transmission.
Otherwise, it sends a NACK asking R for assistance [68]. Figure 3.11 shows a flow
diagram of this protocol. The original proposal of this protocol, made in [22], suf-
fered from error propagation as it did not adopt decision-based relaying. Recently,
[69], [70], and [73] combined IR and SR into a selection incremental relaying (SIR)
protocol.

3.4.2.4 Selection Incremental Relaying

By combining decision-based relaying and decision-based assistance, SIR achieves
a good balance between throughput reduction and performance gain. Figure 3.12
shows a flow diagram of this protocol, which encompasses the previous three pro-
tocols as special cases. In particular, when $\lambda_d = \infty$ and $\lambda_r = 0$, SIR reduces to fixed

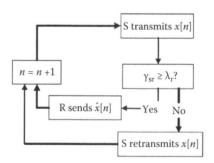

FIGURE 3.10 Flow diagram of SR.

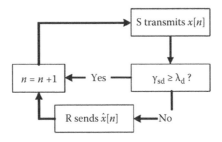

FIGURE 3.11 Flow diagram of IR.

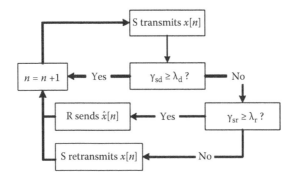

FIGURE 3.12 Flow diagram of SIR.

relaying (FR), while when $\lambda_d = \infty$ and $0<\lambda_r<\infty$, SIR reduces to SR, and finally when $0<\lambda_d<\infty$ and $\lambda_r = 0$, SIR reduces to IR.

3.5 SELECTIVE COOPERATIVE SPECTRUM SENSING STRATEGIES

This section presents three selective CSS strategies that reduce the duration of the reporting phase of CSS and the CCI experienced by a potentially active PU. These strategies are the dual-threshold selective CSS (DTCSS) strategy, the maximum cooperative spectrum sensing (MCSS) strategy, and the maximum–minimum CSS (MMCSS) strategy.

The performance of these strategies is analyzed in terms of the achievable detection probability, false alarm probability, and the average number of retransmissions. It is shown that, while these strategies reduce the CCI effect and the duration of the reporting phase, they achieve a performance comparable to, and sometimes superior to, the performance of the conventional CSS strategy studied in Section 3.2.

3.5.1 DUAL-THRESHOLD SELECTIVE CSS STRATEGY

Consider the K terminal CRN shown in Figure 3.7. At the end of the sensing period, the local decisions d_k's are forwarded to the FC in a round-robin manner to make the final decision D. However, to reduce the CCI effect to a potentially active PU, a modified version of the energy detector is used. In particular, the local energy estimates φ_k are compared against two decision thresholds λ_L and λ_U, where $\lambda_L \leq \lambda_U$ such that

$$d_k = \begin{cases} 0, & \phi_k \leq \lambda_L \\ 1, & \phi_k \geq \lambda_U \\ \text{No decision}, & \lambda_L < \phi_k > \lambda_U \end{cases} \tag{3.13}$$

After this comparison, only terminals with local decisions, that is, $d_k = 0$ or 1, report to the FC while those with no-decision remain silent [71]. The width of the no-decision region, $\lambda_U - \lambda_L$, is inversely proportional to the amount of CCI affecting an active PU. Hence, the wider this region is, the smaller is the amount of CCI. However, this also reduces the number of reports transmitted to the FC, which lowers the detection capability. Hence, the width of this region should be chosen to strike a balance between these two conflicting objectives.

In general, we set λ_U and λ_L to be functions of the Neyman–Pearson threshold value, λ such that $\lambda_L = f_L(\lambda)$ and $\lambda_U = f_U(\lambda)$ are two arbitrary functions that satisfy $\lambda_L \in (0,\lambda]$ and $\lambda_U \in [\lambda,\infty]$ for arbitrary λ, respectively.

If we denote the set of terminals with local decisions by C_r, then the FC receives a total of $K_r \leq K$ local decisions, where K_r is the number of elements in C_r. These local decisions will be sent over imperfect reporting channels, and hence will be subject to nonnegligible decoding errors. To reduce the impact of these errors, the FC eliminates the reports whose instantaneous SINR, or SNR, falls below a preset threshold of τ_{th}. While this threshold can be arbitrarily chosen to meet a certain performance level, its value is lower bounded by the outage threshold of $2^Q - 1$ where Q is the spectral efficiency expressed in bits per second per Hertz (bps/Hz).

At the end of the reporting phase, the FC will have a total of K_d reliably decoded local decisions. These decisions form the decoding set C_d, which is a subset of C_r. With the aid of these K_d reports, the FC calculates the decision metric Θ as

$$\Theta = \sum_{l=0}^{K_d+1} \hat{d}_k \tag{3.14}$$

and compares it to the decision threshold Θ_{th} to make the final decision as

$$\begin{aligned} H_1^{FC} : & \quad \Theta \geq \Theta_{th} \\ H_0^{FC} : & \quad \Theta < \Theta_{th} \end{aligned} \tag{3.15}$$

The performance of this scheme may be evaluated at both, the terminal level and FC level. In particular, the following performance metrics are important to know:

- Detection and false alarm probabilities of R_k
- The average number of reporting terminals in the presence and the absence of a PU

- The detection and false alarm probabilities of the FC
- The average number of retransmission requests in the presence and the absence of a PU

According to the decision rule, the probabilities of detection and false alarm are given by

$$P_{d,DT,k} = \Pr\left[\phi_k \geq \lambda_U \middle| H_1^{PU}\right] \quad \text{and} \quad P_{fa,DT,k} = \Pr\left[\phi_k \geq \lambda_U \middle| H_0^{PU}\right], \text{respectively}$$

Using the CDFs, these two probabilities become

$$P_{d,DT,k} = 1 - F_{\phi_k|H_1^{PU}}(\lambda_U) \quad \text{and} \quad P_{fa,DT,k} = 1 - F_{\phi_k|H_0^{PU}}(\lambda_U) \tag{3.16}$$

where F_{ϕ_k} denotes the cumulative distribution function (CDF) of ϕ_k

The average number of reporting terminals under H_1^{PU} and H_0^{PU} can be determined as follows.

Since $R_k \in C_r$ only if $\phi_k \geq \lambda_U$ or $\phi_k \leq \lambda_L$, consequently we can write

$$\Pr\left[R_k \in C_r \middle| H_1^{PU}\right] = 1 + F_{\phi_k|H_1^{PU}}(\lambda_L) - F_{\phi_k|H_1^{PU}}(\lambda_U) \tag{3.17}$$

$$\Pr\left[R_k \in C_r \middle| H_0^{PU}\right] = 1 + F_{\phi_k|H_0^{PU}}(\lambda_L) - F_{\phi_k|H_0^{PU}}(\lambda_U) \tag{3.18}$$

With the aid of these probabilities, the average number of reporting terminals is calculated as

$$\hat{K}_{r,H_1^{PU}} = \sum_{k=1}^{K} \Pr\left[R_k \in C_r \middle| H_1^{PU}\right] \tag{3.19}$$

$$\hat{K}_{r,H_0^{PU}} = \sum_{k=1}^{K} \Pr\left[R_k \in C_r \middle| H_0^{PU}\right] \tag{3.20}$$

Observe that by setting $\lambda_U = \lambda_L = \lambda$, the width of the no-decision region vanishes, and hence $\Pr\left[R_k \in C_r \middle| H_1^{PU}\right] = \Pr\left[R_k \in C_r \middle| H_0^{PU}\right] = 1$, which makes $\hat{K}_{r,H_1^{PU}} = \hat{K}_{r,H_0^{PU}} = K$.

3.5.1.1 Simulations and Illustrations

Here, we will study the detection and false alarm probabilities at the terminal level.

Figure 3.13 shows the impact of increasing λ_U on $P_{d,DT,k}$ and $P_{fa,DT,k}$. In this figure, we consider the case of a PU located at D_{ref} distance units from an arbitrary terminal and transmitting a 5 dB signal over a 6 MHz channel. By gradually increasing Δ where $\lambda_U = (1 + \Delta)\lambda$, the SU degrades its $P_{fa,DT}$ and $P_{d,DT}$. However, this degradation

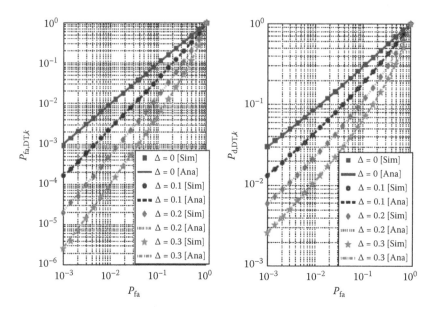

FIGURE 3.13 $P_{fa,DTCSS}$ and $P_{d,DTCSS}$ for different values of Δ.

is reflected in a reduction in the number of reporting terminals as Figure 3.14 shows. In this figure, we consider $K = 10$ terminals randomly distributed within a circle with radius $2D_{ref}$ centered at the PU. As the figure shows, increasing Δ reduces the average number of reporting terminals, and hence reduces the effect of CCI on the PU.

Next, let us look at the performance at the FC side. Figure 3.15 shows $P_{fa,DTCSS}$ and $P_{d,DTCSS}$ for the $K = 10$ terminals network. However, this time, the PU uses -5 dB while the FC uses $\Theta_{th} = 1$, $\tau_{th} = 1$, and the K terminals use a 5 dB signal. As the figure shows, increasing Δ reduces $P_{fa,DTCSS}$, which is a desirable feature. However, it also reduces $P_{d,DTCSS}$, which is undesirable, which is intuitively right since $K_d \leq K$.

Let $M_{1,DTCSS}$ and $M_{0,DTCSS}$ denote the average numbers of retransmissions under H_1^{PU} and H_0^{PU}, respectively.

Finally, we consider $M_{0,DTCSS}$ and $M_{1,DTCSS}$. Figure 3.16 shows that increasing Δ results in an increase in both metrics. In this figure, a $K = 5$ terminal network has been considered with a PU transmitting a -5 dB signal. The network terminals are randomly distributed within [0, 2Dref] distance units from the PU. The reports are sent using a 5 dB signal, and the FC uses $\tau_{th} = 1$ and $\Theta_{th} = 1$.

3.5.2   MAXIMUM COOPERATIVE SPECTRUM SENSING STRATEGY

When the number of reporting terminals is large, the duration of the reporting period becomes intolerably long, which results in a nonnegligible level of CCI experienced by an active PU. This can be avoided if we choose a subgroup of terminals to represent the entire network. For instance, we can allow the terminals with $d_k = 1$ to report to the FC, while other terminals remain silent. Alternatively, we can choose

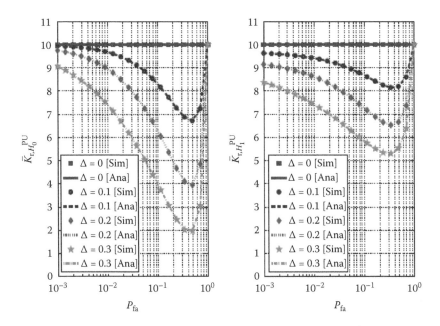

FIGURE 3.14 Average number of reporting terminals as functions of P_{fa}.

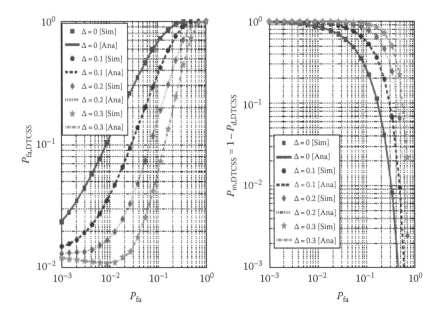

FIGURE 3.15 $P_{fa,DTCSS}$ and $P_{d,DTCSS}$ for the $K = 10$ terminals network.

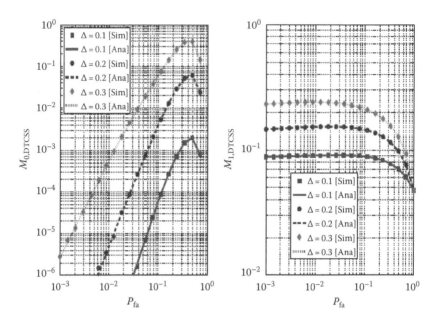

FIGURE 3.16   Average number of retransmissions.

the terminals with $d_k = 0$ to report. A third option is to let the terminals with the highest N energy estimates to report to the FC. However, in all these scenarios, the selection criterion is made in a distributed manner, which means that the duration of the reporting period should be kept at K time slots even if some of these slots are not used. Attempting to reduce this duration results in a nonzero collision probability, which would only worsen the CCI experienced by an active PU. Fortunately, we can still exploit this notion by allowing a single terminal to report to the FC. Obviously, this achieves the minimum reporting time. This terminal can be chosen to have the maximum energy estimate, the minimum energy estimate, the median energy estimate, or be randomly chosen. Of these four options, the maximum energy estimate criterion is the most appealing. Choosing the one with the minimum energy estimate will reduce the PU protection as it will choose H_0^{PU} most of the time even if the PU is active especially for well-populated networks. On the other hand, choosing based on the median energy estimate cannot be made in a distributed manner while random selection wastes the cooperation gain. Consequently, we adopt the maximum energy estimate criterion to achieve a minimum reporting time, a minimum CCI effect, and a desirable level of performance, especially in PU protection. At the end of the sensing period, the terminal with the highest energy estimate sends its local decision to the FC. This selection is made by letting every terminal set a timer, which is inversely proportional to its energy estimate such that the timer corresponding to the highest estimate expires first. Consequently, if R_* denotes the selected terminal, then we can write

$$R_* = \arg_k \max \left\{ \phi_1, \phi_2, \ldots, \phi_k, \ldots \phi_K \right\} \tag{3.21}$$

where the corresponding energy estimate is denoted by ϕ_* and the local decision is d_*. At the FC side, the decoded version of d_*, denoted by $\hat{d}_*$, is used to calculate the decision threshold Θ as $\Theta = \hat{d}_0 + \hat{d}_*$.

Finally, the FC makes the final decision following the decision rule:

$$H_1^{\text{FC}} : \quad \Theta \geq \Theta_{\text{th}}$$
$$H_0^{\text{FC}} : \quad \Theta < \Theta_{\text{th}} \tag{3.22}$$

To reduce the effect of decoding errors, the FC examines the quality of the reporting channel against a preset threshold, τ_{th}. If the instantaneous SINR or SNR exceeds this threshold, the FC makes a final decision using Θ; otherwise, it asks R_* to retransmit.

In the following sections, we shall look at the performance of this strategy. In doing so, we shall look at the same performance metrics used in the DTCSS case except for the average number of reporting terminals, which is one.

Since R_* has $\phi_* = \max_k\{\phi_1,\phi_2,...,\phi_K\}$, then the corresponding detection and false alarm probabilities are defined as $P_{\text{d,MCSS,*}} = \Pr\left[\phi_* \geq \lambda \big| H_1^{\text{PU}}\right]$ and $P_{\text{fa,MCSS,*}} = \Pr\left[\phi_* \geq \lambda \big| H_0^{\text{PU}}\right]$, respectively.

These two probabilities can be rewritten as

$$P_{\text{d,MCSS,*}} = \Pr\left[\phi_* \geq \lambda \big| H_1^{\text{PU}}\right] = 1 - \prod_{k=1}^{K} \Pr\left[\phi_k < \lambda \big| H_1^{\text{PU}}\right] = 1 - \prod_{k=1}^{K}(1 - P_{\text{d},k}) \tag{3.23}$$

$$P_{\text{fa,MCSS,*}} = \Pr\left[\phi_* \geq \lambda \big| H_0^{\text{PU}}\right] = 1 - \prod_{k=1}^{K} \Pr\left[\phi_k < \lambda \big| H_0^{\text{PU}}\right]$$
$$= 1 - \prod_{k=1}^{K}(1 - P_{\text{fa},k}) = 1 - (1 - P_{\text{fa}})^K \tag{3.24}$$

Intuitively, as K increases, $P_{\text{d,MCSS}}$ exponentially approaches 1; however, at the same time, $P_{\text{fa,MCSS}}$ will approach 1, which is undesirable feature. However, since these two events occur at two different rates, a desirable performance can be achieved by properly controlling the probability of false alarm of the individual stations, P_{fa}.

The detection and false alarm performance of R_* are proportional to K, as shown in Figure 3.17. The figure also shows that the gain achieved in $P_{\text{d,MCSS}}$ is accompanied with a loss in $P_{\text{fa,MCSS}}$. However, the latter can be mitigated by reducing the operating P_{fa}. The performance at the FC side is shown in Figure 3.18 for $K = 15$, 10, 5, and 2. In this figure, we show the detection and false alarm probabilities as functions of P_{fa}. The network terminals are randomly distributed within [0 2Dref] of the PU and within [0 Dref] of the FC. The PU transmits at −5 dB, the terminals report using 5 dB, and the FC uses $\tau_{\text{th}} = 1$ and $\Theta_{\text{th}} = 1$. Figure 3.19 shows the average numbers of retransmission requests for this network. As the figure shows, increasing K reduces the average number of retransmission requests.

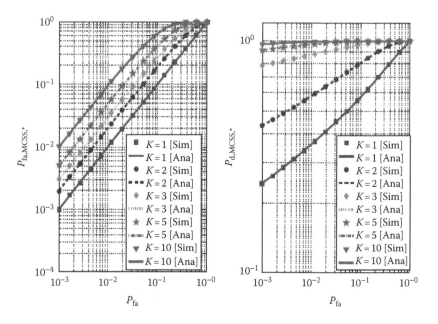

FIGURE 3.17   MCSS detection and false alarm performance for R_*.

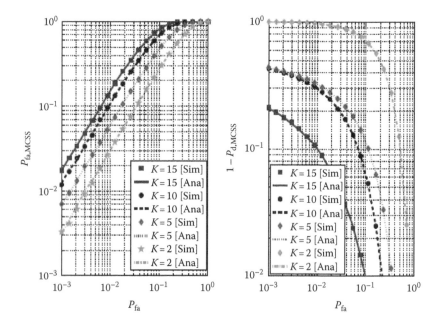

FIGURE 3.18   MCSS false alarm and misdetection performance for different numbers of network terminals.

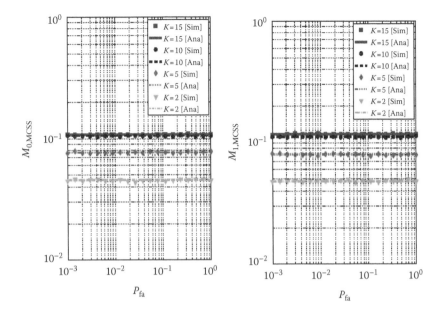

FIGURE 3.19 MCSS average numbers of retransmission requests.

3.5.3 MAXIMUM–MINIMUM CSS STRATEGY

The MCSS strategy does not account for the quality of the reporting channels, assuming that it is most likely changing during the CSS period. While this assumption can be justified for networks with mobile terminals, like those based on the Ecma-392 standard [35], it is not necessarily true for networks with immobile terminals. In fact, it can be argued that such networks have quasi-static reporting channels with coherence times larger than the duration of the entire CSS process. This applies, for instance, to wireless regional area networks (WRANs) based on the IEEE 802.22 standard [72]. In this case, the MCSS strategy can be modified to account for this additional information. In particular, R_* can be chosen according to the max–min criterion proposed in [62] as

$$R_* = \arg_k \max \min \{\phi_k, \psi_k\} \tag{3.25}$$

However, this strategy suffers from two shortcomings. First, the reporting channels under H_1^{PU} are not reciprocal, since the interference effect of the PU at R_k is different from that at the FC. Consequently, using ψ_k in this case does not reflect the status of the reporting channel at the FC side. Second, since the range of ϕ_k is generally higher than that of ψ_k, the inner part of the selection strategy selects ψ_k most of the time. To overcome these shortcomings, the following modified selection criterion is considered here:

$$R_* = \arg_k \max \min \{\Phi_k, \Psi_k\} \tag{3.26}$$

where

$$\Phi_k = \frac{\phi_k}{\lambda} \quad \text{and} \quad \Psi_k = \frac{\psi_k}{\tau_{th}} \qquad (3.27)$$

This criterion achieves reciprocity [61] by only considering the SNR part, ψ_k, while it achieves fairness by normalizing the two variables by their corresponding thresholds, λ and τ_{th}. At the end of the sensing period, every terminal sets a timer inversely proportional to the minimum of Φ_k and Ψ_k such that the timer corresponding to the maximum value expires first. When this happens, the chosen terminal R_* forwards its local report, d_*, to the FC. After this point, the FC follows the same procedure of the MCSS strategy.

Once again, let us now look at the performance of this strategy using the detection and false alarm probabilities of R_* and the FC as well as the average number of retransmission requests.

Figure 3.20 illustrates the detection and false alarm probabilities of R_* for different values of K. It can be seen that MMCSS achieves performance comparable to that of the MCSS strategy. However, this strategy surpasses the MCSS in terms of the average number of retransmission requests as shall be illustrated shortly.

The terminals are randomly distributed within [0 Dref] from the PU which uses −5 dB. The performance at the FC is shown in Figure 3.21. This figure illustrates the gain achieved through increasing K in terms of the detection and false alarm probabilities while Figure 3.22 shows the gains in the average number of retransmission requests.

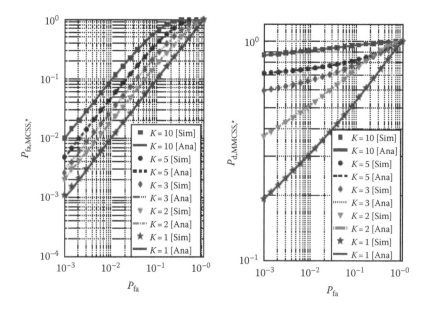

FIGURE 3.20 MMCSS detection and false alarm probabilities of R_* for different values of K.

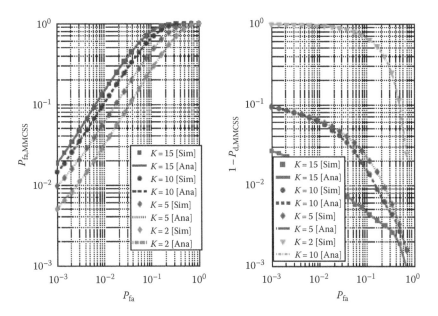

FIGURE 3.21 MMCSS false alarm and misdetection performance for different numbers of network terminals.

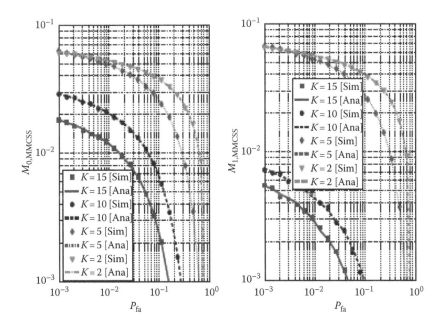

FIGURE 3.22 MMCSS average numbers of retransmission requests.

3.5.4 COMPARISON AND DISCUSSION

Having analyzed the performance of the three strategies, let us now have an insight about the relative performance, and hence the achieved gains of MCSS and MMCSS over the DTCSS strategy. We consider a network with $K = 8$ terminals distributed within [0 2D$_{ref}$] distance units from a PU that uses *0 dB* and within [0 D$_{ref}$] from a FC that uses $\tau_{th} = 1$. These terminals report using 5 dB. Figure 3.23 shows that the MCSS and MMCSS strategies outperform the DTCSS in terms of detection probability when $\Theta_{th} = 1$. But when $\Theta_{th} = 2$, the DTCSS becomes better. It also shows that the MCSS and MMCSS strategies achieve higher false alarm probabilities compared to the DTCSS strategy. However, for very small values of P_{fa}, decoding errors dominate the performance of DTCSS and cause an error floor, which gives the MCSS and MMCSS an additional advantage.

Now, we look at the average number of retransmission requests for $\Theta_{th} = 1,2$. As Figure 3.24 shows, the MMCSS and the MCSS strategies are not affected by this increase, while the DTCSS suffers from an increase, especially when P_{fa} is larger than 10%.

3.5.5 CONCLUSION

This section presented three selective CSS strategies to reduce the duration of the reporting phase and, hence, minimize the CCI effect at a potentially active PU. The performance of these strategies was discussed in terms of the detection and false alarm probabilities, the average number of reporting terminals, and the average

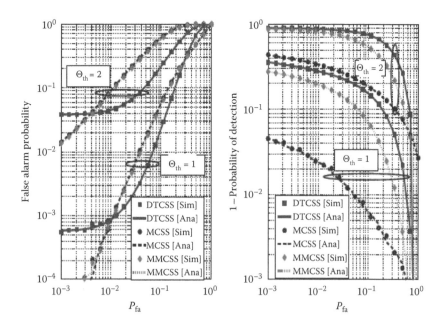

FIGURE 3.23 Detection and false alarm probabilities for DTCSS, MCSS, and MMCSS.

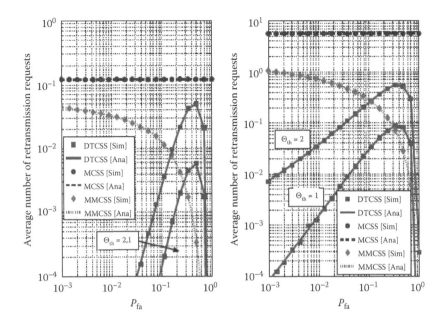

FIGURE 3.24 Average number of retransmission requests of DTCSS, MCSS, and MMCSS.

number of retransmission requests. Simulation results revealed that the MMCSS strategy achieves the best performance among the three in all measures. This performance superiority is achieved while requiring a single reporting terminal with a potential for retransmission request.

3.6 CONCLUSION

This chapter covered cooperative spectrum acquisition techniques. The basic principles and mechanisms of CSS and combining have been presented. The multiband case has been discussed, highlighting its advantages over single-case paradigms. The chapter provided an in-depth description of a number of cooperative spectrum acquisition techniques including decision-based CSS using energy detection, HARQ, and cooperative diversity (FR, SR, IR, and SIR). Selective CSS strategies have been covered. These strategies reduce the CCI effect and the duration of the reporting phase compared to the conventional CSS strategy, while achieving a comparable or a better performance in terms of detection and false alarm probability. The chapter has shown that cooperation is essential to enhance the performance of CR sensing and acquisition techniques.

REFERENCES

1. I. F. Akyildiz, B. F. Lo, and R. Balakrishnan, Cooperative spectrum sensing in cognitive radio networks: A survey, *Physics Communications*, 4 (1), 40–62, March 2011.
2. K. Letaief and W. Zhang, Cooperative communications for cognitive radio networks, *Proceedings of the IEEE*, 97 (5), 878–893, May 2009.

3. D. Cabric, A. Tkachenko, and R. Brodersen, Spectrum sensing measurements of pilot, energy, and collaborative detection, in *Proceedings of the IEEE Military Communications Conference (MILCOM'06)*, Washington, DC, October 2006, pp. 1–7.

4. T. Yucek and H. Arslan, A survey of spectrum sensing algorithms for cognitive radio applications, *IEEE Communications Surveys and Tutorials*, 11 (1), 116–130, January 2009.

5. Z. Quan, S. Cui, A. Sayed, and H. Poor, Optimal multiband joint detection for spectrum sensing in cognitive radio networks, *IEEE Transactions on Signal Processing*, 57 (3), 1128–1140, March 2009.

6. Z. Tian, Compressed wideband sensing in cooperative cognitive radio networks, in *Proceedings of the IEEE Global Telecommunication Conference (GLOBECOM'08)*, New Orleans, LA, December 2008, pp. 1–5.

7. Z. Fanzi, C. Li, and Z. Tian, Distributed compressive spectrum sensing in cooperative multihop cognitive networks, *IEEE Journal of Selected Topics in Signal Processing*, 5 (1), 37–48, February 2011.

8. Y. Wang, A. Pandharipande, Y. L. Polo, and G. Leus, Distributed compressive wideband spectrum sensing, in *Information Theory and Applications Workshop*, La Jolla, CA, February 2009, pp. 2337–2340.

9. N. Shankar, C. Cordeiro, and K. Challapali, Spectrum agile radios: Utilization and sensing architectures, in *Proceedings of the IEEE International Symposium on New Frontiers in Dynamic Spectrum Access Networks (DySPAN'05)*, Baltimore, MD, November 2005, pp. 160–169.

10. W. Zhang, R. Mallik, and K. Letaief, Optimization of cooperative spectrum sensing with energy detection in cognitive radio networks, *IEEE Transactions on Wireless Communications*, 8 (12), 5761–5766, December 2009.

11. J. Ma, G. Zhao, and Y. Li, Soft combination and detection for cooperative spectrum sensing in cognitive radio networks, *IEEE Transactions on Wireless Communications*, 7 (11), 4502–4507, November 2008.

12. E. Visotsky, S. Kuffner, and R. Peterson, On collaborative detection of TV transmissions in support of dynamic spectrum sharing, in *Proceedings of the IEEE International Symposium on New Frontiers in Dynamic Spectrum Access Networks (DySPAN'05)*, Baltimore, MD, November 2005, pp. 338–345.

13. S. Mishra, A. Sahai, and R. Brodersen, Cooperative sensing among cognitive radios, in *Proceedings of the IEEE International Conference on Communications (ICC'06)*, Istanbul, Turkey, vol. 4, June 2006, pp. 1658–1663.

14. Z. Quan, S. Cui, A. Sayed, and H. Poor, Spatial-spectral joint detection for wideband spectrum sensing in cognitive radio networks, in *Proceedings of the IEEE International Conference on Acoustics, Speech and Signal Processing (ICASSP'08)*, Las Vegas, NV, April 2008, pp. 2793–2796.

15. R. Fan and H. Jiang, Optimal multi-channel cooperative sensing in cognitive radio networks, *IEEE Transactions on Wireless Communications*, 9 (3), 1128–1138, March 2010.

16. R. Fan, H. Jiang, Q. Guo, and Z. Zhang, Joint optimal cooperative sensing and resource allocation in multichannel cognitive radio networks, *IEEE Transactions on Vehicular Technology*, 60 (2), 722–729, February 2011.

17. M. Derakhtian, F. Izedi, A. Sheikhi, and M. Neinavaie, Cooperative wideband spectrum sensing for cognitive radio networks in fading channels, *IET Signal Processing*, 6 (3), 227–238, May 2012.

18. Z. Song, Z. Zhou, X. Sun, and Z. Qin, Cooperative spectrum sensing for multiband under noise uncertainty in cognitive radio networks, in *Proceedings of the IEEE International Conference on Communications Workshops (ICC'10)*, Cape Town, South Africa, May 2010, pp. 1–5.

19. S.-J. Kim and G. Giannakis, Sequential and cooperative sensing for multi-channel cognitive radios, *IEEE Transactions on Signal Processing*, 58 (8), 4239–4253, August 2010.
20. G. Hattab and M. Ibnkahla, Multiband cognitive radio: Great promises for future radio access, *Proceedings of the IEEE*, 102 (3), 282–306, March 2014.
21. Federal Communications Commission (FCC), Second memorandum opinion and order in the matter of unlicensed operation in the TV broadcast bands, additional spectrum for unlicensed devices below 900 MHz and in the 3 GHz band, Technical Report, Washington, DC, ET Docket No. 10–174, September 2010.
22. J. Laneman, D. Tse, and G. Wornell, Cooperative diversity in wireless networks: Efficient protocols and outage behavior, *IEEE Transactions on Information Theory*, 50 (12), 3062–3080, December 2004.
23. Z. Quan, S. Cui, H. Poor, and A. Sayed, Collaborative wideband sensing for cognitive radios, *IEEE Signal Processing Magazine*, 25 (6), 60–73, November 2008.
24. K. Ben Letaief and W. Zhang, Cooperative communications for cognitive radio networks, *Proceedings of the IEEE*, 97 (5), 878–893, May 2009.
25. A. El-Mougy, M. Ibnkahla, G. Hattab, and W. Ejaz, Reconfigurable dynamic networks, *The Proceedings of the IEEE* (submitted for publication), 2014.
26. B. F. Lo, A survey of common control channel design in cognitive radio networks, *Physical Communication*, 4 (1), 26–39, 2011.
27. W. Zhang and K. Letaief, Cooperative spectrum sensing with transmit and relay diversity in cognitive radio networks—[transaction letters], *IEEE Transactions on Wireless Communications*, 7 (12), 4761–4766, December 2008.
28. C. Sun, W. Zhang, and K. Letaief, Cluster-based cooperative spectrum sensing in cognitive radio systems, in *Proceedings of the IEEE International Conference on Communications (ICC)*, Glasgow, Scotland, June 24–28, 2007, pp. 2511–2515.
29. J. Lee, Y. Kim, S. Sohn, and J. Kim, Weighted-cooperative spectrum sensing scheme using clustering in cognitive radio systems, in *10th International Conference on Advanced Communication Technology (ICACT)*, Phoenix Park, Korea, February 2008, pp. 786–790.
30. K. Smitha and A. Vinod, Cluster based cooperative spectrum sensing using location information for cognitive radios under reduced bandwidth, in *IEEE 54th International Midwest Symposium on Circuits and Systems (MWSCAS)*, Seoul, Korea, August 2011, pp. 1–4.
31. M. Ben Ghorbel, H. Nam, and M. Alouini, Cluster-based spectrum sensing for cognitive radios with imperfect channel to cluster-head, in *IEEE Wireless Communications and Networking Conference (WCNC)*, Paris, France, April 2012, pp. 709–713.
32. Y. Zou, Y.-D. Yao, and B. Zheng, A selective-relay based cooperative spectrum sensing scheme without dedicated reporting channels in cognitive radio networks, *IEEE Transactions on Wireless Communications*, 10 (4), 1188–1198, April 2011.
33. K. Umebayashi, J. Lehtomaki, T. Yazawa, and Y. Suzuki, Efficient decision fusion for cooperative spectrum sensing based on OR-rule, *IEEE Transactions on Wireless Communications*, 11 (7), 2585–2595, July 2012.
34. C. Stevenson, G. Chouinard, Z. Lei, W. Hu, S. Shellhammer, and W. Caldwell, IEEE 802.22: The first cognitive radio wireless regional area network standard, *IEEE Communications Magazine*, 47 (1), 130–138, January 2009.
35. ECMA International, MAC and PHY for operation in TV white space, Technical Report, Geneva, Switzerland, December 2009.
36. D. Chase, Code combining—A maximum-likelihood decoding approach for combining an arbitrary number of noisy packets, *IEEE Transactions on Communications*, 33 (5), 385–393, May 1985.

37. E. Dahlman, S. Parkvall, and J. Sköld, *4G LTE/LTE-Advanced for Mobile Broadband*, 1st ed., Academic Press, Burlington, MA, 2011.

38. A. Sendonaris, E. Erkip, and B. Aazhang, User cooperation diversity: Part I. system description, *IEEE Transactions on Communications*, 51 (11), 1927–1938, November 2003.

39. A. Sendonaris, E. Erkip, and B. Aazhang, User cooperation diversity: Part II. Implementation aspects and performance analysis, *IEEE Transactions on Communications*, 51 (11), 1939–1948, November 2003.

40. M. Dohler and Y. Li, Cooperative Communications: Hardware, Channel and PHY, 1st ed., John Wily & Sons, Hoboken, NJ, 2010.

41. H. Suraweera, H. Garg, and A. Nallanathan, Performance analysis of two hop amplify-and-forward systems with interference at the relay, *IEEE Communications Letters*, 14 (8), 692–694, August 2010.

42. F. S. Al-Qahtani, T. Q. Duong, C. Zhong, K. A. Qaraqe, and H. Alnuweiri, Performance analysis of dual-hop AF systems with interference in Nakagami-m fading channels, *IEEE Signal Processing Letters*, 18 (8), 454–457, August 2011.

43. N. Milosevic, Z. Nikolic, and B. Dimitrijevic, Performance analysis of dual hop relay link in Nakagami-m fading channel with interference at relay, in *21st International Conference Radioelektronika (RADIOELEKTRONIKA)*, Brno, Czech Republic, April 2011, pp. 1–4.

44. C. Zhong, S. Jin, and K.-K. Wong, Dual-hop systems with noisy relay and interference-limited destination, *IEEE Transactions on Communications*, 58 (3), 764–768, March 2010.

45. S. Ikki and S. Aïssa, Performance analysis of dual-hop relaying systems in the presence of co-channel interference, in *GLOBECOM 2010, 2010 IEEE Global Telecommunications Conference*, Miami, FL, December 2010, pp. 1–5.

46. D. Lee and J. H. Lee, Outage probability for dual-hop relaying systems with multiple interferers over Rayleigh fading channels, *IEEE Transactions on Vehicular Technology*, 60 (1), 333–338, January 2011.

47. H. Suraweera, D. Michalopoulos, R. Schober, G. Karagiannidis, and A. Nallanathan, Fixed gain amplify-and-forward relaying with co-channel interference, in *IEEE International Conference on Communications (ICC)*, Kyoto, Japan, June 2011, pp. 1–6.

48. H. Suraweera, D. Michalopoulos, and C. Yuen, Performance analysis of fixed gain relay systems with a single interferer in Nakagami-m fading channels, *IEEE Transactions on Vehicular Technology*, 61 (3), 1457–1463, March 2012.

49. N. Beaulieu and K. Hemachandra, Exact performance analysis of multihop relaying systems operating in co-channel interference using the generalized transformed characteristic function, in *Australasian Telecommunication Networks and Applications Conference (ATNAC)*, Melbourne, Victoria, Australia, November 2011, pp. 1–6.

50. T. Soithong, V. Aalo, G. Efthymoglou, and C. Chayawan, Performance of multihop relay systems in a Rayleigh fading environment with co-channel interference, in *IEEE Global Telecommunications Conference (GLOBECOM)*, Houston, TX, December 2011, pp. 1–6.

51. S. Ikki and S. Aissa, Effects of co-channel interference on the error probability performance of multi-hop relaying networks, in *IEEE Global Telecommunications Conference (GLOBECOM)*, Houston, TX, December 2011, pp. 1–5.

52. T. Soithong, V. Aalo, G. Efthymoglou, and C. Chayawan, Outage analysis of multihop relay systems in interference-limited Nakagami-m fading channels, *IEEE Transactions on Vehicular Technology*, 61 (3), 1451–1457, March 2012.

53. D. Lee and J. H. Lee, Outage probability for opportunistic relaying on multicell environments, in *IEEE 69th Vehicular Technology Conference (VTC-Spring)*, Barcelona, Spain, April 2009, pp. 1–5.

54. Q. Yang and K. Kwak, Outage performance of cooperative relaying with dissimilar Nakagami-m interferers in Nakagami-m fading, *IET Communications*, 3 (7), 1179–1185, July 2009.

55. J. Si, Z. Li, and Z. Liu, Outage probability of opportunistic relaying in Rayleigh fading channels with multiple interferers, *IEEE Signal Processing Letters*, 17 (5), 445–448, May 2010.

56. D. Lee and J. H. Lee, Outage probability of amplify-and-forward opportunistic relaying with multiple interferers over Rayleigh fading channels, in *IEEE 73rd Vehicular Technology Conference (VTC-Spring)*, Budapest, Hungary, May 2011, pp. 1–5.

57. D. Lee and J. H. Lee, Outage probability of decode-and-forward opportunistic relaying in a multicell environment, *IEEE Transactions on Vehicular Technology*, 60 (4), 1925–1930, May 2011.

58. H. Yu, I.-H. Lee, and G. L. Stuber, Outage probability of decode-and-forward cooperative relaying systems with co-channel interference, *IEEE Transactions on Wireless Communications*, 11 (1), 266–274, January 2012.

59. S.-I. Kim and J. Heo, Outage probability of interference-limited amplify-and-forward relaying with partial relay selection, in *IEEE 73rd Vehicular Technology Conference (VTC-Spring)*, Budapest, Hungary, May 2011, pp. 1–5.

60. S. Ikki and S. Aissa, Impact of imperfect channel estimation and co-channel interference on regenerative cooperative networks, *IEEE Wireless Communications Letters*, 1 (5), 436–439, October 2012.

61. I. Krikidis, J. Thompson, S. Mclaughlin, and N. Goertz, Max-min relay selection for legacy amplify-and-forward systems with interference, *IEEE Transactions on Wireless Communications*, 8 (6), 3016–3027, June 2009.

62. A. Bletsas, A. Khisti, D. Reed, and A. Lippman, A simple cooperative diversity method based on network path selection, *IEEE Journal on Selected Areas in Communications*, 24 (3), 659–672, March 2006.

63. S. Il Kim and J. Heo, An efficient relay selection strategy for interference limited relaying networks, in *IEEE 21st International Symposium on Personal Indoor and Mobile Radio Communications (PIMRC)*, Istanbul, Turkey, September 2010, pp. 476–481.

64. H. Ryu, J. Lee, and C. Kang, Relay selection scheme for orthogonal amplify and forward relay-enhanced cellular system in a multi-cell environment, in *IEEE 71st Vehicular Technology Conference (VTC-Spring)*, Taipei, Taiwan, May 2010, pp. 1–5.

65. F. Onat, A. Adinoyi, Y. Fan, H. Yanikomeroglu, J. Thompson, and I. Marsland, Threshold selection for SNR-based selective digital relaying in cooperative wireless networks, *IEEE Transactions on Wireless Communications*, 7 (11), 4226–4237, November 2008.

66. N. C. Beaulieu and J. Hu, A closed-form expression for the outage probability of decode-and-forward relaying in dissimilar Rayleigh fading channels, *IEEE Communications Letters*, 10 (12), 813–815, December 2006.

67. T. Duong and H.-J. Zepernick, On the performance gain of hybrid decode-amplify-forward cooperative communications, *EURASIP Journal on Wireless Communications and Networking*, 2009 (1), 479463, 2009.

68. S. Ikki and M. Ahmed, Performance analysis of incremental-relaying cooperative-diversity networks over Rayleigh fading channels, *IET Communications*, 5 (3), 337–349, February 2011.

69. Q. Zhou and F. Lau, Two incremental relaying protocols for cooperative networks, *IET Communications*, 2 (10), 1272–1278, November 2008.

70. J. Ran, W. Yafeng, D. Yang, and W. Xiang, A novel selection incremental relaying strategy for cooperative networks, in *IEEE Wireless Communications and Networking Conference (WCNC)*, Cancun, Mexico, March 2011.

71. C. Sun, W. Zhang, and K. Letaief, Cooperative spectrum sensing for cognitive radios under bandwidth constraints, in *Proceedings of IEEE WCNC*, Hong Kong, China, March 11–15, 2007.

72. IEEE draft standard for information technology–telecommunications and information exchange between systems: Local and Metropolitan Area Networks: Specific requirements—Part 22.1: Standard to enhance harmful interference protection for low power licensed devices operating in the TV broadcast bands, IEEE Unapproved Draft Std P802.22.1/D6, February 2009.

73. K. Tourki, H.-C. Yang, and M.-S. Alouini, Error-rate performance analysis of incremental decode-and-forward opportunistic relaying, *IEEE Transactions on Communications*, 59 (6), 1519–1524, June 2011.

74. A. Abu-Alkheir, Cooperative cognitive radio networks: Spectrum acquisition and co-channel interference effect, PhD dissertation, Department of Electrical and Computer Engineering, Queen's University, Kingston, Ontario, Canada, February 2013.

75. A. Abu-Alkheir and M. Ibnkahla, A selective reporting strategy for decision-based cooperative spectrum sensing, *IEEE Communications Letters* (submitted for publication), 2014.

76. A. Abu-Alkheir and M. Ibnkahla, Outage performance of incremental relaying in a spectrum sharing environment, *IEEE Wireless Communications Letters* (submitted for publication), 2014.

77. G. Hattab and M. Ibnkahla, Multiband cognitive radio: Great promises for future radio access (long version), Internal Report, Queen's University, WISIP Laboratory, Kingston, Ontario, Canada, December 2013.

4 Cooperative Spectrum Acquisition in the Presence of Interference

4.1 INTRODUCTION

The dynamic spectrum sharing nature of cognitive radio networks (CNRs) leads to nonnegligible levels of cochannel interference (CCI). This CCI, caused by undetected primary users (PUs) or by other CRNs, can result in significant performance losses. Cooperative spectrum acquisition is investigated in this chapter where different sources of interference are considered. Chase-combining hybrid automatic repeat request (HARQ) and regenerative cooperative diversity are explored in the context of spectrum interweave. The chapter also covers cooperative spectrum acquisition in the context of spectrum underlay where interference is considered as a design parameter. The chapter is organized as follows. Section 4.2 focuses on chase-combining HARQ based on the results of [1–6]. Section 4.3 is devoted to regenerative cooperative diversity in the presence of interference based on [1–6]. Section 4.4 covers spectrum underlay where interference level is a design parameter based on the results presented in [7].

4.2 CHASE-COMBINING HARQ

Consider a Two-Terminal CRN where a source S communicates with a destination D in the presence of N sources of CCI, I_1, I_2,..., I_N as shown in Figure 4.1 [2]. To guarantee reliable communication, S uses a chase-combining HARQ protocol with a maximum number of retransmissions of $K - 1$ [8]. This protocol operates as follows. After receiving the first transmission at time n, the destination examines the signal to interference and noise ratio (SINR), denoted ψ_0, against a preset threshold τ. If $\psi_0 \geq \tau$, then D sends an ACK asking S to transmit a new message in the following time slot. Otherwise, it sends a negative acknowledgment (NACK) asking S for a retransmission. After receiving this retransmission, D combines the two replicas using MRC and examines the combined SINR, denoted ψ_1 against τ. If failure persists, the process is repeated until the condition is met or the maximum number of retransmissions is reached.

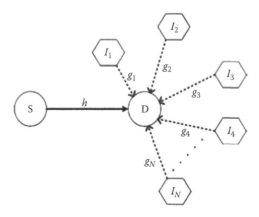

FIGURE 4.1 A two-terminal network with N sources of CCI.

Let $x(n)$ be the transmitted signal during the nth time slot with energy E_x, then $y(n)$ can be written as

$$y(n) = h(n)x(n) + \sum_{i=1}^{N} g_i(n)s_i(n) + w_r(n) \tag{4.1}$$

where

$h(n)$ and $g_i(n)$ are the coefficients of the S–D and I_i–D channels, respectively
$s_i(n)$ is the transmitted signal of I_i with transmission energy E_i
$w_r(n)$ is a zero-mean AWGN with variance σ^2

Since D is unaware of the presence of the CCI, it uses coherent detection to decode the transmitted signal $x(n)$. Assuming perfect knowledge of the channel gain, $h(n)$, D multiplies $y(n)$ by the conjugate of $h(n)$, that is, $h^*(n)$, which yields

$$h^*(n)y(n) = |h(n)|^2 x(n) + h^*(n) \sum_{i=1}^{N} g_i(n)s_i(n) + h^*(n)w_r(n) \tag{4.2}$$

Consequently, by taking the energies of the signal part and the interference plus noise part, the SINR can be expressed as

$$\psi_0 = \frac{E_x |h(n)|^2}{\sum_{i=1}^{N} |g_i(n)|^2 E_i + \sigma^2} = \frac{\gamma_0}{\sum_{i=1}^{N} \zeta_{i,0} + 1} \tag{4.3}$$

where γ_0 and $\zeta_{i,0}$ are the signal to noise ratio (SNR) and the interference to noise ratio (INR) of the S–D and the I_i–D links at the end of the first transmission attempt of $x(n)$, respectively.

After calculating ψ_0, D sends an ACK or a NACK depending on the outcome of the comparison

$$\text{ACK} : \psi_0 \geq \tau$$
$$\text{NACK} : \psi_0 < \tau$$

(4.4)

When a NACK is sent, S retransmits $x(n)$ while D combines $y(n+1)$ with $y(n)$ using MRC. In this case, D perceives a combined SINR of

$$\Psi_1 = \psi_0 + \psi_1$$

(4.5)

where ψ_1 is the SINR of the retransmitted signal, given by

$$\psi_1 = \frac{\gamma_1}{\sum_{i=1}^{N} \zeta_{i,1} + 1}$$

(4.6)

where

$$\gamma_1 = \frac{E_x |h(n+1)|^2}{\sigma^2}$$

(4.7)

and

$$\zeta_{i,1} = \frac{E_i |g_i(n+1)|^2}{\sigma^2}$$

(4.8)

The comparison process in (4.4) is then repeated using Ψ_1.

If failure persists, S keeps retransmitting $x(n)$, a total of $K-1$ times before discarding it.

In general, the SINR of the kth transmission attempt, $k = 0, 1, \ldots, K-1$, can be written as

$$\psi_k = \frac{\gamma_k}{\sum_{i=1}^{N} \zeta_{i,k} + 1}, \quad k = 0, 1, \ldots, K-1$$

(4.9)

4.2.1 PERFORMANCE RESULTS

4.2.1.1 Outage Probability

Here, we look at the performance of this protocol in the presence of arbitrary numbers of CCI sources. In Figure 4.2, we consider the case where $N=3$ sources of

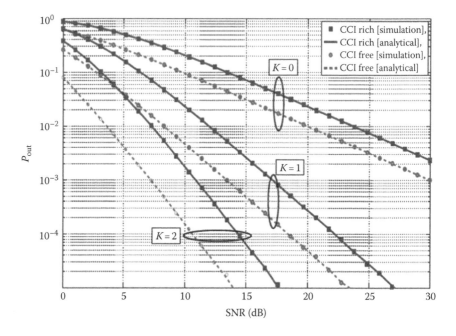

FIGURE 4.2 P_{out} for HARQ as a function of NSR for $K = 0, 1, 2$.

CCI are randomly distributed in the range [$Dref$, $2Dref$], where $Dref$ is the distance between S and D. Each of these CCI sources uses a transmit SNR of 0 dB. In this figure, P_{out} is compared in the presence and in the absence of CCI. It is clear that CCI results in a significant performance loss, regardless of the value of K. However, a larger K still enhances the performance compared to a smaller value.

Now the impact of increasing N for a particular K is examined. Figure 4.3 shows the case of $K = 2$, with N gradually increasing from 0 to 5, assuming a transmit INR for all CCI sources of 5 dB. As can be seen, increasing N worsens the outage performance even when the terminals are located more than $Dref$ distance units from D.

4.2.1.2 Error Probability

In terms of bit error probability (BER), the impact of increasing K is different. As Figure 4.4 shows, increasing K beyond a certain point does not cause significant enhancement to the performance. In this figure, the case of binary phase shift keying (BPSK) modulation is considered with $N = 5$ sources of CCI distributed in the range [$0.5Dref$, $1.5Dref$] and transmitting at 5 dB. Increasing K from 0 to 3 created a performance gain of more than 5 dB, while increasing it from 3 to 6 has made very marginal impact on the performance. This is because the ratio of signals requiring a large number of retransmissions is very small compared to those requiring smaller numbers. Hence, the error probability performance reaches a level of saturation after a certain K.

In Figure 4.5, the impact of increasing N on the performance is illustrated for $K = 1$. Similar to the outage probability case, the figure shows that increasing N even when the transmission SNR is −5 dB causes a significant performance loss.

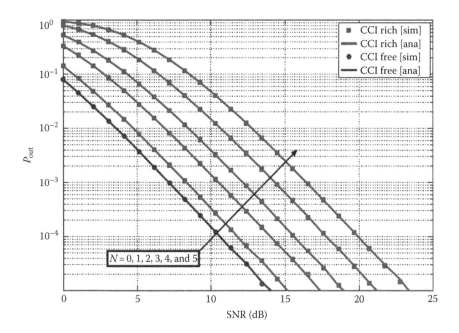

FIGURE 4.3 P_{out} for HARQ as a function of NSR for $K = 2$ and $N = 0,1,\ldots,5$.

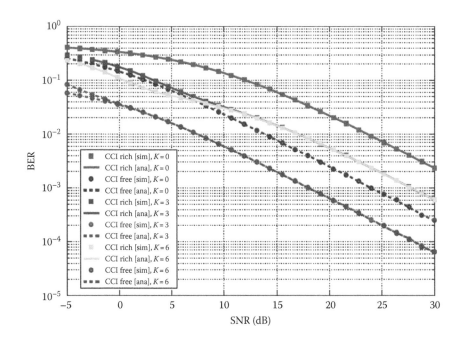

FIGURE 4.4 BER for HARQ as a function of NSR using BPSK modulation for $K=0$, 3, 6 and $N=5$.

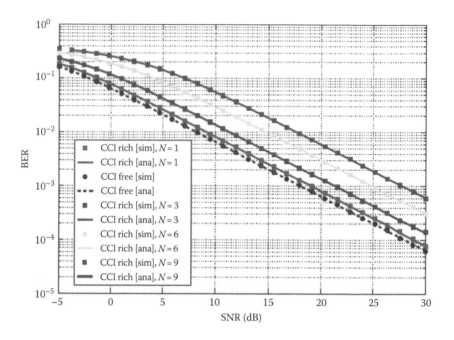

FIGURE 4.5 BER for HARQ as a function of SNR using BPSK modulation for $K = 1$ and $N = 0, 1, 3, 6, 9$.

4.3 REGENERATIVE COOPERATIVE DIVERSITY

Here, a three-terminal relay network is considered with N sources of interference (Figure 4.6) [2].

The systems consist of a source S, a relay R, and a destination D operating in the presence of N dissimilar sources of CCI, $I_1, I_2, \ldots, I_N$. The transmitted signals of S and I_i, denoted by $x(n)$ and $s_i(n)$, have transmission energies of E_x and E_i, respectively, while the AWGN at R and D has zero means and variances of σ_r^2 and σ_d^2, respectively.

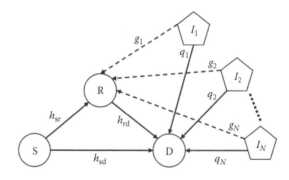

FIGURE 4.6 Three-terminal relay network with N sources of CCI.

The coefficients of the S–R, S–D, R–D, I_k–R, and I_j–D channels are denoted by h_{sr}, h_{sd}, h_{rd}, g_i, and q_i, and are assumed to follow a flat Rayleigh fading model. The receiving terminals are assumed to have no information about g_i and q_i and to have perfect knowledge of h_{sr}, h_{sd}, and h_{rd}.

The transmission of an arbitrary signal $x(n)$ can consume up to two consecutive time slots where the first one is always used by S to broadcast $x(n)$. At the end of this time slot, D and R receive $y_{sd,1}$ and $y_{sr,1}$ given by

$$y_{sd,1} = h_{sd,1}x(n) + \sum_{i=1}^{N} q_{i,1}s_i(n) + w_{d,1} \tag{4.10}$$

$$y_{sr,1} = h_{sr,1}x(n) + \sum_{i=1}^{N} g_{i,1}s_i(n) + w_{r,1} \tag{4.11}$$

where the subscript 1 denotes the first time slot. Using coherent detection, these terminals perceive SINRs of $\psi_{sd,1}$ and ψ_{sr} given by

$$\psi_{sd,1} = \frac{E_x \left| h_{sd,1} \right|^2}{\sum_{i=1}^{N} E_i \left| q_{i,1} \right|^2 + \sigma_d^2} = \frac{\gamma_{sd,1}}{\sum_{i=1}^{N} \zeta_{si,1} + 1} \tag{4.12}$$

$$\psi_{sr} = \frac{E_x \left| h_{sr,1} \right|^2}{\sum_{i=1}^{N} E_i \left| g_{i,1} \right|^2 + \sigma_r^2} = \frac{\gamma_{sr}}{\sum_{i=1}^{N} \rho_i + 1} \tag{4.13}$$

where

$$\gamma_{sd,1} = \frac{E_x \left| h_{sd,1} \right|^2}{\sigma_d^2}, \quad \gamma_{sr} = \frac{E_x \left| h_{sr} \right|^2}{\sigma_r^2}, \quad \zeta_{si,1} = \frac{E_i \left| q_{i,1} \right|^2}{\sigma_d^2}, \quad \rho_i = \frac{E_i \left| g_i \right|^2}{\sigma_r^2}$$

The second time slot can be used to either transmit a new signal, $x(n+1)$ or assist D depending on the employed protocol. Assistance can take the form of a regenerated replica of $x(n)$, denoted by $\hat{x}(n)$, or a retransmission of $x(n)$. Consequently, D receives either $y_{rd,2}$ or $y_{sd,2}$ given by

$$y_{rd,2} = h_{rd,2}\hat{x}(n) + \sum_{i=1}^{N} q_{i,2}s_i(n+1) + w_{d,2} \tag{4.14}$$

$$y_{sd,2} = h_{sd,2}x(n) + \sum_{i=1}^{N} g_{i,2}s_i(n) + w_{d,2} \tag{4.15}$$

Using coherent detection, these two signals contribute additional SINR of either ψ_{rd} or $\psi_{sd,2}$ defined as

$$\psi_{rd} = \frac{\gamma_{rd}}{\sum_{i=1}^{N} \zeta_{ri,2} + 1} \tag{4.16}$$

$$\psi_{sd,2} = \frac{\gamma_{sd,2}}{\sum_{i=1}^{N} \zeta_{si,2} + 1} \tag{4.17}$$

where

$$\gamma_{rd} = \frac{E_x |h_{rd}|^2}{\sigma_d^2}, \quad \gamma_{sd,2} = \frac{E_x |h_{sd,2}|^2}{\sigma_d^2}, \quad \zeta_{ri,2} = \frac{E_i |q_{i,2}|^2}{\sigma_d^2}, \quad \zeta_{si,2} = \frac{E_i |g_{i,2}|^2}{\sigma_d^2}$$

After receiving assistance, D combines the two replicas and makes a final decision.

4.3.1 AVERAGE SPECTRAL EFFICIENCY

From the previous chapter, it is easy to notice that fixed relaying (FR) and selection relaying (SR) relaying protocols operate at a constant spectral efficiency of $0.5Q$ bps/Hz, and incremental relaying (IR) and selective IR (SIR) relaying protocols operate at a spectral efficiency of Q when $\psi_{sd,1} \geq \lambda_d$ and $0.5Q$ when $\psi_{sd,1} < \lambda_d$, where λ_d is the detection threshold.

Consequently, here the average operating spectral efficiency, $\bar{Q}$, is expected to be in the range $[0.5Q, Q]$.

4.3.2 OUTAGE PROBABILITY

It can be obtained knowing that an event of outage occurs when one or more links fail to support the point-to-point spectral efficiency Q.

4.3.3 ERROR PROBABILITY

According to the etiquettes of the FR protocol, D decodes the signal after receiving the two replicas. However, since R also decodes the signal before forwarding it, D may be subject to error propagation. Consequently, the error probability at D can be written as

$$P_{e,FR} = \hat{P}_{e,r} \hat{P}_{e,prop} + \left(1 - \hat{P}_{e,r}\right) \hat{P}_{e,d} \tag{4.18}$$

where $\hat{P}_{e,r}$, $\hat{P}_{e,prop}$, and $\hat{P}_{e,d}$ are the error probability at R, the error propagation probability, and the error probability at D at the end of the second time slot, respectively.

On the other hand, SR protocol imposes a condition on ψ_{sr}, λ_r, below which S retransmits.

Consequently, this protocol experiences an error probability of

$$P_{e,SR} = \left(1 - F_{\psi_{sr}}(\lambda_r)\right)\left[P_{e,r}\,\hat{P}_{e,prop} + \left(1 - P_{e,r}\right)\hat{P}_{e,d}\right] + F_{\psi_{sr}}(\lambda_r)\hat{P}_{e,3} \qquad (4.19)$$

where $P_{e,r}$ is the error probability at R, while $\hat{P}_{e,3}$ is the error probability at D at the end of the second time slot when assistance comes from S. $F_{\psi_{sr}}$ indicates the cumulative distribution function (CDF) function of ψ_{sr}.

To exploit a good S–D link condition, SIR allows assistance only when $\psi_{sd,1} < \lambda_d$. Consequently, it experiences an error probability of

$$P_{e,SIR} = \left(1 - F_{\psi_{sd,1}}(\lambda_d)\right)P_{e,1} + F_{\psi_{sd,1}}(\lambda_d)\left(1 - F_{\psi_{sr}}(\lambda_r)\right)\left[\hat{P}_{e,r}\,P_{e,prop} + \left(1 - \hat{P}_{e,r}\right)P_{e,d}\right]$$

$$+ F_{\psi_{sd,1}}(\lambda_d)F_{\psi_{sr}}(\lambda_r)P_{e,3} \qquad (4.20)$$

where $P_{e,d}$ is the error probability at D at the end of the second time slot when assistance comes from R. Finally, the error probability of IR can be obtained by setting $\lambda_r = 0$ in (4.19), that is,

$$P_{e,SIR} = \left(1 - F_{\psi_{sd,1}}(\lambda_d)\right)P_{e,1} + F_{\psi_{sd,1}}(\lambda_d)\left[\hat{P}_{e,r}\,P_{e,prop} + \left(1 - \hat{P}_{e,r}\right)P_{e,d}\right] \qquad (4.21)$$

4.3.4 Simulation Results

Figure 4.7 shows the average spectral efficiency, $\bar{Q}$, in the presence or absence of CCI where the path loss exponent was taken equal to 4. The following simulation parameters were considered: $D_{sr} = 0.5D_{sd}$, $Q = 4$ bps/Hz subject to the transmissions of two CCI sources, located at $D_{1,d} = D_{1,r} = D_{sd}$, $D_{2,d} = D_{2,r} = 1.5D_{sd}$ and using transmission INRs of −5 and 0 dB, respectively.

A number of interesting observations can be drawn from this computer simulation. First, as expected, CCI degrades the quality of the directly received signal at D and, hence, requires additional assistance, which reduces $\bar{Q}$. Second, when λ_d is chosen larger than ψ_{th}, a lower $\bar{Q}$ is achieved even in the absence of CCI.

Next, we look at the outage probabilities using $\lambda_d = \lambda_r = \psi_{th}$. In Figure 4.8, it is shown that CCI equally affects all four protocols. It also shows that SIR and SR, which have identical outage probability, outperform IR and FR.

Similar conclusions can be drawn for the error probability. In particular, consider a network with $N = 2$ sources of CCI transmitting at 5 and 10 dB and located at $D_{1,d} = 1.5D_{sd}$, $D_{2,d} = 2D_{sd}$, and $D_{1,r} = D_{sd}$, $D_{2,r} = 1.5D_{sd}$, respectively. The network has $D_{sr} = D_{rd} = 0.5D_{sd}$ and operates at $Q = 2$ bps/Hz. The error probability of this network is shown in Figure 4.9.

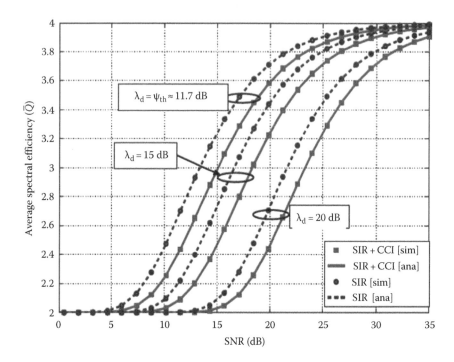

FIGURE 4.7 Average spectral efficiency as a function of the transmission SNR in the presence and absence of CCI effect.

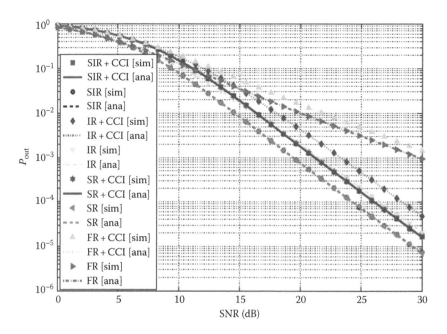

FIGURE 4.8 Outage probabilities in the presence and absence of CCI effect.

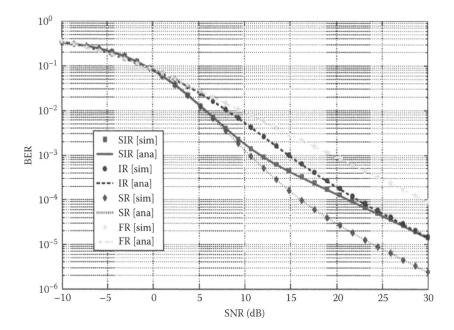

FIGURE 4.9   BER performance in the presence of CCI.

It is shown that the selectivity feature of SR along with its sustained diversity gain allows it to outperform the remaining protocols. The figure also shows that $P_{e,SIR}$ converges to $P_{e,IR}$ at high SNR. This should not be surprising for the two protocols that will eventually be dominated by the direct transmission.

Figure 4.10 shows $P_{e,SIR}$ and the three constituting components: the direct transmission, the relay-assisted transmission, and the retransmission-assisted case. The results indicate that each component dominates the performance over a particular range of SNRs. In particular, while retransmission assisted dominates the performance for very low SNR, the relay-assisted transmission dominates for medium-to-high values. Over these two ranges, SIR enjoys diversity gain. However, this gain is gradually lost as SNR increases because the direct transmission becomes the dominant factor.

4.3.5   CONCLUSIONS

This section studied the impact of CCI on the performance of chase-combining HARQ, DF FR, SR, IR, and SIR. Through deriving closed-form expressions for the various performance metrics, it was possible to quantify the impact of this unavoidable channel impairment on the performance of these protocols. It was shown that their performance suffers from a nonnegligible loss when the CCI sources are within the transmission range of the source. It was also shown that these losses can be mitigated through increasing the transmission SNR of the source terminal.

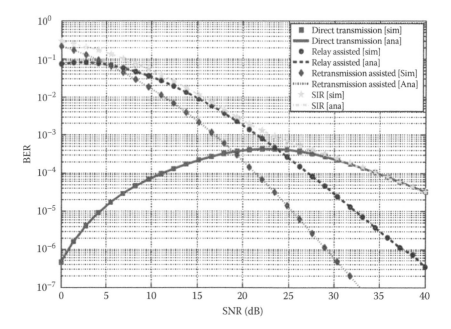

FIGURE 4.10 Total BER for SIR and its three components (direct transmission, relay-assisted transmission, and retransmission assisted).

4.4   SPECTRUM UNDERLAY

Underlay spectrum sharing is one of the proposed access paradigms in CR. It allows SUs to coexist with PUs at the same time, using the same band and in the same region. This is performed through transmit power control. The SU terminal adjusts its transmission power in order to guarantee an acceptable interference to PUs.

The most important issue in underlay access is not to harm the PU link. To do so, a perfect channel state information (CSI) of all channels is required, especially in a full interference environment. However, this arises privacy concerns for the PU. Thus, a suitable scenario in this case would be to consider different levels of knowledge of these channels at the SU stations.

This section discusses an efficient relaying scheme for underlay spectrum sharing. With the introduction of relaying schemes, a considerable enhancement in the performance is expected. In particular, IR is suitable for the context of underlay because it provides a spatial diversity gain. The performance of this scheme is studied in terms of outage and error probabilities. A peak transmit power is imposed at the SU. Furthermore, this section considers the interference caused by the PU on the SU (which has been ignored in previous studies, see, e.g., [9]).

4.4.1   UNDERLAY ACCESS

The main advantage of the underlay paradigm is the simultaneous access to the spectrum by PUs and SUs. Basically, the SU is allowed to transmit if the interference caused to the PU does not exceed a given threshold. The potential interference at the PU will be limited regardless of the state of the channel of the primary–primary link.

For instance, in the case where the primary–primary channel is good, this will prohibit the SU from transmitting with higher rate. In the case where the primary–primary channel is bad, with the presence of the SU, the received SINR will be also bad. Therefore, another criterion is needed to optimize the overall resources.

A good criterion would be to satisfy a minimum SINR at the PU [9,10]. In this case, the performance will be better for both PU and SU. In particular, if the primary–primary link is good, the interference will be relaxed to allow the SU transmit with higher power. In the case where the primary–primary channel is bad, the SUs transmitted power will be reduced to keep the SINR at the accepted level.

Therefore, in this section, we will assume that the SU will transmit when the PU transmits.

Here, the capacity of the SU is studied for a number of scenarios concerning the level of knowledge of the propagation channel characteristics based on the results of [7]. The best performance corresponds to the full knowledge of all channels while the poorest corresponds to a statistical knowledge.

The system has a Z-shaped configuration, as illustrated in Figure 4.11. The channels are assumed to experience Rayleigh fading.

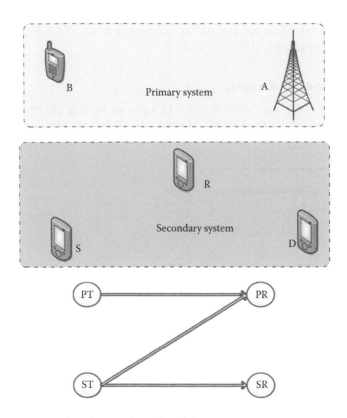

FIGURE 4.11   The Z-interference channel model.

We denote by Ω_p, Ω_s, and Ω_{sp} the average channel gains of the primary link, secondary link, and the secondary–primary link.

4.4.2 RELAYING USING DECODE AND FORWARD (DF)

Among the valuable works in this context is the work of Bao and Bac [11]. Under an interference constraint, the authors studied the performance of relaying in underlay spectrum sharing paradigm. Fixed relaying, selective relaying as well as IR were investigated in terms of outage probability and bit error rate. Results show that IR outperforms the other relaying schemes. However, the link between the primary transmitter and the secondary receiver was assumed to be interference free. Researchers, in the literature, often use this assumption motivated by the fact that the primary transmitter is usually located far from the receiver. But, in dense networks or even in normal cases, this is not usually the case. Furthermore, a perfect knowledge of all channels was assumed, which is not always the case in practice.

4.4.3 RELAYING USING AF

The amplify and forward scheme attracted less consideration in the literature. In [12], the dual-hop AF was investigated where the amplifying gain depends not only on the source–relay link but also on the interference to the PU. Selection relaying was also studied recently [13].

4.4.4 INCREMENTAL RELAYING

The flowchart for IR is shown in Figure 4.12. Here, an SU pair and PU pair concurrently use the spectrum (Figure 4.13). The PU pair consists of a PU transmitter A and

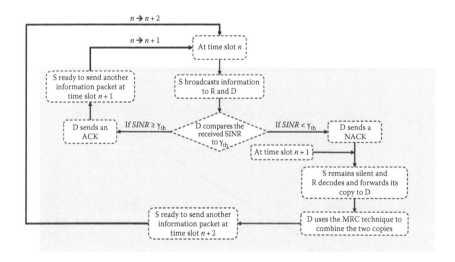

FIGURE 4.12 Flowchart of the IR technique.

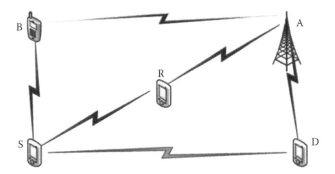

FIGURE 4.13 Illustration of the transmission during the first time slot.

a PU receiver B, while the SU pair includes an SU source S, an SU relay R, and an SU destination D. All signals are assumed to be affected by independent nonidentical Rayleigh fading channels.

4.4.4.1 Time Slots

Following the etiquette of IR, the communication between the source and the relay may require two time slots.

4.4.4.1.1 First Time Slot

During a given time slot n, the secondary user S broadcasts its information to the relay R and the secondary destination D with a transmission power $P_{t,s}(n)$. However, due to the presence of the communication of the PU pair (primary transmitter A transmits to primary receiver B), an additive interference will be received at R and D in addition to the scattered version of the secondary user's signal.

Mathematically, the received signals by D and R are written as

$$y_{s,d}(n) = h_{s,d}(n)\sqrt{P_{t,s}(n)}x_s(n) + h_{a,d}(n)\sqrt{P_p}\,x_a(n) + w_d(n) \tag{4.22}$$

$$y_{s,r}(n) = h_{s,r}(n)\sqrt{P_{t,s}(n)}\,x_s(n) + h_{a,r}(n)\sqrt{P_p}\,x_a(n) + w_r(n) \tag{4.23}$$

where
   $x_s(n)$ and $x_a(n)$ are the transmitted signals with unit energy
   $h_{s,d}(n)$ is the fading channel gain between S and D
   $h_{s,r}(n)$ is the fading channel gain between S and R
   $h_{a,d}(n)$ is the fading channel gain between A and D
   $h_{a,r}(n)$ is the fading channel gain between A and R during the time slot n
   $P_{t,s}(n)$ is the transmission power sent by S during the time slot n
   P_p is the transmission power sent by A, which is assumed to be fixed during the
      whole communication
   $w_d(n)$ and $w_r(n)$ are AWGN noises with zero mean and variances σ_d^2 and σ_r^2,
      respectively

D is assumed to have a perfect knowledge of channel $h_{s,d}(n)$. So, it employs a coherent detection by multiplying the received signal by the conjugate value of the channel gain $h_{s,d}(n)$.

The SINR at D can be expressed as

$$\gamma_{s,d}(n) = \frac{P_{s,r}(n)g_{s,d}(n)}{P_p g_{a,d}(n) + \sigma_d^2} \tag{4.24}$$

where

$$g_{s,d}(n) = \left| h_{s,d}(n) \right|^2$$

4.4.4.1.2   Second Time Slot

Figure 4.14 describes the communication during the second time slot $n + 1$ if needed. In fact, at the end of the first time slot n, D compares the received SINR $\gamma_{s,d}(n)$ given in (4.24) to a given threshold γ_{th}, which represents the minimum received SINR for which D could, successfully, decode the received signal. If $\gamma_{s,d}(n) \geq \gamma_{th}$, D sends positive feedback (ACK) to indicate that the signal was successfully decoded. After receiving the ACK, S transmits another information signal during time slot $n+1$ and R remains silent. However, if $\gamma_{s,d}(n) < \gamma_{th}$, D sends a NACK asking for assistance. In this case, R sends a regenerative replica of the signal in time slot $n+1$. Hence, the received signal at D is

$$y_{r,d}(n+1) = h_{r,d}(n+1)\sqrt{P_{s,r}(n+1)}x_r(n+1)$$

$$+ h_{a,d}(n+1)\sqrt{P_p}\,x_a(n+1) + w_d(n+1) \tag{4.25}$$

where $x_r(n+1)$ is the transmitted symbol at the time slot $n+1$ (assumed to have unit energy).

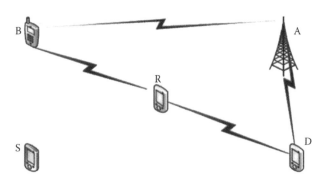

FIGURE 4.14   Illustration of the transmission during the second time slot.

4.4.4.2 MRC Combination at the Receiver

An MRC combiner is used at the receiver. Applying this combining rule to the system, D combines the two received signals over the two time slots n and $n+1$. Thus, in MRC combining, the total SINR at D is given by

$$\gamma_{s,d} = \gamma_{s,d}(n) + \gamma_{r,d}(n+1) = \frac{P_{s,r}(n)g_{s,d}(n)}{P_p g_{a,d}(n) + \sigma_d^2} + \frac{P_{s,r}(n+1)g_{s,d}(n+1)}{P_p g_{a,d}(n+1) + \sigma_d^2} \qquad (4.26)$$

4.4.4.3 Constraints

This section describes the different constraints to be considered. These constraints are mainly the SINR threshold at the secondary receiver, the peak transmit power at the secondary transmitter, and the threshold at the secondary receiver, which will control the relay's assistance.

4.4.4.3.1 SINR Constraints at the Primary Receiver

For the primary receiver B, the received signals at time slots n and $n+1$ are

$$y_{a,b}(n) = h_{a,b}(n)\sqrt{P_p}x_a(n) + h_{s,b}(n)\sqrt{P_{t,s}(n)}x_s(n) + w_b(n) \qquad (4.27)$$

$$y_{a,b}(n+1) = h_{a,b}(n+1)\sqrt{P_p}x_a(n+1) + h_{r,b}(n+1)\sqrt{P_{t,r}(n+1)}x_r(n+1) + w_b(n+1) \qquad (4.28)$$

where

w_b is an AWGN with zero mean and variance σ_b^2
$h_{a,b}$, $h_{s,b}$, and $h_{r,b}$ are the channel gains between A and B, S and B, and R and B, respectively

The path loss exponent n is assumed the same for all links. To reflect the effect of the path loss, the SNRs at R and D are assumed to be related to those at S (noted as SNR) as

$$\Omega_{s,d} = \left(\frac{d_{\text{ref}}}{d_{s,d}}\right)^n \text{SNR} \qquad (4.29)$$

$$\Omega_{s,r} = \left(\frac{d_{\text{ref}}}{d_{s,r}}\right)^n \text{SNR} \qquad (4.30)$$

$$\Omega_{r,d} = \left(\frac{d_{\text{ref}}}{d_{r,d}}\right)^n \text{SNR} \qquad (4.31)$$

where d_{ref}, $d_{\mathrm{s,r}}$, and $d_{\mathrm{r,d}}$ are a reference distance, the distance between S and R, and the distance between R and D, respectively.

Similarly, we define

$$\Omega_{\mathrm{a,b}} = \left(\frac{d_{\mathrm{ref}}}{d_{\mathrm{a,b}}}\right)^{n} \mathrm{SNR}_{\mathrm{a,b}} \tag{4.32}$$

$$\Omega_{\mathrm{s,b}} = \left(\frac{d_{\mathrm{ref}}}{d_{\mathrm{s,b}}}\right)^{n} \mathrm{SNR}_{\mathrm{s,b}} \tag{4.33}$$

$$\Omega_{\mathrm{r,d}} = \left(\frac{d_{\mathrm{ref}}}{d_{\mathrm{r,b}}}\right)^{n} \mathrm{SNR}_{\mathrm{r,b}} \tag{4.34}$$

The primary receiver B suffers from the presence of the CCI. To limit the interference level, we propose that during the two time slots, the received SINR, $\gamma_{\mathrm{b}}(n)$ and $\gamma_{\mathrm{b}}(n+1)$, should be kept above a certain threshold γ_{T}. This threshold corresponds to a minimum spectral efficiency R_1 expressed in bit/s/Hz, such that

$$\gamma_{\mathrm{T}} = 2^{R_1} - 1. \tag{4.35}$$

Mathematically, we have the following conditions:

$$\gamma_{\mathrm{b}}(n) = \frac{P_{\mathrm{p}} g_{\mathrm{a,b}}(n)}{P_{\mathrm{s,d}}(n) g_{\mathrm{s,b}}(n) + \sigma_{\mathrm{b}}^2} \geq \gamma_{\mathrm{T}} \tag{4.36}$$

$$\gamma_{\mathrm{b}}(n+1) = \frac{P_{\mathrm{p}} g_{\mathrm{a,b}}(n+1)}{P_{\mathrm{r,d}}(n) g_{\mathrm{r,b}}(n+1) + \sigma_{\mathrm{b}}^2} \geq \gamma_{\mathrm{T}} \tag{4.37}$$

where

$$g_{\mathrm{a,b}} = \left|h_{\mathrm{a,b}}\right|^2$$

$$g_{\mathrm{r,b}} = \left|h_{\mathrm{r,b}}\right|^2$$

To achieve this, S and R should choose their transmission powers such that the following equations are satisfied:

$$P_{\mathrm{s,d}}(n) \leq \frac{P_{\mathrm{p}} g_{\mathrm{a,b}}(n) - \sigma_{\mathrm{b}}^2 \gamma_{\mathrm{T}}}{\gamma_{\mathrm{T}} g_{\mathrm{s,b}}(n)} \tag{4.38}$$

$$P_{r,d}(n+1) \leq \frac{P_p g_{a,b}(n+1) - \sigma_d^2 \gamma_T}{\gamma_T g_{r,b}(n+1)} \qquad (4.39)$$

We will assume perfect CSI knowledge at the transmitters (i.e., $h_{s,r}(n)$ is known at R, and $h_{s,d}(n)$ and $h_{r,d}(n+1)$ are known at D for coherent detection).

With the presence of the CCI, the primary receiver could experience an outage event with a probability $P_{out,min}$ that we define as follows:

$$\Pr\left(\gamma_b(n) < \gamma_T\right) = P_{out,min} \qquad (4.40)$$

$$\Pr\left(\gamma_b(n+1) < \gamma_T\right) = P_{out,min} \qquad (4.41)$$

where $P_{out,min}$ is assumed to be small.

4.4.4.3.2   Peak Transmit Power

According to expressions (4.38) and (4.39), the transmission powers $P_{t,s}(n)$ and $P_{t,r}(n+1)$ could be very high if the channel gain at the primary system, $g_{a,b}$, is very strong and the secondary–primary channel gains $g_{s,b}(n)$ and $g_{r,b}(n+1)$ are very weak. Hence, we will impose a maximum transmission power P_m that is not exceeded by the source and the relay [7,14] (as power limitation is required in most standards).

4.4.4.3.3   Minimum Received SINR at the Secondary Receiver

This constraint controls the relaying process. The received SINR during the first time slot n is compared to a threshold γ_{th}. This threshold is fixed according to a minimum required spectrum efficiency R_0. The threshold is related to the spectrum efficiency as $\gamma_{th} = 2^{R_0} - 1$

The value of γ_{th} has to strike a balance between the error probability and the spectrum efficiency. In fact, having a large value of γ_{th} will reduce the error probability, since in this case, D will ask more often for assistance. But, by doing so, the communication will require two time slots instead of one.

By analyzing the extreme values of γ_{th}, we notice that a minimum value of 0 can eliminate the use of the relay (in this case, we can have the best spectrum efficiency). On the other side, an infinite value of γ_{th} corresponds to the case where the relay always assists the secondary receiver.

4.4.4.4   Channel Knowledge

In underlay spectrum sharing, the challenge that faces the secondary network is to have accurate CSI knowledge about the secondary–primary and the primary–primary channels.

4.4.4.4.1   Primary–Primary Channel

It is difficult to estimate the primary–primary channel; therefore, only a statistical knowledge of the channel gain will be assumed, which is more practical in this case.

4.4.4.4.2 *Secondary–Primary Channels*

In practice, it is difficult to estimate the channel gain between the secondary and primary systems due to the absence of direct communications between them. Based on that, we will draw two scenarios of knowledge of the CSI, which are as follows:

> *Scenario* 1: The secondary network has only a statistical knowledge of the secondary–primary channels that affect the primary receiver, that is, $\Omega_{s,b}$ and $\Omega_{r,b}$ as well as the two channels that affect the performance at receivers D and R, that is, $\Omega_{a,d}$ and $\Omega_{a,r}$.
> *Scenario* 2: The secondary network has full knowledge of the channel gains that affect B's performance, that is, $h_{s,b}(n)$ and $h_{r,b}(n+1)$, as well as the channel gains affecting D's performance, that is, $h_{a,d}(n)$, $h_{a,d}(n+1)$, and $h_{a,r}(n)$.

4.4.5 SIMULATION RESULTS AND ILLUSTRATIONS

Figure 4.15 shows the impact of R_0 and thus (γ_{th}) on the secondary receivers performance. A high required rate increases the outage probability. The following parameters are considered in the simulation: $d_{s,b} = d_{r,b} = d_{a,r} = d_{a,d} = d_{ref}$ and the average SNR at the primary channel $\Omega_{a,b} = 25\,dB$, the average SNR of the secondary–primary channels is equal to 0 dB, the spectrum efficiency at B is $R_1 = 3$.

A comparison between the two scenarios is illustrated in Figure 4.16. It can be seen that, for the same parameters, the second scenario outperforms the first. In the simulation, the following parameters were used: $d_{s,b} = d_{r,b} = d_{a,r} = d_{a,d} = d_{a,b} = d_{ref}$, the average SNR at the primary channel $\Omega_{a,b} = 25\,dB$, the average SNR of the secondary–primary channels is equal to 0 dB, the spectrum efficiency at B is $R_1 = 2$ bit/s/Hz,

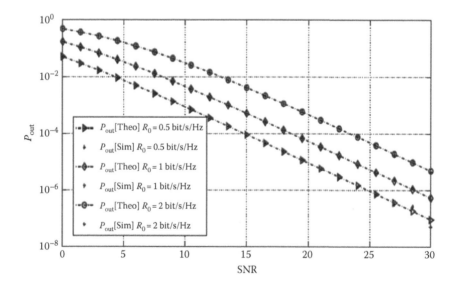

FIGURE 4.15 Outage probability for different rates at the secondary receiver (second scenario), secondary–primary SNR equals 0 dB.

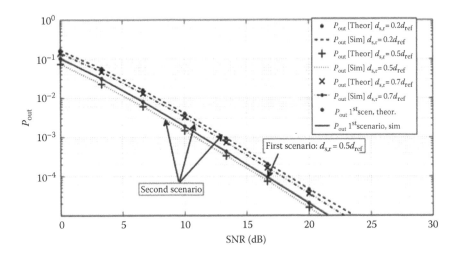

FIGURE 4.16 Comparison between the first and second scenarios for IR protocol, secondary–primary SNR equals 0 dB.

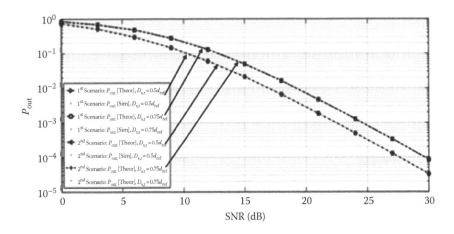

FIGURE 4.17 Comparison between the first and second scenario for IR protocol, secondary–primary SNR equals 10 dB.

$R_0 = 0.5$ bit/s/Hz. Note that for the first scenario $d_{s,r} = 0.5d_{ref}$, for the second scenario $d_{s,r}$ took several values ranging from $0.2d_{ref}$ to $0.7d_{ref}$.

Figure 4.17 shows that for strong average SNR at the primary receiver, the first scenario achieves the same performance as the second scenario when the average SNR of the secondary–primary channel is equal to 10 dB. The used parameters are as follows: $d_{s,b} = d_{r,b} = d_{a,r} = d_{a,d} = d_{ref}$, $d_{a,b} = 0.5d_{ref}$ and the average SNR at the primary channel $\Omega_{a,b} = 25$ dB, the average SNR of the secondary–primary channels is equal to 10 dB, the spectrum efficiency at B is $R_1 = 1$ bit/s/Hz, $R_0 = 1$ bit/s/Hz. Note that for the second scenario, $d_{s,r}$ took the following values: $0.5d_{ref}$ and $0.75d_{ref}$.

4.4.6   ERROR PROBABILITY

According to the IR protocol described earlier, two modes of operation could be distinguished. The first mode corresponds to the normal operation where the received SINR from the direct link is above the threshold γ_{th}. This is denoted by M_{dir}. The second mode corresponds to the case when the destination asks for the assistance from the relay and combines two received signals. This is denoted by M_{div}.

The average error probability can be expressed as

$$P_e = \Pr\left[M_{dir}\right]P_{e,dir} + \Pr\left[M_{div}\right]P_{e,div} \tag{4.42}$$

where

$\Pr[M_{dir}] = \Pr[\gamma_{s,d}(n) \geq \gamma_{th}]$
$\Pr[M_{div}] = \Pr[\gamma_{s,d}(n) < \gamma_{th}]$
$P_{e,dir}$ is the error probability under M_{dir}
$P_{e,div}$ is the error probability under M_{div}

Moreover, $P_{e,dir}$ can be expressed as

$$P_{e,div} = P_{e|r}\left[1 - P_{e|d}\right] + P_{e|d}\left[[1 - P_{e|r}]\right] \tag{4.43}$$

where

$P_{e|r}$ is the error probability at the relay
$P_{e|d}$ is the error probability after performing MRC combining [7]

Figure 4.18 presents the average error probability for binary frequency shift keying (BFSK) and BPSK modulations. The first scenario was considered for mathematical

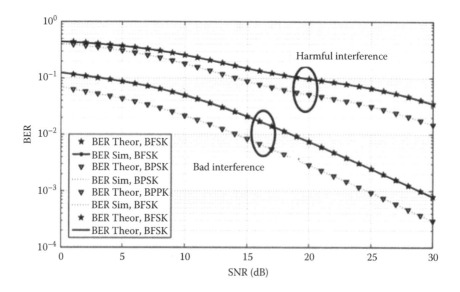

FIGURE 4.18   BER for different modulation schemes.

convenience (a statistical knowledge of the primary–secondary channels and the primary–primary channel). The different distances were taken equal to d_{ref}. For the *harmful interference* case, the different SNRs were taken equal to 15 dB. In this case, the channel gains between primary and secondary systems have high SNRs. Thus, the secondary system will send with lower power, which will result in a poor performance. For the *bad interference* case, the different SNRs were taken equal to 0 dB. The channel gains between primary and secondary systems were so low that the secondary transmitter could send with much higher power, which resulted in better performance. Note that the performance of BPSK is better than BFSK.

4.5 CONCLUSION

This chapter investigated cooperative spectrum acquisition in the presence of interference. The system included a transmitting and receiving SU and PU pairs in addition to other sources of interference. A number of relaying stations have been used to enhance performance. It has been shown that interference can highly affect the system's performance and that the design parameters are different from the interference-free case.

REFERENCES

1. A. Alkheir and M. Ibnkahla, Selective cooperative spectrum sensing in cognitive radio networks, in *Proceedings of the IEEE Global Telecommunications Conference (GLOBECOM'11)*, December 2011, pp. 1–5.
2. A. Abu-Alkheir, Cooperative cognitive radio networks: Spectrum acquisition and co-channel interference effect, PhD dissertation, Department of Electrical and Computer Engineering, Queen's University, Kingston, Ontario, Canada, February 2013.
3. A. Abu-Alkheir and M. Ibnkahla, An accurate approximation of the exponential integral function using a sum of exponentials, *IEEE Communications Letters*, 17 (7), 1364–1367, July 2013.
4. A. Abu-Alkheir and M. Ibnkahla, Impact of co-channel interference and imperfect channel estimation on decode and forward selection incremental relaying, *IEEE Transactions on Wireless Communications* (submitted for publication), 2014.
5. A. Abu-Alkheir and M. Ibnkahla, A selective reporting strategy for decision-based cooperative spectrum sensing, *IEEE Communications Letters* (submitted for publication), 2014.
6. A. Abu-Alkheir and M. Ibnkahla, Outage performance of incremental relaying in a spectrum sharing environment, *IEEE Wireless Communications Letters* (submitted for publication), 2014.
7. B. Khalfi, Performance analysis of underlay spectrum sharing in cooperative diversity networks, MSc dissertation, SupCom, Tunisia, and Internal Report, WISIP Lab, Queen's University, Kingston, Ontario, Canada, August 2012.
8. D. Chase, Code combining—A maximum-likelihood decoding approach for combining an arbitrary number of noisy packets, *IEEE Transactions on Communications*, 33 (5), 385–393, May 1985.
9. Z. Rezki and M.-S. Alouini, Ergodic capacity of cognitive radio under imperfect channel-state information, *IEEE Transactions on Vehicular Technology*, 61 (5), 2108–2119, June 2012.
10. P. Dmochowski, H. Suraweera, P. Smith, and M. Shafi, Impact of channel knowledge on cognitive radio system capacity, in *Vehicular Technology Conference Fall (VTC 2010-Fall)*, September 2010, pp. 1–5.

11. V. N. Q. Bao and D. H. Bac, A unified framework for performance analysis of DF cognitive relay networks under interference constraints, in *International Conference on ICT Convergence (ICTC)*, September 2011, pp. 537–542.
12. L. P. Tuyen and V. N. Q. Bao, Outage performance analysis of dual-hop AF relaying system with underlay spectrum sharing, in *14th International Conference on Advanced Communication Technology (ICACT 2012)*, February 2012, pp. 481–486.
13. D. Li, Outage probability of cognitive radio networks with relay selection, *IET Communications*, 5 (18), 2730–2735, 2011.
14. M. Khojastepour and B. Aazhang, The capacity of average and peak power constrained fading channels with channel side information, in *IEEE Wireless Communications and Networking Conference*, vol. 1, March 2004, pp. 77–82.

5 Spectrum Sensing
Performance Measures and Design Trade-Offs

5.1 INTRODUCTION

The performance measures in cognitive radio networks are very important as they dictate the performance of the overall network [1,2]. In this chapter, we categorize the performance measures of spectrum sensing techniques into two broad categories. In the first one, we analyze the main metrics that test the performance of spectrum sensing; in the second part, we focus on testing the throughput of multiband cognitive radio networks (MB-CNRs). We discuss the different trade-offs that need to be taken into account during the design and operating phases of sensing techniques based on the findings of [1,2].

5.2 RECEIVER OPERATING CHARACTERISTICS

The *receiver operating characteristic* (ROC) is probably the most common performance metric in spectrum sensing. It is simply a plot of the probability of detection, P_D, versus the probability of false alarm, P_{FA}. Both single-band and multiband spectrum sensing cases are considered.

5.2.1 SINGLE BAND

For single-band CR, P_D is simply the probability that the secondary user (SU) correctly detects the primary user (PU) when it is present in a given band (true positive). Hence, for the test given in Chapter 2 Equation 2.2, P_D is expressed as

$$P_D = \Pr(T(y) \geq \lambda | H_1). \tag{5.1}$$

On the other hand, P_{FA} is the probability that the SU incorrectly decides the presence of PUs albeit they are actually idle (false positive). This is expressed as

$$P_{FA} = \Pr(T(y) \geq \lambda | H_0). \tag{5.2}$$

It is desirable to have high probability of detection and low probability of false alarm. The former guarantees minimal interference with the PU, and the latter guarantees throughput improvements for the secondary users. Nevertheless, a trade-off between these two is inevitable. For example, if the SU has a full knowledge of the

PU's transmitted signal, then x is deterministic, and hence, for the given model in Chapter 2 Equation 2.1, we have

$$H_0: \quad T(y) \text{ is } N\left(0, \sigma^2 \|x\|^2\right)$$

$$H_1: \quad T(y) \text{ is } N\left(\|x\|^2, \sigma^2 \|x\|^2\right)$$

(5.3)

With direct computations of (5.1) and (5.2), we have [3, ch. III]

$$P_{FA} = Q\left(\frac{\lambda}{\sqrt{\sigma^2 \|x\|^2}}\right)$$

(5.4a)

$$P_D = Q\left(\frac{\lambda - \|x\|^2}{\sqrt{\sigma^2 \|x\|^2}}\right)$$

(5.4b)

where $Q(.)$ is the complementary distribution function (CDF) of the standard Gaussian.

With direct computation of (5.4a) and (5.4b), we have [3, ch. III]

$$P_D = Q\left(Q^{-1}(P_{FA}) - \sqrt{N}\gamma\right)$$

(5.5)

where the SNR $\gamma = \|x\|^2/\sigma^2$.

On the other hand, when the SU does not have prior knowledge about x, we can assume it follows a Gaussian distribution (i.e., $N(0, \|x\|^2)$). If the energy detector is used, we have

$$H_0: \quad T(y) \text{ is } \chi_N^2$$

$$H_1: \quad T(y) \text{ is } \frac{\|x\|^2 + \sigma^2}{\sigma^2} \chi_N^2$$

(5.6)

where χ_N^2 denotes a central chi-square distribution with N degrees of freedom. It can be shown that [3, ch. III]

$$P_{FA} = \frac{\Gamma\left(\dfrac{N}{2}, \dfrac{\lambda}{2}\right)}{\Gamma\left(\dfrac{N}{2}\right)}$$

(5.7a)

$$P_{\mathrm{D}} = \frac{\Gamma\left(\dfrac{N}{2}, \dfrac{\lambda\sigma^2}{2(\|x\|^2 + \sigma^2)}\right)}{\Gamma\left(\dfrac{N}{2}\right)} \tag{5.7b}$$

where $\Gamma(.)$ and $\Gamma(.,.)$ are the complete and incomplete gamma functions, respectively [4].

Using the central limit theorem, it can be shown that [3, ch. III]

$$P_{\mathrm{D}} = Q\left(\frac{1}{\sqrt{2\gamma + 1}}\left(Q^{-1}(P_{\mathrm{FA}}) - \sqrt{N}\gamma\right)\right) \tag{5.8}$$

Figure 5.1 illustrates the performance of three main detectors. The coherent detector, the energy detector, and a feature detector that exploits the second-order statistics of an OFDM signal under two different scenarios. In the first scenario, the SU has knowledge about the number of useful symbols in the OFDM block, and in the second scenario, it has an additional knowledge about the cyclic prefix (CP) (see [5] for derivations). We observe that the best performance is attained by the coherent detector since we have full knowledge of x. Second, the feature detector has an excellent performance when we exploit more features such as the CP. Third, the energy detector has a poor performance at the low SNR region. This region is of significant importance because the PU signals might be weak at the SU receiver, and to reliably detect them, it must be equipped with a detector that performs well under low SNR conditions.

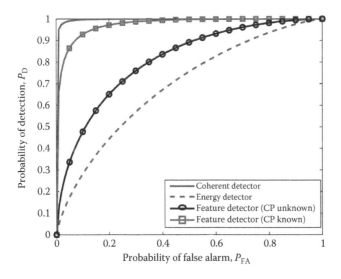

FIGURE 5.1 ROC curves for different detectors in single-band framework ($SNR = -10$ dB).

5.2.2 COOPERATIVE SPECTRUM SENSING

In cooperative spectrum sensing, let $P_D^{(i)}$ and $P_{FA}^{(i)}$ denote the probability of detection and false alarm of the ith SU, respectively. In hard combining, if we have a k out of K rule, then the overall detection and false alarm probabilities are, respectively,

$$Q_D = \sum_{q=k}^{K} \binom{K}{q} \left\{ \prod_{i=1}^{q} P_D^{(i)} \times \prod_{j=1}^{K-q} \left(1 - P_D^{(j)}\right) \right\} \quad (5.9a)$$

$$Q_{FA} = \sum_{q=k}^{K} \binom{K}{q} \left\{ \prod_{i=1}^{q} P_{FA}^{(i)} \times \prod_{j=1}^{K-q} \left(1 - P_{FA}^{(j)}\right) \right\} \quad (5.9b)$$

For example, for OR-logic rule, where $k = 1$, we have

$$Q_D^{OR} = \prod_{i=1}^{K} \left(1 - P_D^{(i)}\right) \quad (5.10)$$

For AND-logic rule, where $k = K$, we have

$$Q_D^{AND} = \prod_{i=1}^{K} P_D^{(i)} \quad (5.11)$$

Q_{FA} can be obtained for these special cases in a similar fashion.

Figure 5.2 shows the ROC curves of an energy detector with different number of SUs. We assume that all SUs have identical P_{FA} and P_D. It is observed that increasing the number of cooperating SUs improves the performance. Also, the OR-logic rule achieves a better detection performance compared to the AND-logic rule.

For soft combining, the detection and false alarm probabilities must be derived explicitly for a given model. For instance, the probability of detection of T in Chapter 3 Equation 3.3 is

$$P_D = \Pr\left(\sum_{k=1}^{K} c_k T_k(y) > \lambda \middle| H_1 \right) \quad (5.12)$$

assuming a total of K SUs (i.e. the summation starts with 1 and not 0). Thus, once we find the probability distribution of T, we can use the preceding expression to find P_D, and a similar procedure is required to find P_{FA}.

5.2.3 MULTIBAND COGNITIVE RADIO

Unlike single-band spectrum sensing, there is no unified definition for the detection and false alarm probabilities when multiple bands are considered. Intuitively, one

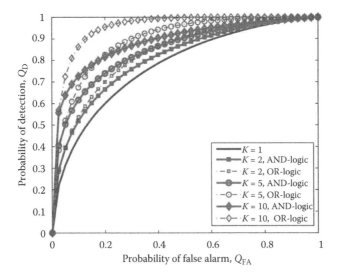

FIGURE 5.2 ROC curves with different number of cooperating SUs ($SNR = -10$ dB).

can calculate these probabilities for each subband individually. That is, we compute $P_{D,m}$ for $m = 1, 2, ..., M$. A naive approach is to average the performance of all bands together, that is,

$$P_D^M = \frac{1}{M} \sum_{m=1}^{M} P_{D,m} \tag{5.13}$$

Nevertheless, averaging may lead to incorrect insights especially when there are outliers (i.e., if a single channel has a very low detection probability, the average detection probability is most probably going to be low). A more robust approach could be to use a normalized weighting average such as

$$P_D^M = \sum_{m=1}^{M} a_m P_{D,m} \tag{5.14}$$

where a_m is a weighting factor for subband, m, subject to

$$\sum_{m=1}^{M} a_m = 1 \tag{5.15}$$

These weights could reflect the sensitivity or the importance of a channel. For instance, if a channel does not require tight interference requirements, one can assign a low weight to that channel such that the performance report of that channel does not have significant impact on the overall performance. Clearly, when the weights are equal, (5.14) reduces to (5.13).

In [6], another definition of these probabilities is proposed such that

$$P_M^D = \Pr\left(\text{at least one band detected as occupied}\,\middle|\,H_1^M\right) \tag{5.16a}$$

$$P_M^{FA} = \Pr\left(\text{at least one band detected as occupied}\,\middle|\,H_0^M\right) \tag{5.16b}$$

where H_1^M and H_0^M denote the events that all channels are busy and all channels are vacant, respectively. In [7], the false alarm probability is defined as the probability of deciding that all the channels are busy even though there exists at least one vacant channel. If we let $H_0^{(i)}$ denote that there exist at least i free channels for $i = 1, \ldots, M$, then the probability of false alarm can be rewritten as

$$P_M^{FA} = \Pr\left(H_1^M\,\middle|\,\bigcup_{i=1}^{M} H_0^{(i)}\right) \tag{5.17}$$

which can be also expressed as

$$P_M^{FA} = \frac{\sum_{i=1}^{M} \Pr\left(H_1^M \bigcap H_0^{(i)}\right)}{\Pr\left(\bigcup_{i=1}^{M} H_0^{(i)}\right)} \tag{5.18}$$

While the previous measures assume perfect knowledge of the subchannel boundaries, this is not the case in edge detection (i.e., in wavelet detection) where the performance of estimating these edges is also necessary. The authors in [8] propose new measures to test the performance of the edge detection algorithms. If we denote $N_B = M + 1$ to be the actual number of subchannels' boundaries in a system and $\hat{N}_B$ to be the number of boundaries that are correctly detected, then the probability of miss detecting an actual boundary is

$$P_{ME} = \frac{N_B - \hat{N}_B}{N_B} \tag{5.19}$$

and the probability of detecting a false edge is given by

$$P_{FE} = \frac{N_T - \hat{N}_B}{N_{FFT} - N_B} \tag{5.20}$$

where

N_T is the total number of all detected edges including both the actual subchannels' edges and the false edges

N_{FFT} is the utilized fast Fourier transform (FFT) size

Clearly, increasing the size of FFT will improve the detection performance. Finally, the average edge detection error probability is defined as

$$P_E = \frac{P_{ME} + P_{FE}}{2} \tag{5.21}$$

Alternatively, in [6], the authors define the band occupancy error (BOE) as

$$BOE = \frac{\sum_{m=1}^{M} |BOD_a(m) - BOD_e(m)|^2}{\sum_{m=1}^{M} |BOD_a(m)|^2} \tag{5.22}$$

where the subscripts "a" and "e" indicate the actual and the estimated band occupancy degree (BOD), respectively, which is expressed as

$$BOD(m) \begin{cases} \gamma_m, & \text{subchannel } m \text{ is occupied} \\ 0, & \text{otherwise} \end{cases} \tag{5.23}$$

where γ_m is the SNR at band m.

A very similar procedure is also proposed for individual subcarriers within each band. Note that a simpler BOD is achieved if we replace γ_m by 1. That is, 0 and 1 represent an unoccupied and occupied subchannel, respectively.

In addition to the aforementioned metrics, several modified versions can be used such as the complementary ROC where the probability of miss detection $P_M = 1 - P_D$ is plotted versus P_{FA} or $Q_M = 1 - Q_D$ versus Q_{FA} in case of using cooperative communications. Other intuitive performance metrics are based on plotting these probabilities versus SNR.

5.3 THROUGHPUT PERFORMANCE MEASURES

Multiband cognitive radio networks promise to tangibly improve the aggregate throughput of the network, and perhaps to guarantee, to a certain extent, Quality of Service (QoS) for SUs. Therefore, it is no surprise to use throughput as a common performance measure in this paradigm. We consider a general transmission paradigm where the SU accesses the band whether the PU is absent (interweave paradigm) or the PU is present (underlay paradigm). In the latter, power adaptation is mandatory to protect the PUs. Such combination of the two is commonly known as sensing-based spectrum sharing [9]. Furthermore, we assume imperfect sensing. Under these two assumptions, there are four possible scenarios for SUs transmission:

- The SU correctly decides that the PU is absent, and thus, it transmits at power ρ_s^0 with probability $1 - P_{FA,m}$.
- The SU incorrectly decides that the PU is absent, and thus, it transmits at power ρ_s^0 with probability $1 - P_{D,m}$.
- The SU correctly decides that the PU is present, and thus, it transmits at power $\rho_s^1 < \rho_s^0$ with probability $P_{D,m}$.
- The SU incorrectly decides that the PU is present, and thus, it transmits at ρ_s^1 with probability $P_{FA,m}$.

Let r_{ij} be the transmission rate of the SU given that it decides H_i when H_j is the true hypothesis. Thus, we have

$$r_{00} = B\log\left(1 + \frac{\rho_s^0}{\sigma^2}\right) \tag{5.24}$$

$$r_{01} = B\log\left(1 + \frac{\rho_s^0}{I + \sigma^2}\right) \tag{5.25}$$

$$r_{11} = B\log\left(1 + \frac{\rho_s^1}{I + \sigma^2}\right) \tag{5.26}$$

$$r_{10} = B\log\left(1 + \frac{\rho_s^1}{\sigma^2}\right) \tag{5.27}$$

where
  B is the subchannel bandwidth
  I is the interference due to the PU transmission when it is present

If we assume that the SU accesses one subchannel out of M, then the average throughput of the network is given by

$$R = \sum_{m=1}^{l} r_{00,m}(1 - P_{FA,m})p(H_0) + r_{01,m}(1 - P_{D,m})p(H_1) + r_{11,m}P_{D,m}p(H_1) + r_{10,m}P_{FA,m}p(H_0)$$

$$\tag{5.28}$$

where
  l is the number of channels that are being used by the CRN
  $p(H_i)$ is the probability that H_i occurs

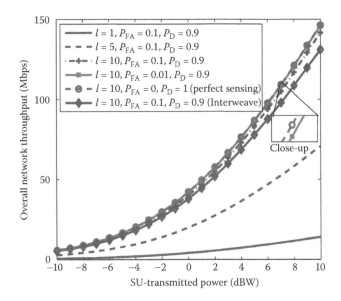

FIGURE 5.3 Throughput performance with different number of channels and probabilities of false alarm and detection, $\rho_s^l = 0.4\rho_s^0$, $\sigma^2 = 1$, $I = -20$ dB, $p(H_0) = 1 - p(H_1) = 0.7$.

We remark the following. First, under perfect sensing conditions, r_{01} and r_{10} become 0. Also, under the interweave paradigm (i.e., the SU only transmits when the PU is absent), the contributions of r_{11} and r_{10} become 0. The throughput performance using MB-CRNs is illustrated in Figure 5.3. The subchannels are assumed to have a bandwidth of 6 MHz, each with the same SNR conditions. The power allocation is uniformly distributed among the l channels.

It can be concluded from Figure 5.3 that if the SU accesses more channels (i.e., l increases), the throughput would increase as MB-CRN promises. Also, for a tighter P_{FA}, the throughput is further improved since the data interruptions are less frequent. Finally, sensing-based spectrum sharing (or hybrid access) gives better throughput in comparison with interweave access, because it allows the SU to coexist with the PU.

5.4 FUNDAMENTAL LIMITS AND TRADE-OFFS

Like any wireless communication system, cognitive radio networks have fundamental limits and many trade-offs among its design parameters. Since these networks are basically extensions of SB-CRNs and may incorporate cooperative communications, and the design of the network's parameters becomes even more challenging. The main design parameters of MB-CRNs include sensing time, network's throughput, detection reliability, number of cooperating SUs, number of sensed channels, power control, channel assignment, fairness, etc.

We consider here the general case of multiband cognitive radio. In general, the design procedure is as follows. For a set of parameters, the designer wants to choose the best values that maximize some function such as the throughput, or minimize another function such as interference to PUs. Mathematically, this can be formulated as an optimization problem, which can have the form [10]

$$\text{maximize } f(o)$$
$$\text{subject to } g_i(o), \quad i = 1, 2, \ldots, q \tag{5.29}$$

where
  $f(o)$ is the objective function
  o is the optimization variable
  $\{g_i(o)\}$ are constraint functions bounded by $\{b_i\}$

For example, a typical optimization problem in MB-CRN is where the objective function is the throughput, the optimization variable is the sensing duration, and the constraint functions include transmit power constraints and interference constraints. In this section, we will illustrate the design trade-offs and discuss some of the techniques that could provide improvements to these parameters.

5.4.1 SENSING TIME OPTIMIZATION

One of the key parameters in spectrum sensing is the sensing duration, which strongly impacts the network's throughput. To illustrate this, Figure 5.5a shows a MAC frame structure that has been widely adopted for cognitive networks. The T seconds frame consists of two slots: a sensing slot, τ, and a transmission slot, $T - \tau$, if assuming the SU accesses the subchannel. Thus, when sensing duration is considered, the throughput becomes

$$C = \frac{T - \tau}{T} R \tag{5.30}$$

Clearly, increasing τ will shorten the transmission slot, and thus, the throughput of the SU decreases, as shown in Figure 5.4a. However, longer sensing improves the detection performance since $N = \tau f_s$, where f_s is the sampling frequency (i.e., the SU receiver collects more samples for its test statistics), Figure 5.4b.

Liang et al. have studied this trade-off in [11] for SB-CR. Mathematically, the optimization problem can be expressed as

$$\max C(\tau)$$
$$\text{subject to } P_D(\lambda) \geq \beta, 0 < \tau < T \tag{5.31}$$

where β is called the target detection probability.

Other variations arise in the literature where additional constraints are used. These constraints can be, for example, on the probability of false alarm, or the detector's thresholds can be jointly optimized with sensing time (e.g., $o = \{\lambda, \tau\}$

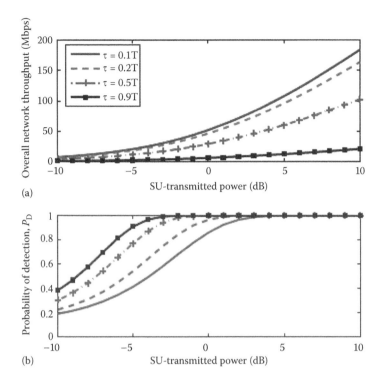

(a)

(b)

FIGURE 5.4 Impact of sensing time on throughput performance. (a) Throughput vs. SU transmitted power. (b) Probability of detection vs. transmitted power.

which is the case in the multiband sensing-time-adaptive joint detector (MSJD) detector).

In the following, we will discuss some of the techniques that can be used to balance the sensing time-throughput trade-off.

5.4.1.1 MAC Frame Structures

The authors in [11] have proposed a modified frame structure as shown in Figure 5.5b where the sensing duration τ is decomposed into S equal sensing slots. It is shown that increasing the number of these slots would reduce the optimum sensing time, and thus, the throughput is improved, and interestingly, the probability of false alarm is also reduced. This work is extended for MB serial sensing in [12] where the authors analyze two different frame structures: multiband slotted frame, where each subchannel is allocated a fixed number of slots, and multiband continuous frame, where each subchannel has an arbitrary length duration bounded by the total sensing time as shown in Figure 5.5c (each subchannel here is allocated two slotted frames) and Figure 5.5d, respectively. In the practical sense, the slotted frame is easier to implement, yet it is demonstrated that solving for the optimum time is less complex if we use the continuous frame. It is also demonstrated that increasing the number of channels for sensing improves the throughput.

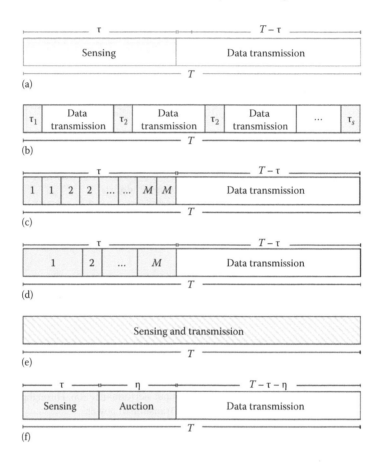

FIGURE 5.5 (a) A conventional MAC frame in CR; (b) the structure of a slotted frame; (c) a multiband slotted frame; (d) a multiband arbitrary-length slotted frame; (e) a novel frame where transmission and sensing occur at the same time; (f) an auction-based frame.

In [13], a novel frame structure is proposed where sensing and transmission are carried out simultaneously, as shown in Figure 5.5e, using decoding in parallel with spectrum sensing. The basic principle is as follows. The SU initially senses the spectrum and starts transmitting when a band is vacant. However, it transmits both its data and the spectrum occupancy to the base station. The base station decodes the received signal, extracts the SU data, and uses the rest of the signal to analyze the spectrum occupancy using any of the spectrum sensing techniques. Since both decoding and sensing are done in parallel, the advantages are as follows: longer sensing duration that enhances the detection performance, longer transmission rate that enhances the throughput, better PU protection since the sensing is always carried on, and finally, sensing time optimization becomes unnecessary, since it is done for the entire frame duration. The drawbacks, however, are twofold. First, it is demonstrated in [14] that decoding errors compromise the performance. Thus, it is shown that there is a lifetime, of which this novel frame is superior to the conventional frame, and beyond it, the novel frame loses its advantages due to the accumulation of decoding errors over consecutive frames and the dependence of the SU data transmission on the sensing results of the previous frames. Second, the existence of other SUs in the region requires the receiver to successfully decode multiple signals from multiple sources; thus, it becomes more challenging to detect the mixed signals.

5.4.1.2   Dual Radio

A dual-radio architecture is proposed in [15] where the receiver has one dedicated radio for sensing and one for transmission. This would improve the spectrum sensing, and more importantly, it would provide higher throughput. Obviously, the drawback is the higher cost, higher power consumption, and higher receiver complexity.

5.4.1.3   Adaptive Sensing Algorithms

Several works have investigated the impact of using adaptive sensing time on the networks' throughput [16–19]. Using dynamic sensing durations reduces the required number of samples, particularly when the subchannel has high SNR, and thus, the overall network's throughput is significantly improved [16]. In [17], Datla et al. propose an adaptive algorithm where longer sensing durations are used for channels with high idle probabilities because they are more rewarding in terms of throughput. Each time a channel is declared to be occupied, the required sensing duration for that channel is reduced in future sensing. The drawback of this algorithm is that it misses few available channels especially those which are declared occupied in previous sensing results. A two-phase adaptive algorithm is proposed in [18]. In the first phase, the SU iteratively explores the possible set of idle channels such that after each iteration, the number of candidate channels is exponentially decreased by excluding those with low idle probability. By the second phase, the SU would have a very small subset of candidate channels, and it will allocate the sampling budget accordingly to perform fine sensing. This is extended in [19] such that the goal is now to detect all available idle subchannels instead of a subset of them. The algorithm significantly improves the throughput at the low SNR regime. At high SNR regime, it is shown that the nonadaptive algorithm becomes optimal, where each subchannel has an equal sampling budget. Yang et al. present a QoS aware low complexity scheme where the SU may access some channels without spectrum sensing (i.e., $\tau = 0$) [20]. These channels have either high idle probabilities or high tolerable interference limits. It is demonstrated that this algorithm improves the throughput. Yet, further analysis is required on how to obtain the idle probabilities of these channels and to quantify the risk of accessing a spectrum without spectrum sensing.

5.4.1.4   Sequential Probability Ratio Tests

In [21], parallel sequential probability ratio tests (SPRTs) are used to optimize the number of samples required for spectrum sensing (i.e., $o = N$) and hence the sensing duration. Compared to fixed-sample size (FSS) detectors, parallel SPRTs significantly reduce the number of samples due to two gains: the gain of using the SPRT and the gain of simultaneously sensing multiple bands (parallel sensing). However, a key challenge in the SPRT bank is that each detector may yield a different sample size since, in general, the sample size is a random variable that depends on the observed data, and hence, the overall sensing time would be dictated by the largest sensing delay among the parallel detectors. Caromi et al. have investigated both parallel and serial SPRTs under limited and unlimited sensing duration [22]. Since the SPRT has no upper limit on the number of samples required to achieve a decision, the authors propose several truncated SPRTs to limit N. In serial sensing, N_{opt} increases as the number of sensing channels increases. On the contrary, for parallel sensing, when the number of channels increases, N_{opt} decreases. On the other hand, the channel sensing order is studied in [23]. It is demonstrated that the intuitive sensing order, where the SU sequentially senses the channels with higher to lower idle probabilities, is not always optimum. Specifically, if nonadaptive transmission rate is used, the intuitive order is optimal, but this optimality is lost when adaptive transmission is employed.

5.4.1.5　Number of Cooperating Users

Here, the optimization problem is

$$\max C(\tau, k)$$

$$\text{subject to } 0 < k \le K, 0 < \tau < T \tag{5.32}$$

where we want to jointly optimize the number of SUs with sensing duration.

Soft combining and hard combining are analyzed in [11]. It is demonstrated that increasing the number of total SUs, K, reduces the optimum sensing time, and hence, the throughput is improved. This is because when we have more SUs, the probability of detection increases. Thus, for a predefined target probability, we can reduce the sensing duration as we increase the number of SUs. Also, it is demonstrated that the majority voting rule has the best performance among other hard-combining rules, since it provides the lowest optimum sensing time and the highest achievable throughput. Nevertheless, the gain saturates if the number of SUs keeps increasing. In [24], Peh et al. investigate the optimum number of cooperating SUs, k_{opt}, to maximize the throughput under hard-combining schemes. It is observed that the optimum value depends on the wireless environment, and hence, there is no single voting rule that optimize the throughput for different SNRs. Nevertheless, optimizing k reduces the sensing time and improves the throughput. Interestingly, it is demonstrated that when the channel condition is bad (low SNR), it is more advantageous to allocate more time for sensing (i.e., reduce $T - \tau$), and when it is extremely bad (extremely low SNR), then it is more advantageous to allocate more time for transmission than sensing. The reason is that at such regions, P_{FA} is very high, regardless of τ, and thus, it is more beneficial to increase $T - \tau$. Soft combining is analyzed in [12] where it is shown that increasing the number of SUs improves the network's throughput with diminishing gain as k increases. In [25], soft combining and hard combining are investigated for multiband detection. It is demonstrated that soft combining provides better throughput, but hard combining causes less interference to the PUs since it has short overhead.

While increasing the number of cooperating SUs improves the reliability of detection and reduces τ_{opt}, it incurs a long delay due to the time required to collect the information from all of SUs. To tackle this issue, the SUs can simultaneously send their decisions on orthogonal frequency bands [26], yet this requires larger bandwidth. Thus, another solution is sought in [31] where the authors derive the least required number of SUs to achieve a target performance. On the other hand, censoring is discussed in [27] to limit the number of cooperating SUs. The basic principle is that the SU only cooperates if its detection result is considered useful. This technique reduces the number of cooperating users as well as save the total power budget. Similarly, selective-based cooperative spectrum sensing is presented in [28–30] where the proposed algorithms jointly reduce the overhead and mitigate the false reports sent by unreliable SUs. The basic principle is not only to limit the number of cooperating SUs, but also to admit merely those who have reliable decisions based on several factors such as the SNR of the reporting channel and the quality of sensing. Finally, an optimum voting rule for a fixed value of K is derived in [31] to minimize the sum of probabilities of misdetection and false alarm $Q_M + Q_{FA}$, where $Q_M = 1 - Q_D$. Thus, if we take the derivative of the summation with respect to k and set it to zero, we obtain

$$k_{opt} \approx \left\lceil \frac{K}{1+\varepsilon} \right\rceil \tag{5.33}$$

where

$$\varepsilon = \frac{\ln\left(\dfrac{P_{FA}}{P_D}\right)}{\ln\left(\dfrac{1-P_D}{1-P_{FA}}\right)} \tag{5.34}$$

Thus,
1. If $P_{FA} \approx P_D$, then $\varepsilon \approx 0$, and the majority rule is optimal.
2. If $P_{FA} \ll 1 - P_D$, then $\varepsilon \approx K + 1$, and the OR-logic rule is optimal.
3. If $P_{FA} \gg 1 - P_D$, then $\varepsilon \approx 1$, and the AND-logic rule is optimal.

5.4.2 DIVERSITY AND SAMPLING TRADE-OFFS

Integrating the cooperative communication paradigm with cognitive networks requires a trade-off between the spatial diversity achieved by cooperation and the expensive hardware requirements for sensing a very wideband spectrum. Figure 5.6 shows this trade-off. A multiband framework is assumed where there are M subchannels with a bandwidth of 6 MHz each. Also, sampling is done at the Nyquist rate. It is observed that the sampling cost could be significantly reduced if we compromise the full diversity by allowing each SU to sense only a subset of the M subchannels. For instance, when each subchannel must be monitored simultaneously by only two SUs

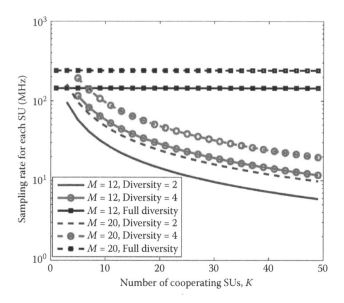

FIGURE 5.6 The sampling cost reduces as the number of cooperating users increases.

(i.e., diversity is 2), then the sampling cost would be less than the case of full diversity where all the SUs must monitor the subchannel. Also, increasing the number of cooperating users will further reduce the sampling cost because the individual SU would have to sense fewer channels. In addition, higher sampling costs are required when M is increased. In Figure 5.7, it is easy to see the inevitable trade-off between diversity and sampling cost. Observe that when there are more SUs, the trade-off's impact becomes less. Wang et al. show that compressive sampling can help reduce this trade-off impact by proposing a cooperative detection algorithm based on rank minimization of the SUs' collected measurement vectors [32].

5.4.3 Power Control and Interference Limits Trade-Offs

Optimum power allocation is vital for improving the network's throughput as well as for protecting the PUs. It becomes even more important when the underlay scheme is used since power adaptation becomes necessary. Many papers have studied joint optimization of power and sensing time to maximize the throughput. Other works include adding transmit power and interference bounds as constraint functions.

5.4.3.1 Average and Peak Transmit Powers

There are two commonly used transmit power constraints on the SUs, namely, the average power, ρ_{avg}, and the peak power, ρ_{peak}. The former is preferred when we want to maintain long-term power budget while the latter is used to limit the peak power for practical considerations including the nonlinearity of power amplifiers [33]. If we assume that the SUs access l channels, then the average power constraint is typically expressed as

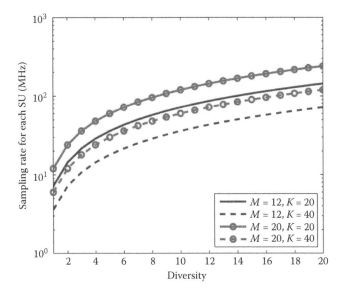

FIGURE 5.7   Sampling cost-diversity trade-off.

$$\sum_{m=1}^{l} E[\rho_m] \leq \rho_{avg} \tag{5.35}$$

while the peak power constraint is expressed as

$$\sum_{m=1}^{l} \rho_m \leq \rho_{peak} \tag{5.36}$$

To understand why the latter is more restrictive, consider the following scenario, where two SUs accessing together l channels, and the power bounds are $\rho_{avg} = \rho_{peak} = 1$ W. For simplicity, assume that the transmit powers are fixed, and no fading conditions are imposed. Then during the power allocation, under the average constraint, both SUs can transmit at any power as long as the average is 1 W. For instance, a possible combination is $\rho_1 = 0.5$ W, $\rho_2 = 1.5$ W, and so on. However, under the peak constraint, both SUs must satisfy $\rho_1 + \rho_2 \leq 1$ W.

Pei et al. study the joint optimization of sensing time and power allocation subject to these power constraints [34]. It is shown that both constraints have a water-filling power control where the SU allocates more power for the subchannels with higher SNR. The difference, however, is that for ρ_{peak}, the allocation strategy depends on the number of vacant channels unlike the allocation strategy for ρ_{avg} where a certain power allocation is assigned, regardless of the activity of other channels. In other words, the power allocation in ρ_{peak} is done after the SU is aware of the spectrum sensing results. It is shown that the throughput is higher when ρ_{avg} is used since it is less restrictive compared to ρ_{peak}. Also, the optimum sensing time, τ_{opt}, is increased when the power budget is increased. Interestingly, the variations of τ_{opt} are small for different powers, and hence, using a fixed sensing time for different power budgets would only slightly degrade the throughput. Nevertheless, a fixed sensing time guarantees sensing synchronization among other SUs and reduces the network's complexity. Water filling is also proposed by Wang in [35] where the throughput is maximized by jointly optimizing the power allocation and the channels to be sensed. In [36], Barbarossa et al. jointly optimize the power allocation with energy detection thresholds where it is shown that the power control is simply a multilevel water-filling procedure with levels depending on the power budget, the channel quality, and the detection reliability. In [37], Zou et al. propose an auction-based power allocation scheme when multiple SUs compete for the network resources. It works as follows. If the SU works in a spectrum interweave scheme, the power allocated to the SU will be proportional to its payment. However, if the SU works in an underlay scheme, the power allocated is going to be constrained to protect the QoS of the PU network. An additional slot is added to the MAC frame to allow the SU's bid for transmit powers, as illustrated in Figure 5.5f. This means that for a fixed T, the throughput would decrease if transmission duration is decreased, or the detection performance is degraded if sensing duration is reduced instead. On the other hand, the sensing time and power control are jointly optimized based on the distance between the SU and the PU [38]. The basic principle is that when the PU is far away, it is better to merely use power control to allow the SU to simultaneously exist with the PU. This is because to detect the weak PU signal, longer sensing time τ is required, which impacts the throughput. However, if the PU is very close to the SU, a short sensing time would be enough to reliably detect the PU. Thus, power control is not required since it is recommended that the SU stops its transmission to protect the PU. Simulation results demonstrate that the proposed technique outperforms adaptive power allocation when the separation distance is long, and it outperforms the adaptive sensing time when the distance is short. Nevertheless, this algorithm is susceptible to inaccurate information about the PUs' locations.

5.4.3.2 Average and Peak Interferences

Interference to the PU happens when the SU simultaneously exists with the PU. This occurs when the SU either mis-detects the PU, or when the SU adapts its power to meet an interference bound in the underlay scheme. In order to protect PUs, an interference bound is set on the optimization problem. There are two common bounds, the average interference constraint I_{avg} and the peak interference constraint, I_{peak}. Mathematically, the former is expressed as

$$\sum_{i=1}^{K^{(m)}} E[\rho_i] \leq I_{\text{avg}} \tag{5.37}$$

where $K^{(m)}$ is the number of SUs transmitting over subchannel m, and the peak interference constraint is

$$\sum_{i=1}^{K^{(m)}} \rho_i \leq I_{\text{peak}} \tag{5.38}$$

Zhang in [39] shows that the throughput of the SU network is higher when the I_{avg} constraint is imposed. The reason is that the average interference constraint, I_{avg}, is generally more flexible compared to peak interference constraint, I_{peak}, given that $I_{\text{avg}} = I_{\text{peak}}$. Surprisingly, while one may expect that the latter provides better protection to the PU network (since the peak bound is very restrict), it is shown otherwise due to an interesting interference diversity phenomenon of the PU. This phenomenon is attributed to the convexity of the throughput function with respect to the interference power. It is shown that the random interference levels that arise in I_{avg} case are more advantageous when the interference powers are fixed as in I_{peak}. In other words, the I_{avg} constraint is not only good for the SUs, but it also provides less throughput losses to the PUs! Furthermore, water-filling power allocation is optimum under I_{avg}, and the truncated channel inversion power allocation is optimum under I_{peak}, which is a fair power allocation scheme that maintains a constant power by inverting the channel fading [40]. In [41], it is recommended I_{avg} be imposed on delay-insensitive PU systems, whereas I_{peak} is preferred when the systems are delay sensitive [41]. Stotas and Nallanathan have analyzed the impact of interference tolerance to PUs on the sensing time [42]. It is shown that a higher average interference bound reduces the optimum sensing time, and thus improves the throughput as expected.

5.4.3.3 Beamforming

Joint beamforming is shown to be powerful to overcome the sensing-throughput trade-off [43]. Particularly, Fattah et al. demonstrate that beamforming reduces the sensing time, improves the throughput, and more importantly, it maintains a good PU protection [44]. Yet, it requires CSI at the SU transmitter and receiver as well as an antenna array. This is more challenging in CRNs, since there are different PU networks, and the PUs may not be willing to feedback the CSI to the SUs. This motivates a new research direction on robust beamforming algorithms against imperfect CSI.

5.4.3.4 Power and Resource Allocation

In [45], joint optimization of power control, channel selection, and rate adaptation is used to maximize the throughput. Two algorithms are proposed. In the first one, the SU selects the best available channels, and then power and rate adaptation are implemented accordingly. To guarantee that each SU selects the best channels, frequent channels' reselections become inevitable, and hence, high overhead is incurred. To reduce the overhead, an alternative algorithm is proposed where the SU selects a channel as long as it can support the least possible transmission rate. Otherwise, an alternative channel is randomly selected. This algorithm has lower throughput, yet it reduces the frequency of channels' reselections. This work advises to use adaptive bandwidth selection for MB-CRNs to further maximize the network's throughput. In [46], joint admission control and power allocation are studied for MB-CRNs to maximize the throughput under different constraints on QoS and power consumption. In [47], a joint optimization of detection thresholds, channel assignment, and power allocation is presented to maximize the throughput. In [48], sensing time and channel selection are jointly optimized to maximize the throughput. Particularly, once the optimum sensing time for each subchannel is selected, the SU selects a subset (say l) from M subchannels, over which the aggregate throughput can be maximized. It is shown that for a fixed M, increasing l reduces the throughput, whereas for a fixed l, increasing M improves the throughput with a diminishing gain. In [49], iterative centralized power and scheduling algorithms are proposed to enhance the throughput.

5.4.4   RESOURCE ALLOCATION TRADE-OFFS

Resource allocation has recently become an active area of research for cognitive radio networks [50]. Such resources include power allocation, channel (or bandwidth) selection, admission control (which links to access), etc. Fairness among users is also an important factor in resource allocation.

5.4.4.1   Bandwidth Selection

One can presume that accessing all available bands would theoretically increase the throughput. However, when an SU accesses all these bands, there is a higher probability that a PU returns to at least one of them, and thus, handoff becomes necessary, which consequently increases the network's overhead and interrupts the SU's transmission. Therefore, optimizing the number of subchannels for spectrum access becomes necessary. Dan et al. investigate the optimum bandwidth (or optimum number of a subset of subchannels, l_{opt}) to maximize network's throughput [51]. The authors investigate both contiguous (CON) and noncontiguous (NCON) channel allocations for delay-sensitive and delay-insensitive traffic. For serial sensing, it is demonstrated that l_{opt} is higher when the idle channel probability is high. Also, NCON has larger l_{opt} since a larger overhead is needed to search for l_{opt} contiguous channels in CON scheme. Also, CON is less sensitive to the channel idle durations compared to NCON. If the idle probability is high, it is recommended to use parallel sensing since l_{opt} becomes significantly higher compared to serial sensing [51]. If the occupancy of the channels is correlated, and the SU has prior knowledge about it, then further improvements can be attained in terms of throughput, and these benefits are observed more in CON since the contiguous channels usually have higher correlation. In addition, when there are multiple SUs in vicinity, channel reconfiguration is important in CON. To see why, imagine there are four consecutive idle bands (1–4) and two SUs in the network (SU1 and SU2). If $l_{opt} = 2$, then if SU1 accesses bands 2 and 3, then SU2 cannot access 1 and 4 since they are noncontiguous. However, if SU1 accesses 1 and 2, then SU2 can be accommodated to access 3 and 4. Clearly, the advantage of channel reconfiguration lies by accommodating more SUs, but the disadvantage is the larger overhead due to the reallocation processes, which incurs transmission interruptions and delays because of setting up new links. Finally, the gains of the reconfiguration scheme are shown to diminish as the number of SUs becomes larger since the losses incurred by evacuating and reconnecting to new channels overwhelm the gains. On the other hand, to enhance bandwidth efficiency, Khambekar et al. propose a novel scheme where the guard interval of OFDM symbol is utilized for spectrum sensing [52].

5.4.4.2 Fairness

Fairness is an important criterion in resource allocation [53]. Obviously, one would like to maximize the network's throughput and allow as many users as possible to use the network's resources. However, allocating some of the resources to SUs, which are in bad channel conditions (e.g., deep fading), would severely impact the network's throughput. Nevertheless, it is also unfair to ignore such SUs. Hence, a trade-off between fairness and the network's performance exists. Two fairness algorithms are discussed in [54], namely, the proportional fairness and max–min fairness. The former is a common scheduling scheme where the resources are allocated based on the SU's channel quality. It strikes a good balance between fairness and throughput, yet it may become unfair when the users experience varying channel conditions as well as it does not guarantee QoS. The latter scheme is based on equal allocations among all SUs. If an SU cannot benefit from its allocated resources, then these resources are fairly distributed among the others. Thus, a certain minimum of QoS can be guaranteed. It is shown that the former is better in terms of throughput. On the other hand, consensus-based protocols are proposed in [53,55] to enhance spectrum sharing fairness among multiple SUs. This work is extended in [56] where the performance of both maximum fairness and uniform fairness, where each SU has an equal resource allocation, is evaluated.

5.5 CONCLUSION

There are several key parameters in cognitive radio networks that need to be carefully designed. Sensing time must be carefully optimized because it is inversely proportional to the throughput. Several MAC frames have been discussed to improve the sensing time-throughput trade-off. Other techniques include adaptive algorithms and SPRTs. Also, beamforming and dual radio can enhance the sensing time and throughput at the expense of additional complexity. In addition, cooperation improves the detection performance by exploiting spatial diversity, and it also reduces the sensing time, and consequently, the throughput is improved. Nevertheless, this requires larger overhead and higher power consumption, and techniques like censoring must be used to limit the number of cooperating users. On the other hand, there is an inevitable trade-off between diversity and sampling cost. Thus, full diversity imposes high sampling cost on the SU receiver, and to reduce the sampling requirements, we need to compromise the diversity. Other important considerations include power and interference control as well as resource allocation.

The main aforementioned limits and trade-offs are summarized in the trade-off map presented in Figure 5.8. Note that describing a relationship between two parameters by good or bad is with respect to the impact of increasing the first parameter on

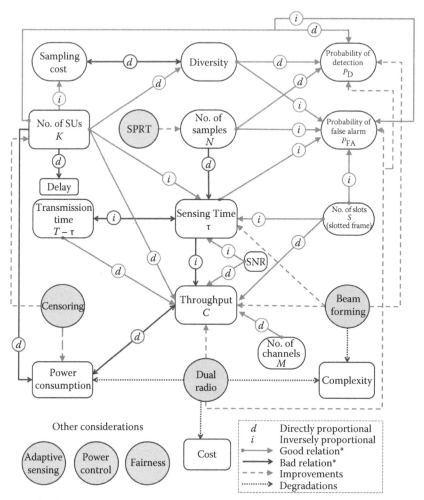

FIGURE 5.8 Trade-off map.

the other one. The following example describes how to use this diagram. Consider the number of collected samples, N. Then if we increase N (i.e., if the SU collects more samples during spectrum sensing), we have three consequences:

1. P_{FA} will decrease. This means N and P_{FA} are inversely proportional. Since this impact of increasing N is good to the system.
2. P_D will increase. This means N and P_D are directly proportional. Since this impact of increasing N is good to the system.
3. τ will increase. This means N and τ are directly proportional. Since this impact of increasing N is bad to our system considering it will reduce the throughput (recall the sensing time-throughput trade-off).

REFERENCES

1. G. Hattab and M. Ibnkahla, Multiband cognitive radio: Great promises for future radio access, *Proceedings of the IEEE*, 102 (3), 282–306, March 2014.
2. G. Hattab and M. Ibnkahla, Multiband cognitive radio: Great promises for future radio access (long version), Internal Report, Queen's University, WISIP Laboratory, Kingston, Ontario, Canada, December 2013.
3. H. Poor, *An Introduction to Signal Detection and Estimation*, Springer, New York, 1994.
4. A. Jeffrey and D. Zwillinger, *Table of Integrals, Series, and Products*, Elsevier Science, Boston, MA, 2000.
5. S. Chaudhari, V. Koivunen, and H. Poor, Autocorrelation-based de-centralized sequential detection of OFDM signals in cognitive radios, *IEEE Transactions on Signal Processing*, 57 (7), 2690–2700, July 2009.
6. Y. Zeng, Y.-C. Liang, and M. W. Chia, Edge based wideband sensing for cognitive radio: Algorithm and performance evaluation, in *Proceedings of the IEEE International Symposium on New Frontiers in Dynamic Spectrum Access Networks (DySPAN'11)*, Aachen, Germany, May 2011, pp. 538–544.
7. Y. Xin, G. Yue, and L. Lai, Efficient channel search algorithms for cognitive radio in a multichannel system, in *Proceedings of the IEEE Global Telecommunications Conference (GLOBECOM'10)*, Miami, FL, December 2010, pp. 1–5.
8. S. El-Khamy, M. El-Mahallawy, and E. Youssef, Improved wideband spectrum sensing techniques using wavelet-based edge detection for cognitive radio, in *Proceedings of the International Conference on Computing, Networking and Communications (ICNC'13)*, San Diego, CA, January 2013, pp. 418–423.
9. X. Kang, Y.-C. Liang, H. Garg, and L. Zhang, Sensing-based spectrum sharing in cognitive radio networks, *IEEE Transactions on Vehicular Technology*, 58 (8), 4649–4654, October 2009.
10. S. Boyd and L. Vandenberghe, *Convex Optimization*, Cambridge University Press, Cambridge, U.K., 2004.
11. Y.-C. Liang, Y. Zeng, E. Peh, and A. T. Hoang, Sensing-throughput tradeoff for cognitive radio networks, *IEEE Transactions on Wireless Communications*, 7 (4), 1326–1337, April 2008.
12. R. Fan and H. Jiang, Optimal multi-channel cooperative sensing in cognitive radio networks, *IEEE Transactions Wireless Communications*, 9 (3), 1128–1138, March 2010.
13. S. Stotas and A. Nallanathan, Overcoming the sensing-throughput tradeoff in cognitive radio networks, in *Proceedings of the IEEE International Conference on Communications (ICC'10)*, Cape Town, South Africa, May 2010, pp. 1–5.
14. L. Tang, Y. Chen, A. Nallanathan, and E. Hines, Spectrum sensing based on recovered secondary frame in the presence of realistic decoding errors, in *Proceedings of the IEEE International Conference on Communications (ICC'12)*, Ottawa, Ontario, Canada, June 2012, pp. 1495–1501.
15. N. Shankar, C. Cordeiro, and K. Challapali, Spectrum agile radios: Utilization and sensing architectures, in *Proceedings of the IEEE International Symposium on New Frontiers in Dynamic Spectrum Access Networks (DySPAN'05)*, Baltimore, MD, November 2005, pp. 160–169.
16. P. Paysarvi-Hoseini and N. Beaulieu, Optimal wideband spectrum sensing framework for cognitive radio systems, *IEEE Transactions on Signal Processing*, 59 (3), 1170–1182, March 2011.
17. D. Datla, R. Rajbanshi, A. M. Wyglinski, and G. Minden, Parametric adaptive spectrum sensing framework for dynamic spectrum access networks, in *Proceedings of the Second IEEE International Symposium on New Frontiers in Dynamic Spectrum Access Networks (DySPAN'07)*, Dublin, Ireland, April, 2007, pp. 482–485.

18. Y. Feng and X. Wang, Adaptive multiband spectrum sensing, *IEEE Wireless Communications Letters*, 1 (2), 121–124, April 2012.

19. Y. Feng, L. Song, and B. Jiao, Adaptive multi-channel spectrum sensing throughput tradeoff, in *Proceedings of the IEEE Conference on Communication Systems (ICCS'12)*, Singapore, November 2012, pp. 250–254.

20. C. Yang, Y. Fu, Y. Zhang, R. Yu, and S. Xie, Optimal wideband mixed access strategy algorithm in cognitive radio networks, in *Proceedings of the IEEE Wireless Communications and Networking Conference (WCNC'12)*, Paris, France, April 2012, pp. 1275–1280.

21. S.-J. Kim, G. Li, and G. Giannakis, Multi-band cognitive radio spectrum sensing for quality-of-service traffic, *IEEE Transactions on Wireless Communications*, 10 (10), 3506–3515, October 2011.

22. R. Caromi, Y. Xin, and L. Lai, Fast multiband spectrum scanning for cognitive radio systems, *IEEE Transactions on Communications*, 61 (1), 63–75, January 2013.

23. H. Jiang, L. Lai, R. Fan, and H. Poor, Optimal selection of channel sensing order in cognitive radio, *IEEE Transactions on Wireless Communications*, 8 (1), 297–307, January 2009.

24. E. Peh, Y.-C. Liang, Y. L. Guan, and Y. Zeng, Optimization of cooperative sensing in cognitive radio networks: A sensing-throughput tradeoff view, *IEEE Transactions on Vehicular Technology*, 58 (9), 5294–5299, November 2009.

25. H. Mu and J. Tugnait, Joint soft-decision cooperative spectrum sensing and power control in multiband cognitive radios, *IEEE Transactions on Signal Processing*, 60 (10), 5334–5346, October 2012.

26. K. Letaief and W. Zhang, Cooperative communications for cognitive radio networks, *Proceedings of the IEEE*, 97 (5), 878–893, May 2009.

27. E. Axell, G. Leus, E. Larsson, and H. Poor, Spectrum sensing for cognitive radio: State-of-the-art and recent advances, *IEEE Signal Processing Magazine*, 29 (3), 101–116, May 2012.

28. A. Alkheir and M. Ibnkahla, Selective cooperative spectrum sensing in cognitive radio networks, in *Proceedings of the IEEE Global Telecommunications Conference (GLOBECOM'11)*, Houston, TX, December 2011, pp.1–5.

29. A. Abu-Alkheir, Cooperative cognitive radio networks: Spectrum acquisition and co-channel interference effect, PhD dissertation, Department of Electrical and Computer Engineering, Queen's University, Kingston, Ontario, Canada, February 2013.

30. A. Alkheir and M. Ibnkahla, An accurate approximation of the exponential integral function using a sum of exponentials, *IEEE Communications Letters*, 17 (7), 1364–1367, July 2013.

31. W. Zhang, R. Mallik, and K. Letaief, Optimization of cooperative spectrum sensing with energy detection in cognitive radio networks, *IEEE Transactions Wireless Communications*, 8 (12), 5761–5766, December 2009.

32. Y. Wang, Z. Tian, and C. Feng, Collecting detection diversity and complexity gains in cooperative spectrum sensing, *IEEE Transactions on Wireless Communications*, 11 (8), 2876–2883, August 2012.

33. X. Kang, Y.-C. Liang, A. Nallanathan, H. Garg, and R. Zhang, Optimal power allocation for fading channels in cognitive radio networks: Ergodic capacity and outage capacity, *IEEE Transactions on Wireless Communications*, 8 (2), 940–950, February 2009.

34. Y. Pei, Y.-C. Liang, K. Teh, and K. H. Li, How much time is needed for wideband spectrum sensing? *IEEE Transactions on Wireless Communications*, 8 (11), 5466–5471, November 2009.

35. X. Wang, Joint sensing-channel selection and power control for cognitive radios, *IEEE Transactions on Wireless Communications*, 10 (3), 958–967, March 2011.

36. S. Barbarossa, S. Sardellitti, and G. Scutari, Joint optimization of detection thresholds and power allocation for opportunistic access in multicarrier cognitive radio networks, in *Proceedings of the Third IEEE International Workshop on Computational Advances in Multi-Sensor Adaptive Processing (CAMSAP'09)*, Dutch Antilles, the Netherlands, December 2009, pp. 404–407.

37. J. Zou, H. Xiong, D. Wang, and C. W. Chen, Optimal power allocation for hybrid overlay/underlay spectrum sharing in multiband cognitive radio networks, *IEEE Transactions on Vehicular Technology*, 62 (4), 1827–1837, May 2013.

38. E. Peh, Y.-C. Liang, and Y. Zeng, Sensing and power control in cognitive radio with location information, in *Proceedings of the IEEE International Conference on Communication Systems (ICCS'12)*, Agadir, Morocco, November 2012, pp. 255–259.

39. R. Zhang, On peak versus average interference power constraints for protecting primary users in cognitive radio networks, *IEEE Transactions on Wireless Communications*, 8 (4), 2112–2120, April 2009.

40. A. Goldsmith, *Wireless Communications*, Cambridge University Press, Cambridge, U.K., 2005.

41. K. Hamdi and K. BenLetaief, Power, sensing time, and throughput tradeoffs in cognitive radio systems: Across-layer approach, in *Proceedings of the IEEE Wireless Communications and Networking Conference (WCNC'09)*, Budapest, Hungary, April 2009, pp. 1–5.

42. S. Stotas and A. Nallanathan, Optimal sensing time and power allocation in multiband cognitive radio networks, *IEEE Transactions on Communications*, 59 (1), 226–235, January 2011.

43. D. Palomar, J. Cioffi, and M.-A. Lagunas, Joint Tx-Rx beamforming design for multicarrier MIMO channels: A unified framework for convex optimization, *IEEE Transactions on Signal Processing*, 51 (9), 2381–2401, September 2003.

44. A. Fattah, M. Matin, and I. Hossain, Joint beamforming and power control to overcome tradeoff between throughput-sensing in cognitive radio networks, in *Proceedings of the IEEE Symposium on Computer Informatics (ISCI'12)*, Penang, Malaysia, March 2012, pp. 150–153.

45. A. Alkheir and M. Ibnkahla, Performance analysis of joint power control, rate adaptation, and channels election strategies for cognitive radio networks, in *Proceedings of the IEEE Global Telecommunications Conference (GLOBECOM'12)*, Anaheim, CA, December 2012, pp. 1126–1131.

46. A. Panwar, P. Bhardwaj, O. Ozdemir, E. Masazade, C. Mohan, P. Varshney, and A. Drozd, On optimization algorithms for the design of multiband cognitive radio networks, in *Proceedings of the 46th Annual Conference on Information Sciences and Systems (CISS'12)*, Princeton, NJ, March 2012, pp.1–6.

47. C. Shi, Y. Wang, T. Wang, and P. Zhang, Joint optimization of detection threshold and throughput in multiband cognitive radio systems, in *Proceedings of the IEEE Consumer Communications And Networking Conference (CCNC'12)*, Las Vegas, NV, January 2012, pp. 849–853.

48. H. Yao, X. Sun, Z. Zhou, L. Tang, and L. Shi, Joint optimization of subchannel selection and spectrum sensing time for multiband cognitive radio networks, in *Proceedings of the International Symposium on Communications and Information Technologies (ISCIT'10)*, Tokyo, Japan, October 2010, pp. 1211–1216.

49. L. Vijayandran, S.-S. Byun, G. Øien, and T. Ekman, Increasing sum rate in multiband cognitive radio networks by centralized power allocation schemes, in *Proceedings of the 20th IEEE International Symposium on Personal, Indoor and Mobile Radio Communications*, Tokyo, Chennai, September 2009, pp. 491–495.

50. R. Zhang, Y.-C. Liang, and S. Cui, Dynamic resource allocation in cognitive radio networks, *IEEE Signal Processing Magazine*, 27 (3), October 2010, pp. 102–114.

51. D. Xu, E. Jung, and X. Liu, Optimal bandwidth selection in multi-channel cognitive radio networks: How much is too much? in *Proceedings of the Third IEEE International Symposium on New Frontiers in Dynamic Spectrum Access Networks (DySPAN'08)*, Chicago, IL, October, 2008, pp. 1–11.

52. N. Khambekar, L. Dong, and V. Chaudhary, Utilizing OFDM guard interval for spectrum sensing, in *Proceedings of the IEEE Wireless Communications and Networking Conference (WCNC'07)*, Hong Kong, China, March 2007, pp. 38–42.

53. P. Hu and M. Ibnkahla, A consensus-based protocol for spectrum sharing fairness in cognitive radio ad hoc and sensor networks, *International Journal of Distributed Sensor Networks*, 2012 (11), 1–12, August 2012.

54. R. Fan, H. Jiang, Q. Guo, and Z. Zhang, Joint optimal cooperative sensing and resource allocation in multichannel cognitive radio networks, *IEEE Transactions on Vehicular Technology*, 60 (2), 722–729, February 2011.

55. P. Hu, Cognitive radio ad hoc networks: A local control approach, PhD dissertation, Department of Electrical and Computer Engineering, Queen's University, Kingston, Ontario, Canada, February 2013.

56. P. Hu and M. Ibnkahla, Consensus-based local control schemes for spectrum sharing in cognitive radio sensor networks, in *Proceedings of the 26th Biennial Symposium on Communications (QBSC'12)*, Kingston, Ontario, Canada, May 2012, pp. 115–118.

6 Spectrum Handoff

6.1 INTRODUCTION

Spectrum maintenance or channel maintenance has been an integral part in modern communication systems. For instance, a seamless handoff from one channel to another is vital for a mobile user. In a broad sense, handoff in conventional systems is initiated when the user's current channel deteriorates, and thus, it looks for a better one. In cognitive radio networks (CRNs), however, the concept is different due to the existence of two types of users, primary users (PUs) and seconday users (SUs). This has given rise to spectrum handoff where the SU initiates a handoff when the PU returns to one of the bands used by the SU. Thus, the spectrum handoff in CRNs depends on the behavior of PUs, and this is a major challenge, since such behavior is random in nature.

In general, if a PU returns to a channel used by an SU, then there are two scenarios. In the first one, the SU remains silent in the channel and postpones its transmission until the PU evacuates the band. In the second one, the SU moves to another channel. Intuitively, the former is inefficient because the SU does not know how long the PU would be active. In the latter, there are two different methods, namely, reactive and proactive handoffs. In reactive handoff, the SU would sense for other available channels when the PU returns. Thus, the SU would waste some time sensing the spectrum again. In proactive handoff, the SU maintains a list of candidate channels to access once the PU returns. This saves time; yet, to obtain a list of available channels, the SU must have accurate data records about the behavior of the PU in order to predict which channels are going to be available for future access.

6.2 SPECTRUM MOBILITY

In CRNs, transmission is performed over a temporarily unused frequency spectrum, which can be claimed at any time by its licensed user. Therefore, the SUs periodically sense the spectrum and move between spectrum holes to complete their transmission. The concept of moving between spectrum bands is called *spectrum mobility*. Spectrum mobility introduces what is called spectrum handoff when the SUs switch their transmission to unused spectrum holes. *Spectrum handoff* can be triggered by any of the following events:

- A PU reclaims its licensed channel that is being used by the SU.
- An SU physically moves to a geographical region where the current spectrum hole may not be available (i.e., another PU or SU is making use of it).
- An SU experiences a link quality degradation such that the QoS requirements cannot be fulfilled.

The ultimate goal of spectrum mobility management is to perform successful and fast spectrum handoff while maintaining the ongoing communication and minimizing interference with PUs. To this end, spectrum mobility management can be realized by two main functionalities:

- *Spectrum handoff:* The SU monitors the radio environment and tries to anticipate handoff triggering events. Once a handoff triggering event occurs, the SU pauses its ongoing transmission and continues it over another suitable channel. This process may involve long delay resulting from searching for a backup channel and RF front-end reconfiguration (e.g., modulation scheme, carrier frequency, and so on).
- *Link management:* As stated earlier, the spectrum handoff process may introduce long delays. Based on the duration of the handoff process, the SU can adapt its upper network protocols to ensure minimum performance degradation.

6.2.1 Relationship with Other Spectrum Management Functions

Spectrum management in CRNs is implemented through four functionalities. These include (1) spectrum sensing, (2) spectrum decision, (3) spectrum sharing, and (4) spectrum mobility. In this section, the relationship between these four functionalities is explained.

Figure 6.1 shows a block diagram of the four spectrum management procedures. Initially, based on the spectrum sensing outcomes, the spectrum decision procedure assigns one of the available channels to an SU. The role of spectrum sharing is to ensure fairness when multiple SUs try to access channels. The spectrum mobility procedure collaborates with the spectrum sensing, spectrum sharing, and spectrum decision functions to detect events that must initiate the spectrum handoff process and selects a suitable channel.

The spectrum mobility function typically includes a subfunction to analyze past PU spectrum usage along with current spectrum sensing results to look for patterns

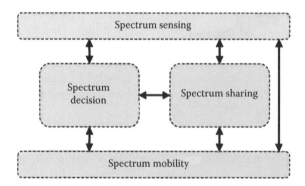

FIGURE 6.1 Spectrum sensing, spectrum decision, spectrum sharing, and spectrum mobility. (Adapted from Wang, L.C. et al., *IEEE Trans. Wireless Commun.*, 18(6), 18, December 2011.)

that help in predicting future spectrum availability. As we will see, an SU can use prediction to anticipate spectrum handoff triggering events.

6.3   SPECTRUM HANDOFF STRATEGIES

Spectrum handoff implementations can be categorized into four strategies [2], depending on the timing of the handoff initiation. As shown in the following sections, implementing a proper handoff strategy plays an important role in obtaining satisfactory performance. Here, we present the details of the four strategies.

6.3.1   NONHANDOFF STRATEGY

In nonhandoff strategy, when a PU reclaims its channel, the SU stops transmission, remains idle, and resumes transmission once the PU vacates the channel. This process may be repeated by the SU as necessary during a single session of data transmission. This strategy is illustrated in Figure 6.3a. Nonhandoff strategy is suitable for infrequent PU transmission and when other licensed bands are overcrowded. One advantage of this scheme is that its implementation does not require sophisticated algorithms. However, nonhandoff strategy introduces high latency because of the time spent by the PU occupying the current channel. Moreover, generally, the SU does not know for how long the PU would use the channel. That is, the latency caused here is typically unpredictable. Thus, the effective data rate will decrease. This unpredictable latency makes this strategy impractical for delay-sensitive applications. Another disadvantage is the unavoidable possibility of the SU's interference with the PU because the detection of the PU's presence is not perfect and the SU will continue transmitting for a period of time until the PU is detected.

6.3.2   REACTIVE HANDOFF STRATEGY

Reactive handoff strategy is shown in Figure 6.3b. Once an SU detects a PU's arrival, it suspends its data transmission and performs spectrum sensing to search for a free channel. Upon succeeding in finding a suitable channel, the SU switches to the new channel and resumes its unfinished transmission. Moreover, a backup channel may be selected in this strategy. The performance of the reactive strategy is highly dependent on the speed and accuracy of the spectrum sensing process.

Figure 6.2 shows two phases for spectrum handoff in this case: link maintenance phase and evaluation phase. The evaluation phase includes spectrum sensing. The link maintenance phase includes the channel handover and switching process as well as reactive search for backup channels.

The reactive handoff strategy suffers from prolonged handoff latency because it takes place only after a handoff request is initiated. In that case, the SU starts performing spectrum sensing, backup channel selection, and channel switching. All of these tasks need time to be fully accomplished and therefore will yield latency. Similar to the nonhandoff strategy, interference with PU cannot be completely avoided due to latency and imperfect spectrum sensing.

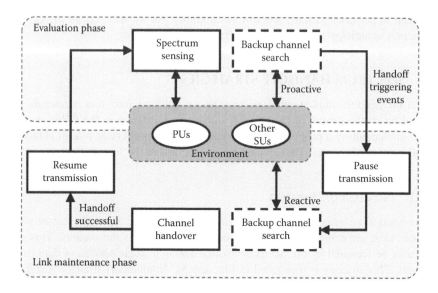

FIGURE 6.2 Spectrum handoff process. (Adapted from Christian, I. et al., *IEEE Comm. Mag.*, 50(6), 114, June 2012.)

6.3.3 PROACTIVE HANDOFF STRATEGY

In proactive handoff, the handoff process is planned and performed before a handoff triggering event occurs. Figure 6.3c illustrates this strategy. First, the SU builds a model for PU spectrum usage based on the history of past PU activity. Moreover, the SU performs spectrum sensing beforehand and prepares a list of backup channels. This list can be exchanged with the receiver. Afterward, the SU predicts future spectrum usage and switches to a suitable channel from the list of backup channels before a handoff request is initiated. Clearly, proactive handoff outperforms the previous two strategies in terms of handoff latency and PU interference avoidance. Also, the backup channel list can be ranked according to some key parameters such as SNR, SINR, delay, and expected lifetime. During the handoff process, the new channel is selected to satisfy QoS requirements.

Figure 6.2 shows the link maintenance phase and evaluation phase for proactive handoff. The evaluation phase includes spectrum sensing and proactive search for backup channels. The link maintenance phase includes the channel handover and switching process.

There are several drawbacks of proactive handoff. First, accurate PU activity modeling plays a fundamental role in this strategy. Inaccurate modeling results in erroneous prediction, which degrades the performance of spectrum handoff. Second, the list of available channels can become outdated. Therefore, there is a chance that the candidate channel has been already occupied at the time of handoff. In particular, in highly dynamic bands such as cellular bands, the list of candidates must be updated more frequently compared to the static TV bands. Third, to implement this strategy, we need sophisticated algorithms.

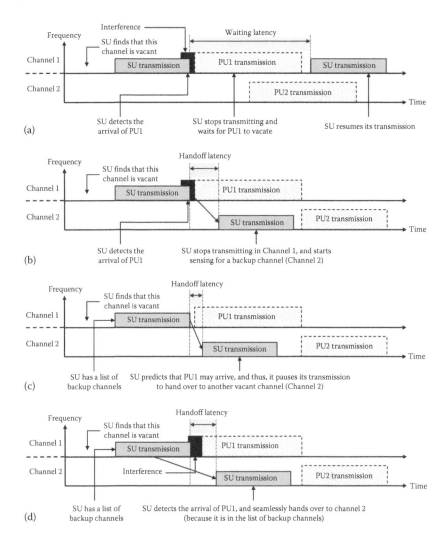

FIGURE 6.3   Spectrum handoff strategies: (a) nonhandoff; (b) reactive handoff; (c) proactive handoff; (d) hybrid handoff. (Adapted from Christian, I. et al., *IEEE Comm. Mag.*, 50(6), 114, June 2012.)

Extensive research has been carried out on proactive spectrum handoff. For instance, the authors of [3] used principles of *renewal theory* to estimate the probability that a channel will be idle in the future. Simulation results show that their approach minimizes interference to PUs by up to 30% and considerably reduces throughput jitter for SUs compared to the reactive handoff strategy. In [4], the authors built a statistical model to approximate future usability of each channel and assign an availability metric to it. The SU uses the estimated availability metric to avoid channels with heavy PU activity. Simulations conducted on a TV-broadcast scenario showed that using this metric, disruptions to PUs are significantly minimized while spectrum utilization of SUs is maximized.

6.3.4 Hybrid Handoff Strategy

This strategy is a combination of reactive and proactive handoff strategies. The backup channel list is prepared proactively while the handoff action is done reactively. Figure 6.3d illustrates this handoff strategy. Hybrid handoff strategy does not need an accurate PU traffic model because switching occurs after the handoff request is initiated. This strategy is a good compromise between reactive and proactive strategies. Short handoff period can be achieved compared to the reactive strategy. However, hybrid handoff may suffer from the problem of outdated backup channel list. In addition, there is an inevitable chance of PU interference resulting from imperfect spectrum sensing.

6.3.5 Comparison

Table 6.1 shows a brief comparison between the four spectrum handoff strategies. Regarding PU interference and handoff latency, proactive handoff has superior

---

TABLE 6.1
Comparison of Spectrum Handoff Strategies

Strategy	Nonhandoff	Reactive	Proactive	Hybrid
Principle	Stay and wait	Reactive sensing Reactive action	Proactive sensing Proactive action	Proactive sensing Reactive action
Advantages	Very low PU interference, low complexity, easy to implement	Moderate complexity	Fastest response Good channel selection	Fast response
Disadvantages	Very high SU interference with PU	Slow response	Outdated available channel selection High computational requirement	Outdated available channel selection High complexity
Handoff latency	High	Medium	Very low	Low
Dependence	PU activity	Spectrum sensing performance	Backup channels Accuracy of PU traffic model	Backup channels
Potential application	Short data transmission Applications that tolerate high delays	Applications that tolerate latency	All PU networks that allow good/accurate theoretical modeling of PU traffic	All

Source: Adapted from Christian, I. et al., *IEEE Comm. Mag.*, 50(6), 114, June 2012.

performance, followed by hybrid handoff, reactive handoff, and nonhandoff, respectively. However, this comes at the expense of increased complexity and hardware requirements to implement the proactive handoff.

6.4 DESIGN REQUIREMENTS FOR SPECTRUM MOBILITY MANAGEMENT

The main role of spectrum mobility management is to provide a smooth and fast handover between channels. This imposes a number of requirements and constraints on the design of spectrum handoff schemes. The following sections provide a brief discussion of design requirements and current issues in spectrum mobility management.

6.4.1 TRANSPORT-LAYER PROTOCOL ADAPTATION

An unsuccessful or very slow handoff will cause route failure, which leads to packet losses. Classical transport-layer protocols are not designed to deal with this kind of packet loss. For instance, TCP interprets increased packet loss as an indication of high congestion in the network. It reacts through congestion mitigation resulting in reduction in throughput. Moreover, after a spectrum handoff, TCP protocol is not supposed to be informed of the new channel characteristics in order to adjust its flow parameters. Therefore, in a CR framework, spectrum mobility management must include mechanisms to provide route failure feedback to TCP in order to avoid interpreting such failure as a result of network congestion [5].

6.4.2 CROSS-LAYER LINK MAINTENANCE AND OPTIMIZATION

Conventional layered architecture is inefficient when considering spectrum mobility issues. Spectrum mobility affects all layers in the protocol stack. Therefore, optimal performance can be achieved by employing cross-layer design methods. This will enable flexible control and information exchange among the different layers.

6.4.3 SEARCH FOR BEST BACKUP CHANNEL

At any time, several available backup channels can be used by an SU. Searching for a suitable backup channel is a challenging task, because it depends on a number of considerations such as channel bandwidth, channel quality, expected PU activity, geographic location, and QoS requirements. Poor channel selection can increase the number of handoffs and reduce the overall throughput of the SU network. Therefore, spectrum mobility management schemes must optimize the selection of backup channel according to some criteria such as expected PU activity, channel quality, bandwidth, fairness requirements, potential interference level, and so on.

6.4.4 CHANNEL CONTENTION DURING SPECTRUM HANDOFF

The appearance of PUs may trigger spectrum handoff among several SUs at the same time. Consequently, several SUs may contend to acquire the same backup channels. This will result in an increased number of unsuccessful handoffs and degrades the overall performance of the CR network. The spectrum mobility management process must include functions to facilitate spectrum access, sharing, and fairness among SUs.

6.4.5 COMMON CONTROL CHANNEL

When an SU transmitter performs a spectrum handoff, the corresponding SU receiver must be informed of the spectrum handoff to adjust its transmitter/receiver parameters and avoid possible PU interference. Performing spectrum handoff without coordinating with other SUs can lead to link failure. Therefore, coordination is very important in spectrum mobility. A common control channel (CCC) can be used to enable handoff-related information exchange among SUs. CCC can also be used to support neighbor discovery, cooperative spectrum sensing, and signaling. The CCC can have local or global coverage and can be in-band or out-of-band. Although global CCC simplifies coordination between SUs, there are several drawbacks in using this approach. For example, it is not practical to assign a dedicated network-wide CCC because the spectrum environment changes continuously. Moreover, the CCC needs to be changed regularly because of the appearance of PUs (which might claim the CCC for their own use).

6.5 PERFORMANCE METRICS

Several metrics have been proposed to measure the performance of spectrum mobility [6,7]. In the following, four performance measures are defined.

1. *Link maintenance probability (or probability of successful handoff)*: Once the PU reclaims its channel, the SU must vacate the channel and initiate a spectrum handoff process. This process may fail or succeed. The link maintenance probability is the probability that the spectrum handoff process is successful, and consequently, the SU communication is maintained.
2. *Number of handoff trials*: This metric is defined as the number of spectrum handoff trials performed by the SU over one session of transmission. The average number of trials over several sessions may give more insight than the actual number of trials per session.
3. *Handoff latency*: During the spectrum handoff process, if there is a suitable channel, the SU will switch to it immediately without delay. On the other hand, if there is no available channel, the SU will wait for a longer period until it acquires a free channel. This waiting time constitutes the main part of the handoff latency. Moreover, there is a delay due to RF front-end reconfiguration and collaboration with other spectrum management functionalities. Furthermore, since the PU detection is not perfect, the SU may continue transmitting for a period of time and causing interference to the PU before evacuating the channel. This delay is also included in the handoff latency. Therefore, handoff latency is an accumulation of individual delays incurred during the handoff process. Reducing the handoff latency is very important to minimize PU interference and to ensure quality for SU communication.

4. *Effective data rate*: During an SU transmission, there may be several interruptions because of spectrum handoff. Effective data rate is defined as the
amount of data transmitted during one session. This parameter is one of the
fundamental performance measures to evaluate spectrum handoff strategies.

The performance measures defined earlier are closely related, and they affect each
other. For example, an increased number of handoff trials result in an increase in the
handoff latency and consequently reduce the overall throughput of the network. This
dependence among performance measures needs to be addressed through making
design trade-offs and a good balance between these metrics.

6.6 MATHEMATICAL MODELS FOR SPECTRUM HANDOFF

Mathematical models can provide insights into the performance of spectrum handoff.
Several studies have employed analytical models with different levels of complexity [1,6–11]. Most of these studies suffer from assumptions that limit the accuracy
of their outcomes. For instance, the modeling in [7] is performed assuming only
a pair of SUs present and interactions with other SUs are ignored. Moreover, it is
assumed that the communicating SUs perfectly detect the presence of PU at the same
time, which is not usually the case in practice. Although all of these studies impose
unrealistic assumptions about the CRN, the obtained results can be used to design
and evaluate spectrum handoff protocols. This section presents two mathematical
models to quantify the impact of PU presence on SU communication and to compare
different spectrum handoff strategies.

6.6.1 PERFORMANCE OF SPECTRUM HANDOFF STRATEGIES

In [7], the authors have developed an analytical model to study the performance
of the nonhandoff strategy, reactive handoff, and another strategy called predetermined channel list strategy. The predetermined channel list strategy (which is different from proactive handoff) is similar to the nonhandoff strategy except that, once
the PU appears in the current transmission, the SU uses a channel randomly from a
preset list of channels.

The PU spectrum usage on each channel was modeled as a Bernoulli process.
Therefore, both the busy and idle periods of the spectrum follow a geometric distribution. The performance is measured in terms of the following criteria:

- The probability that the SU can successfully transmit a packet for a given
number of handoff trials
- The impact of the PU's presence on the effective data rate of the SU

6.6.1.1 System Model

Both PUs and SUs use the slotted transmission system, in which the channel is
divided into time slots. To make the model simple, only two communicating SUs
are considered in the CRN. PUs use a connection-oriented scheme for their transmissions. It is assumed that the PU's presence is detected perfectly by sensing the

connection setup transmission in the PU system. Moreover, the PU's transmission time T_{PU} is fixed and the PU accesses a slot independently with a probability p_{PU}. We denote by p_e the frame outage probability of the SU's transmission in a time slot.

The time required for the SU to perform spectrum sensing is denoted by T_s, and the time required for the SU to reconfigure its RF front end is denoted by T_o. In reactive handoff, it is assumed that after a handoff trial, at least one vacant slot is always available. In proactive handoff, the SU selects its new channel from a predetermined list of target channels. The SU prepares this list before the handoff triggering event by estimating the PU usage model on candidate channels. However, there is a possibility of incorrect decision if the SU selects a channel that is already occupied by a PU. This is modeled as a probability p_s of erroneous channel selection. Identical to the reactive handoff, T_o will be the time required for RF front-end reconfiguration.

The mathematical derivations are reported as derived in [7]: the SU has a payload length (denoted by l_{SU}) equal to 1500 bytes. Furthermore, the SU's data rate (denoted by r_{SU}) is 12 Mbps. Therefore, without any interruption by a PU, the time required for the SU to send its payload is denoted by T_{SU} and is computed as

$$T_{SU} = \frac{l_{SU}}{r_{SU}} = \frac{1500 \times 8 \; \text{bits}}{12 \; \text{Mbps}} = 1 \; \text{ms} \qquad (6.1)$$

Let M be the total number of slots for the SU to send its payload. Given that the slot time is 10 μs, M can be calculated as

$$M = \frac{T_{SU}}{\text{Slot time}} = \frac{1 \; \text{ms}}{10 \; \text{μs}} = 100 \qquad (6.2)$$

The system parameters used in this study are summarized in Table 6.2.

TABLE 6.2
System Parameters

Time Slot	10 μs
PU's transmission time (T_{PU})	2.5 ms
Payload length of an SU's transmission (l_{SU})	1500 bytes
SU's transmission rate (r_{SU})	12 Mbps
Execution time for a spectrum handoff trial (T_o)	1 ms
SU's radio sensing time (T_s)	100 μs–10 ms
Frame error rate for an SU's transmission (p_e)	10^{-2}
Erroneous channel selection probability (p_s)	10^{-3}–10^{-1}
PU's appearance probability in a slot (p_{PU})	10^{-3}–10^{0}
Time required for SU to send its payload without PU's interruption (T_{SU})	1 ms
Number of slots for an SU to send its payload (M)	100

Source: Wang, L. and Anderson, C., On the performance of spectrum handoff for link maintenance in cognitive radio, in *Proceedings of the International Symposium on Wireless Pervasive Computing*, Santorini, Greece, 2008.

6.6.1.2 Link Maintenance Probability

The link maintenance probability p_m is defined as the probability that an SU can send its payload within N handoff trials. Given that the SU's transmission requires M slots, p_m can be written as

$$p_m \triangleq \text{Pr\{the number of handoff trials required for the whole SU's}$$
$$\text{payload is less than } N | \text{an } M\text{-slot transmission\}} \tag{6.3}$$

Let p be the probability that a given slot is not available for an SU. For the nonhandoff and predetermined channel list strategies, the link maintenance probability is given by

$$p_m = \sum_{i=0}^{N} \binom{M-2+i}{i} p^i (1-p)^{M-1} \tag{6.4}$$

where $p_m = 1$ as $N \to \infty$.

In the nonhandoff and predetermined channel list scenarios, the next slot may not be available after a handoff trial due to continuous PU's transmission or incorrect channel selection from the predetermined list. Thus, p_m is always less than one for a limited number of handoff trials.

The nonhandoff and predetermined channel list scenarios differ on the probability p. Given the access probability p_{PU} of PUs and the frame outage probability p_e of the SU's transmission in a slot, the probability p in the nonhandoff scenario can be written as

$$p = p_{PU} + (1 - p_{PU}) p_e \tag{6.5}$$

However, for the predetermined channel list handoff, we must take into account the probability of incorrect channel selection p_s. Thus, in the case of predetermined channel list handoff, the probability p becomes

$$p' = p + (1-p) p_s \tag{6.6}$$

Consider the reactive spectrum handoff case, the link maintenance probability p_m is different from the aforementioned two cases. Because target channel selection is done after performing spectrum sensing, at least, one slot is always available after a handoff trial. Therefore, the link maintenance probability p_m can be written as

$$p'_m = \sum_{i=0}^{N} \binom{M-1}{i} p^i (1-p)^{M-1-i} \tag{6.7}$$

and for $N=M$, $p'_m = 1$.

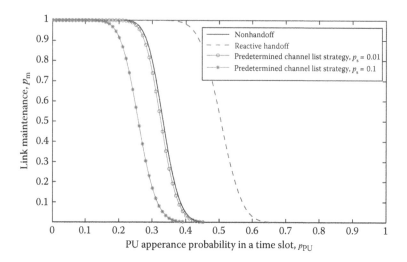

FIGURE 6.4 Link maintenance probability of three handoff strategies.

It is assumed that the outcome of spectrum sensing is always correct. Thus, the probability of erroneous channel selection can be ignored in the calculation of the probability of slot availability p.

Figure 6.4 illustrates the link maintenance of the three handoff strategies. Here, it is assumed that $N = M/2$. It is observed that reactive handoff has the best link maintenance probability. This is expected because it is assumed that there is at least one slot available after a handoff trial. In addition, the link maintenance of nonhandoff is better than that of the predetermined channel list strategy. This is due to the impact of incorrect channel selection in the predetermined channel list handoff. As p_s decreases, the link maintenance probability of the predetermined channel list handoff strategy improves. It should be noted that the link maintenance probability of the predetermined channel list scheme is upper bounded by that of the nonhandoff strategy, with equality if $p_s = 0$. This model shows that it is better not to perform handoff, rather than *blindly* selecting a channel from a predetermined list.

Note that in practice, spectrum sensing is not perfect, and due to the false alarm probability, one must expect lower link maintenance for the nonhandoff strategy. Finally, the link maintenance deteriorates as the PU's appearance probability increases regardless of the handoff strategy, which is expected. For example, the link maintenance is expected to be low in channels allocated for cellular bands, for instance, due to the high activity of the PU compared to TV spectrum channels. To mitigate this, the number of handoff trials N must be increased.

6.6.1.3 Effective Data Rate

Given l_{SU} the payload length of an SU, the effective data rate R_{SU} for SUs can be defined as

$$R_{SU} = \frac{l_{SU}}{E[t_{SU}]} \tag{6.8}$$

where $E[t_{SU}]$ is the average transmission time of an SU. $E[t_{SU}]$ depends on the used spectrum handoff strategy, as demonstrated next.

In the case of nonhandoff scenario, the SU will be delayed by T_{PU} each time a PU appears in the channel. Thus, referring to the derivations of p_m in (6.4), the average transmission time of an SU adopting the nonhandoff strategy can be expressed as

$$E\left[t_{SU}^{NHO}\right] = \sum_{i=0}^{N} \binom{M-2+i}{i} p^i (1-p)^{M-1} (iT_{PU} + T_{SU}) \tag{6.9}$$

where T_{SU} is given by (6.1). As $N \rightarrow \infty$, the average transmission time converges to

$$E\left[t_{SU}^{NHO}\right] = T_{SU} + T_{PU} \frac{(M-1)p}{(1-p)} \tag{6.10}$$

For the predetermined channel list handoff case, the SU's transmission is delayed by the execution time T_o. Similar to (6.9), the average transmission time becomes

$$E\left[t_{SU}^{PHO}\right] = \sum_{i=0}^{N} \binom{M-2+i}{i} p'^i (1-p')^{M-1} (iT_o + T_{SU}) \tag{6.11}$$

As $N \rightarrow \infty$, the average transmission time converges to

$$E\left[t_{SU}^{PHO}\right] = T_{SU} + T_o \frac{(M-1)p'}{(1-p')} \tag{6.12}$$

Finally, for the reactive spectrum handoff strategy, for each handoff trial, the SU needs spectrum sensing time T_s plus execution time T_o. Therefore, the average transmission time can be written as

$$E\left[t_{SU}^{RHO}\right] = \sum_{i=0}^{N} \binom{M-1}{i} p^i (1-p)^{M-1-i} (i(T_s + T_o) + T_{SU}) \tag{6.13}$$

When $N=M$, the transmission time in (6.13) can be expressed as

$$E\left[t_{SU}^{RHO}\right] = T_{SU} + (T_s + T_o)(M-1)p \tag{6.14}$$

Simulation results [7]:

We consider the effective data rate when the SU can always succeed in maintaining its link, that is, $p_m = 1$. Figure 6.5a and b illustrates the impact of the PU's traffic load on the effective data rate of the SU. In Figure 6.5a, we observe that the effective SU's data rate in the predetermined channel list handoff case is higher than that of

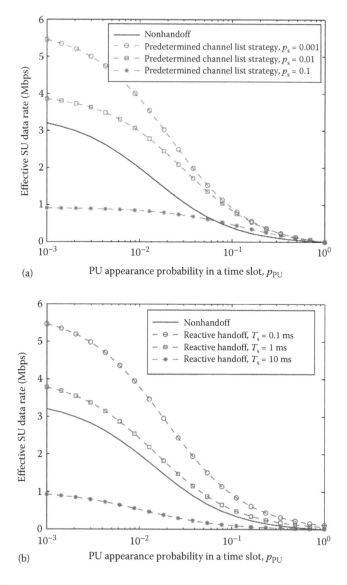

FIGURE 6.5 Impact of the PU's traffic load on the effective data rate of SUs: (a) comparison between nonhandoff and predetermined channel list handoff and (b) comparison between nonhandoff and reactive handoff.

the nonhandoff scheme when p_s is reasonably low. However, for $p_s = 0.1$, the nonhandoff scheme provides better effective data rates for low PU appearance probability. Figure 6.5b shows the impact of sensing time on the reactive handoff strategy. For T_s 0.1 ms, the reactive handoff strategy outperforms the nonhandoff strategy, whereas the performance of the nonhandoff strategy is better when the sensing time is long (e.g., T_s 10 ms).

6.6.1.4   Design of System Parameters for Spectrum Handoff Schemes

In this section, we discuss the design parameters for the predetermined channel list and reactive spectrum handoff schemes [7]. Figure 6.6a shows the required number of handoff trials N and the maximum erroneous channel selection probability p_s for several given effective data rates (contour curves). Figure 6.6b shows the required number of handoff trials N and the maximum spectrum sensing time T_s for several given effective data rates (contour curves). Here, the system satisfies the criterion the link maintenance probability $p_m > 0.9$. In Figure 6.6a and b, the

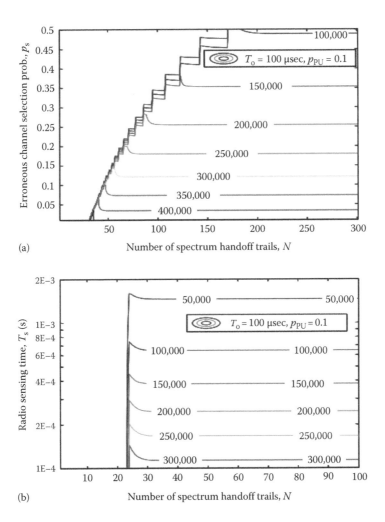

FIGURE 6.6   Contour of the effective data rate for the system parameters (a) p_s and N in the predetermined channel list handoff scheme and (b) T_s and N in the reactive handoff scheme. (From Wang, L. and Anderson, C., On the performance of spectrum handoff for link maintenance in cognitive radio, in *Proceedings of the International Symposium on Wireless Pervasive Computing*, Santorini, Greece, 2008.)

handoff execution time is set to $T_o = 100$ μs and the probability of PU appearance is $p_{PU} = 0.1$. In both figures, we observe that for every given effective rate, the required number of handoff trials increases as p_s or T_s increases. However, when the required number of handoff trials is greater than a certain limit, the maximum allowable p_s or T_s decreases. This behavior can be explained as follows. As the maximum allowable p_s or T_s increases, more handoff trials are required to increase the probability of link maintenance p_m. On the other hand, too many handoff trials imply longer transmission time. Therefore, the maximum allowable p_s or T_s must decrease to achieve the average transmission time required for the given effective data rate [7].

6.6.2 Time Relationship Model of Spectrum Handoff

The spectrum holes were modeled in [8] as a sequence of numbered channels, which are always available but have random holding time for each hole. Additionally, the SU's arrival and departure are both modeled as Poisson's random processes. Then, the spectrum handoff probability is defined as the probability of having n spectrum handoff trials within one service call duration. This model illustrates the impact of the holding time of spectrum holes on the SU's traffic. It is assumed that SU's location does not change during one service call duration.

6.6.2.1 Model

1. *SU's call duration model:* The service call arrival and departure of cognitive user are both Poisson's random process. Thus, the service call duration has a negative exponential distribution. The probability density function of SU call duration is modeled as

$$g(x) = \begin{cases} \mu e^{-\mu x} & x \geq 0 \\ 0 & x < 0 \end{cases} \tag{6.15}$$

where $1/\mu$ is the mean time of SU call duration.

2. *Spectrum holes model:* Spectrum holes used by an SU in one call duration are numbered as 0, 1, 2, and so on. The holding time X_i of each spectrum hole is assumed to be a general continuous random variable, and X_i $(i = 0,1,...)$ has identically independent distribution. The probability density function is denoted by $f(x)$. We can consider also the special case when the arrival and departure of spectrum holes also obey Poisson's distribution. Assuming the mean value of each hole's holding time is $1/\lambda$, the density and distribution functions of X_i $(i = 0,1,...)$ are, respectively,

$$f(x) = \begin{cases} \lambda e^{-\lambda x} & x \geq 0 \\ 0 & x < 0 \end{cases} \tag{6.16}$$

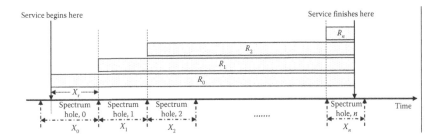

FIGURE 6.7 Time relationship during one service call holding time. (Adapted from Liu, H.-J. et al., Study on the performance of spectrum mobility in cognitive wireless network, in *Proceedings of the IEEE International Conference on Communications Systems*, Singapore, 2008.)

The mean value and variance should meet the following requirements:

$$\begin{cases} E[X_i] = \dfrac{1}{\lambda} < \infty \\[2mm] \mathrm{Var}[X_i] < \infty \end{cases} \qquad i = 0,1,2,\ldots \tag{6.17}$$

When a service call begins, the first spectrum hole accessed is numbered as 0, and subsequent spectrum holes accessed by the SU are numbered as 1, 2, and so on.

3. *Spectrum handoff probability:* Figure 6.7 shows a graph of the spectrum holes experienced by an SU during one service call duration. X_r denotes the residual time of spectrum hole 0 after the beginning of a service call. X_i denotes the holding time of spectrum hole i occupied by the SU. R_0 denotes the entire service call duration time of the SU. R_i denotes the time period from the beginning of spectrum hole i occupied by the SU to the end of the current service call.

In Figure 6.7, there are n times spectrum handoff within a service call duration time R_0. The probability of n times spectrum handoffs occurring within the time period R_0 is denoted by P_n. The probability of having no spectrum handoff within R_0 is denoted by P_0. Thus, P_0 is equal to the probability of $R_0 \le X_r$.

From Figure 6.7, we can see that probability of having n spectrum handoff trials occurring within R_0 equals the probability of the following events happening simultaneously:

a. The residual time of spectrum hole 0 provides X_r less than the current call duration time R_0.

b. The SU switches to spectrum holes 1, 2, ..., $(n-1)$ before current service call ends.

c. The SU switches to spectrum hole n after current service call ends.

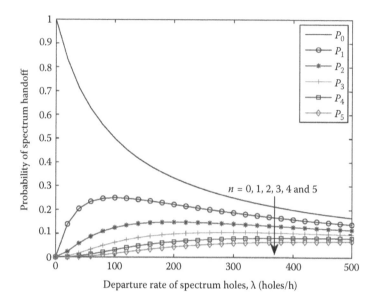

FIGURE 6.8 Impact of λ on spectrum handoff probability.

6.6.2.2 Impact of λ and μ on Spectrum Handoff Probability

The details of the analysis can be found in [8]. Here, we present some of the simulation results to see the impact of the different parameters on the handoff performance.

Let $\mu = 100$ calls/h and $\lambda = 0 \sim 500$ spectrum holes/hour. As Figure 6.8 shows, the probability of spectrum handoff increases as λ increases. This is because as the value of λ increases, the moving rate of spectrum holes becomes faster and the average holding time of each spectrum hole decreases, which means that the SU will experience more spectrum holes during one service call.

Figure 6.9 shows the case when λ is fixed at 100 spectrum holes/hour, and let $\mu = 0$–1000 calls/h. It can be noticed that the spectrum handoff probability increases sharply with respect to μ before it stabilizes after some limit. This is because as μ increases, the average duration time of each service call is shorter. Additionally, the moving rate of spectrum holes is relatively slower when compared with the departure rate of SUs. Therefore, the number of spectrum holes accessed during one service call is smaller, leading to lower number of spectrum handoff trials.

6.7 MB-CRN HANDOFF

The need of frequent handoff can be reduced by using multiband cognitive radio networks (MB-CNRs) [12]. For example, Figure 6.10 shows that at time $t = T_1$, there are three SUs. SU1 and SU3 use single-band detection. The former uses reactive handoff, whereas the latter maintains a list of candidate channels (one channel in this scenario). SU2, however, uses multiband sensing and accesses two consecutive channels.

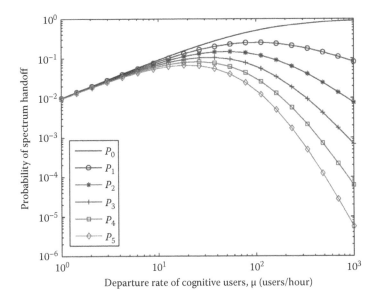

FIGURE 6.9 Impact of μ on spectrum handoff probability.

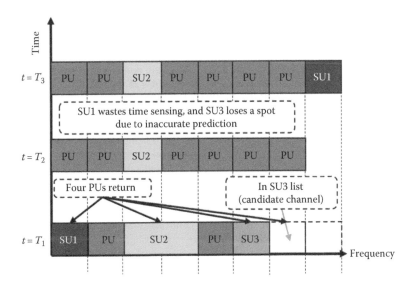

FIGURE 6.10 Illustration of handoff in MB-CRN.

At time $t = T_2$, four PUs return to their licensed bands, as shown in Figure 6.10. Thus, SU1 performs spectrum sensing to check for other available bands. SU3 does not perform sensing because it already has a list of candidate channels. However, due to the inaccurate prediction or information, a PU returns to SU3's candidate channel. Thus, SU3 would not seamlessly handover to that channel. On the other hand, SU2 would stop transmitting on one of its bands and reallocate the power of that band to the other one. Therefore, transmission interruption is not enforced on SU2 even though the throughput is going to be degraded since the bandwidth becomes smaller.

At time $t = T_3$, SU1 finds another channel after spectrum sensing, and thus, it has enforced data interruption, SU2 does not have data interruptions, and SU3 does not find a channel since it has inaccurate data records about the spectrum occupancy. This simple illustration demonstrates that not only multiband sensing is beneficial, but also multiband access is useful, since it reduces the handoff frequency, and consequently, the communication overhead and network load are reduced.

6.8   CONCLUSION

This chapter covered handoff strategies in the context of CRNs. Three main strategies have been discussed: nonhandoff, reactive, and proactive. Some mathematical tools have been provided for the analytical study of the different handoff strategies. Moreover, the chapter discussed the handoff process in the context of MB-CRN, which is expected to allow less frequent and more efficient handoffs.

REFERENCES

1. L. C. Wang, C. W. Wang, and K.-T. Feng, A queueing-theoretical framework for QoS-enhanced spectrum management in cognitive radio networks, *IEEE Transactions on Wireless Communications*, 18 (6), 18–26, December 2011.
2. I. Christian, S. Moh, I. Chung, and J. Lee, Spectrum mobility in cognitive radio networks, *IEEE Communications Magazine*, 50 (6), 114–121, June 2012.
3. L. Yang, L. Cao, and Z. H., Proactive channel access in dynamic spectrum network, in *Proceedings of the Cognitive Radio Oriented Wireless Networks and Communications (CROWNCOM)*, Orlando, FL, 2007.
4. K. Acharya, P. Aravinda, S. Singh, and H. Zheng, Reliable open spectrum communications through proactive spectrum access, in *Proceedings of the First International Workshop on Technology and Policy for Accessing Spectrum*, Boston, MA, 2006.
5. V. Balogun, Challenges of spectrum handoff in cognitive radio networks, *Pacific Journal of Science and Technology*, 11 (2), 304–314, November 2010.
6. Y. Zhang, Spectrum handoff in cognitive radio networks: opportunistic and negotiated situations, in *Proceedings of the International Conference on Communications*, Dresden, Germany, 2009.
7. L. Wang and C. Anderson, On the performance of spectrum handoff for link maintenance in cognitive radio, in *Proceedings of the International Symposium on Wireless Pervasive Computing*, Santorini, Greece, 2008.
8. H.-J. Liu, Z.-X. Wang, S.-F. Li, and M. Yi, Study on the performance of spectrum mobility in cognitive wireless network, in *Proceedings of the IEEE International Conference on Communications Systems*, Singapore, 2008.
9. C. Wang and L. Wang, Modeling and analysis for proactive-decision spectrum handoff in cognitive radio networks, in *Proceedings of the IEEE International Conference on Communications*, Dresden, Germany, 2009.

10. L. C. Wang and C. W. Wang, Spectrum handoff for cognitive radio networks: Reactive-sensing or proactive-sensing?, in *Proceedings of the IEEE International Conference on Performance, Computing and Communications*, Dalian, China, 2008.
11. C. W. Wang, L. C. Wang, and F. Adachi, Modeling and analysis for reactive-decision spectrum handoff in cognitive radio networks, in *IEEE GLOBCOM*, Miami, FL, 2010.
12. G. Hattab and M. Ibnkahla, Multiband cognitive radio: great promises for future radio access, *Proceedings of the IEEE*, 102 (3), 282–306, March 2014.

7 MAC Protocols in Cognitive Radio Networks

7.1 INTRODUCTION

This chapter covers medium access control (MAC) protocols in the context of cognitive radio networks (CRNs). In particular, it addresses cooperative and noncooperative MAC-layer protocols that can be used in the absence of a common control channel (CCC). A special focus is given to fairness protocols and mobility support. The reader is expected to have some basic knowledge of MAC protocols of classical networks (see, e.g., [15]).

7.1.1 FUNCTIONALITY OF MAC PROTOCOL IN SPECTRUM ACCESS

In the open systems interconnection (OSI) model, the MAC layer is a sublayer of the data link layer. It provides addressing and channel access control mechanisms to enable several terminals to communicate and have access to the communication medium. In particular, the channel access control mechanisms provided by the MAC layer enable several stations connected to share the same physical (PHY) medium.

The cognitive radio (CR) concept has evolved as the potential technology to solve the problem of spectrum scarcity and limited spectrum utilization. The CRN allows its users to access the vacant portions of the spectrum in the absence of primary users (PUs) to ensure that no harm is caused to PUs. This necessitates adapting the system parameters to the dynamically changing radio environment, learning about the spectrum occupancy, making decisions on the quality of the available channels, etc. This motivates the research on CR MAC protocols, with an aim of controlling the spectrum sensing process, deciding about the channel occupancy, sharing the spectrum among CR users, maintaining tolerable interference levels to PUs, and ensuring fairness among CR users. Figure 7.1 gives a generic architecture of the layered protocol stack in the context of CRNs [17,18] and the possible role of the different layers.

The shared nature of the wireless channel necessitates coordination among CR users. In this respect, spectrum sharing provides the capability to maintain the Quality of Service (QoS) of SUs without causing harmful interference to PUs. This can be performed by coordinating secondary users' (SUs') access to the medium as well as adaptively allocating resources to them (such as power, bandwidth, and so on) [1].

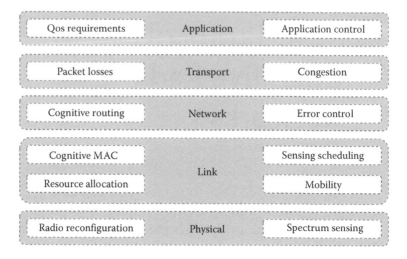

FIGURE 7.1   A generic architecture of the CR protocol stack including five layers.

7.1.2 DIFFERENCE BETWEEN TRADITIONAL MAC AND CR MAC

Studies of CR MAC protocols [1–6] report several differences between classical MAC and cognitive MAC protocols, in particular:

- The carrier sense mechanism at the MAC layer may not reveal complete information regarding the channel occupancy, because it is difficult to distinguish between the energy radiated by SUs and active PUs. For example, if SU1 accesses an empty channel and SU2 wants to sense that channel (through a carrier sense protocol), it is difficult for SU2 to tell whether the channel is used by an SU or a PU.
- Packets may be simply retransmitted in the event of a collision with other CR users, while the transmission must cease immediately if the packet loss is due to PU activity. To differentiate these causes, the physical layer may support the MAC layer in the implementation of the sensing strategy and identifying the origin of the radiated power.

A general framework of the spectrum functions and the interlayer coupling is shown in Figure 7.2. Based on the information coming from the physical-layer radio frequency (RF) environment, the sensing scheduler at the MAC layer can determine the sensing and transmission times. Whenever data packets need to be sent, access to the medium for transmission is coordinated by the spectrum access function.

7.1.3 CENTRALIZED VERSUS DISTRIBUTED ARCHITECTURES

In the interweave paradigm, SUs search for unoccupied channels and build a list of candidate channels. Since the channel list depends on the sensing performance of

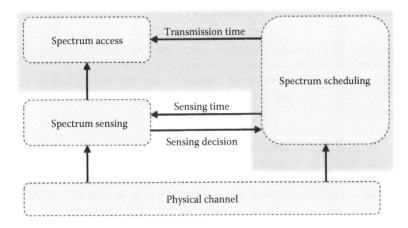

FIGURE 7.2 Spectrum functions at the CR MAC.

SUs, SUs can have different lists. However, for transmission, every communicating SU pair needs to agree on the channels to use. Therefore, careful coordination between SUs to choose and access channels is required. Such coordination can be employed in a centralized or distributed manner.

Centralized-based architectures depend on a fusion center (FC) (also known as a CR base station [CBS], or a sink node). This fusion center coordinates and controls SUs' access (e.g., which user will go for which channel and for how long). Typically, a database is created at the fusion center. Such database could be established via the help of SUs. The SU that wants to access a channel must consult with the fusion center. The main limitation of centralized schemes is the additional overhead incurred due to the mandatory communication between SUs and the fusion center. In addition, PUs' activity is always changing, and thus, periodic updates of the database are vital.

On the contrary, distributed schemes do not rely on a centralized base station (BS). Here, SUs must cooperate with each other to coexist and access the available bands. In particular, each cooperating SU must perform local spectrum sensing and share the results with other SUs. SUs must coordinate with each other for a fair sharing of the available spectrum resources.

7.1.4 Concept of Common Control Channel in CR MAC

In conventional wireless communication systems, control channels are used to facilitate spectrum sharing and spectrum access among multiple users. However, due to the unique challenges of CRNs, coordination between users using a common control channel (CCC) poses many challenges. For instance, is the control channel predetermined, or does it depend on the sensing results? Can the control channel be changed dynamically based on the PU activity?

Two different approaches can be adopted for CCC [5]:

1. *In-band CCC*: The same channel that is used for SU's information data transmission is also used for transmitting control data.
2. *Out-of-band CCC*: A different channel is used for sending control messages. Such a channel could be reserved as CCC either temporarily or permanently.

Clearly, the advantage of the in-band CCC is that CR user switching between two different channels (for information data and control data, respectively) is avoided, unlike the case for the out-of-band CCC (where there are dedicated channels for control data and information data, respectively). Since in in-band CCC an SU sends both data and control messages on the same channel, the throughput of the SU would be reduced. In addition, this scheme is affected by the PU's activity. That is, if a PU suddenly returns, the SU must stop sending these control messages to avoid interference with the PU.

7.1.5 CLASSIFICATION OF MAC PROTOCOLS

CR MAC protocols can be classified into three categories [1]:

1. *Random access protocols*: The advantage of this class of MAC protocols is that many SUs can contend for a given channel. These protocols typically employ carrier sense multiple accesses (CSMA) with collision avoidance (CSMA/CA) scheme. The SU must sense the spectrum, and if it declares that a channel is unoccupied (by both PUs and other SUs), then it can use it.
2. *Time-slotted protocols*: The SUs transmit over dedicated time slots. Here, scheduling and time synchronization are essential.
3. *Hybrid protocols*: These protocols combine the aforementioned schemes. Here, the time is slotted and SUs contend for the channel in the corresponding time slots. Control messages are usually sent over dedicated time slots.

The remainder of this chapter discusses the main issues related to CR MAC protocol design. This is done through the study of some representative protocols that have been recently proposed in the literature. Rather than presenting the basics of MAC protocols, we target some important issues in CR MAC and illustrate them through examples (i.e., study some protocols that have addressed these issues). Section 7.2 provides a description of CSMA-/CS-based protocols and interframe spacing. The section then addresses the case of MAC protocols that do not involve a CCC. Section 7.2 studies a number of representative MAC protocols in the centralized and distributed architectures, respectively. Finally, Sections 7.4 and 7.5 address QoS and mobility support in CR MAC, respectively.

7.2 INTERFRAME SPACING AND MAC CHALLENGES IN THE ABSENCE OF CCC

This section describes interframe spacing in CSMA-/CA-based protocols. We discuss the classes of protocols that do not involve a CCC and in which interframe spacing needs to be carefully managed.

7.2.1 INTERFRAME SPACING IN CSMA-/CA-BASED PROTOCOLS

The design of MAC protocols is more challenging in wireless networks than wired networks. This is due to the fact that in wireless medium the signal strength decays due to fading, which causes the medium characteristics to be highly location dependent. Hence, the traditional CSMA protocol does not work well and this gives rise to the *hidden and exposed terminal problems*. To deal with these problems, IEEE 802.11 protocol supports two modes of operation: the distributed coordination function (DCF) and the point coordination function (PCF) [15]. In DCF, stations act independently and do not have a central control. In PCF framework, an access point controls all activity in its cell.

Random access protocols in wireless LANs may use channel reservation techniques by exchanging short request-to-send (RTS) and clear-to-send (CTS) control packets before the actual data packets are sent [14]. This allows temporary channel reservation for the data packet transmission. In addition, the neighbors of the receivers (receiving CTS or RTS packets) will defer their transmissions for a duration of network allocation vector (NAV). NAV duration is estimated by each potential transmitter as the time required by the current transmitter to finish its transmission and its related acknowledgments (ACKs).

PCF and DCF can coexist within one cell. It works by carefully defining the interframe time interval. This has been defined in [15] as follows. After a frame has been sent, a certain amount of dead time is required before any station may send a frame. Four different intervals are defined, each for a specific purpose. The shortest interval is short interframe spacing (SIFS). It is used to allow the parties in a single dialog to go first. For example, this includes letting a receiver send a CTS as a response to an RTS, letting a receiver send an ACK for a fragment or full data frame, and letting the sender of a fragment burst transmit the next fragment without sending an RTS again. There is only one station that is entitled to respond after an SIFS interval. If it fails to make use of the channel and a PCF interframe spacing (PIFS) time elapses, the BS may send a beacon frame or poll frame. This mechanism allows a station sending a data frame or fragment sequence to finish its frame without anyone else getting in the way, but gives the BS a chance to grab the channel when the previous sender is done without having to compete with eager users. If the BS has nothing to send, and a DCF interframe spacing (DIFS) time elapses, any station may attempt to acquire the channel. The usual contention rules apply. The last time interval, extended interframe spacing (EIFS), is used only by a station that needs to report a bad or unknown frame. Giving this event the lowest priority comes from the fact that the receiver does not know what to do with the frame; therefore, it should wait a substantial time

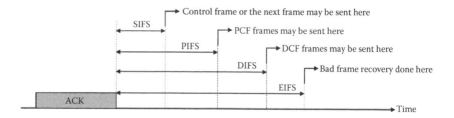

FIGURE 7.3 Interframe spacing in 802.11. (From Tanenbaum, A. and Wetherall, D., *Computer Networks*, 5th ed., Prentice Hall, Upper Saddle River, NJ, October 2010.)

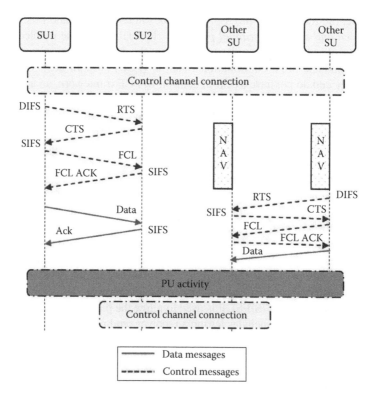

FIGURE 7.4 Generic communication in CR MAC protocol using interframe spacing. (Adapted from Shah, M. et al., An analysis on decentralized adaptive MAC protocols for cognitive radio networks, in *Proceedings of the 18th International Conference on Automation & Computing*, Leicestershire, U.K., September 2012.)

to avoid interference with an ongoing dialog between two stations. This is illustrated in Figures 7.3 and 7.4.

MAC protocols with DCF dedicate one channel for exchanging control messages, and all other channels are used for data exchange. This is basically the CCC mentioned earlier. Figure 7.4 illustrates a generic control and data frame exchanges in

CR MAC. The example includes exchange of RTS/CTS frames and Free Channel List (FCL) information. FCL includes the available channels that may be used by the SU pair for their transmission.

7.2.2 MAC CHALLENGES IN THE ABSENCE OF CCC

There are several MAC protocols that have been designed without CCC [2]. This section describes four of these protocols and discusses their challenges:

- Jamming-based MAC
- Interleaved carrier sense multiple access
- Multichannel Medium Access Control (MMAC)
- Synchronized MAC

1. *Jamming-based MAC (JMAC)* [2]: In this protocol, the channel is subdivided into two subbands: S and R as shown in Figure 7.5. These subbands do not necessarily have equal bandwidth. The former is used for RTS and information data transmissions, and the latter is used for ACK and CTS transmission. The access mechanism in this protocol is as follows. First, the source must send an RTS frame over channel S only if channel R is declared vacant for at least DIFS time duration. However, if channel R is busy, a *Backoff* procedure is initiated. Second, the destination must be listening for RTS frames in channel S. Upon successful reception of the RTS frame, the destination transmits a CTS to the source. Meanwhile, the source must be listening in channel R for CTS frames. Once the source receives the CTS, it starts transmitting the useful data over channel S. Finally, when the destination receives the data, it transmits an ACK frame to the source over channel R. The procedure is illustrated in Figure 7.5.

   Even though the reservation procedure of IEEE 802.11 is similar to that of the JMAC protocol, the medium can be jammed in JMAC as long as required. In addition, the destination in JMAC protocol also jams channel R during the reception of the data on channel S. This is necessary because

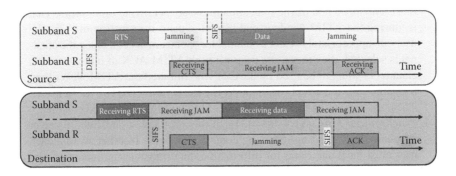

FIGURE 7.5 The access procedure of JMAC protocol.

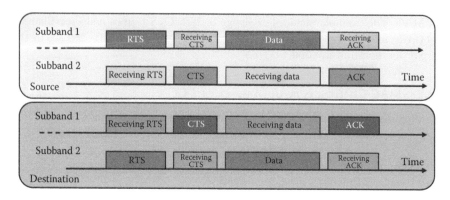

FIGURE 7.6   The access procedure of ICSMAC protocol.

jamming channel R prevents the other users from transmitting RTS frames (recall that an RTS frame is transmitted over channel S only if channel R is vacant for DIFS time period).

2. *Interleaved Carrier Sense Multiple Access (ICSMAC)* [2]: In this protocol, the channel is again divided into two subbands. However, unlike the JMAC protocol, these subbands must have equal bandwidth, and more importantly, the transmission can be initiated in any one of them. Figure 7.6 illustrates the access mechanism of this protocol, where simultaneous transmission is achieved. Particularly, if the source transmits RTS on channel 1, then the data must be transmitted over the same channel, and the destination must transmit CTS and ACK on the same channel as well. To enable simultaneous transmission, the destination must transmit RTS and data over the other channel (channel 2 in this case).

3. *Multichannel Medium Access Control (MMAC)* [2]: In this protocol, each user has a preferred channel list (PCL), which must have up-to-date information about the occupancy of listed channels in vicinity. This list is essential because this protocol allows each user to utilize several channels, where the user dynamically switches between them. Figure 7.7 illustrates the access mechanism of the MMAC protocol. To select a channel, the user must send an ad hoc traffic indication message (ATIM) to other users. That is, during the ATIM window, and using the prior information in PCL, the user can appropriately select a channel for spectrum access. Once the channel is selected, the other users transmit back ATIM-ACK to indicate which channel is going to be used by the user who initiated the ATIM. After that, the same user must reply with an ATIM-Reservation (ATIM-RES) frame. Then, the regular handshake process is used within the selected channel.

4. *Synchronized MAC (SYN-MAC)* [1]: In this protocol, a dedicated receiver is required for sending/receiving control messages. In other words, each user must have one radio for information data transmission/reception and another radio for control message transmission/reception (Figure 7.8) [1].

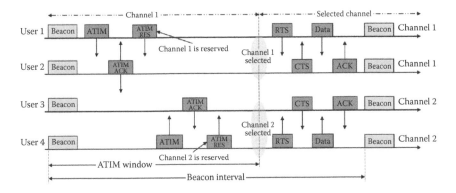

FIGURE 7.7 Access procedure of MMAC protocol. (Adapted from Kosek-Szott, K., *Ad Hoc Networks*, 10(3), 635, May 2012.)

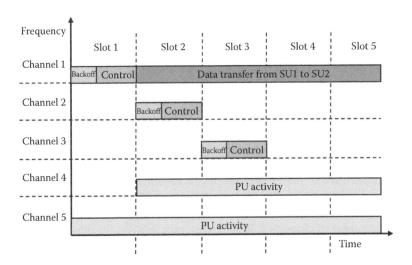

FIGURE 7.8 Control and information data packet exchange in SYN-MAC. (From Cormio, C. and Chowdhury, K., *Ad Hoc Networks*, 7, 1315, 2009.)

The access mechanism is as follows. Time slots are used such that each slot is dedicated for a channel, and control messages can be sent over these time slots. However, information data transmission can occur over any available channel. In other words, frequency hopping is used for control messages, where the user periodically transmits these messages over different frequency bands. For example, in Figure 7.8, there are five channels and five time slots. For each frequency channel, control messages can be sent only on the dedicated time slot. Users send their control data on these time slots after a random backoff period. Assume that SU1 wants to communicate with SU2 and that SU1 has sensed channels 1, 2, and 3 to be available, whereas SU2 has sensed channels 1, 3, and 5 to be available. SU1 picks channel 1 and waits for the beginning of time slot 1 to tune its radio to that channel. After a backoff period, SU1 contends for channel 1. Upon successful contention, data transmission is held between SU1 and SU2.

7.2.3   Network Setup in the Absence of CCC

Both in centralized and in distributed architecture, the network setup problem needs to be addressed. For example, when the PU of a chosen control channel returns, a new control channel needs to be assigned. It is important to know how the first node contacts the control BS and how it would be informed about the chosen control channel for the first time. The authors in [16] present three protocols to address this problem, which are designed without CCC:

1. Exhaustive (EX) protocol
2. Random (RAN) protocol
3. Sequential (SEQ) protocol

To initiate the network, these protocols include a scan and search procedure for the control BS (CBS) and SUs. The primary user traffic rate (PUTR) is defined as the average rate at which the PU changes its state (active/inactive). The channel availability is directly related to PUTR (i.e., a higher PUTR implies that the channel availability for SUs will fluctuate at a higher rate). The total spectrum in which the SUs can operate is divided into a fixed number of channels, N.

It is also assumed that PU's traffic does not vary in one search cycle. The EX protocol assumes that the number of channels N is known, and the channels are searched from lower to higher frequencies by both the CBS and SUs. Unlike the deterministic/ EX approach, the RAN protocol is a probabilistic approach. This is useful in situations where the number of channels N is not known precisely. On the other hand, the SEQ protocol is a modified version of EX protocol which incorporates the multihop case. In this protocol, the total number of channels N is assumed to be known.

7.2.3.1   Basic Access Procedure of the Protocols

- Centralized architecture
  For EX protocol, the system has a CBS equipped with a timer. Initially, the CBS initiates the timer and starts its search from the channel with the lowest frequency. The timer duration is T_s seconds, which is the length of a time slot. When the timer expires, the CBS looks for the next channel. In each time slot, the channel is scanned for the presence of a PU. If the channel is occupied, then the CBS moves to the next channel and restarts the timer again. If the channel is available, a beacon is sent indicating the availability of that channel. The CBS will wait for a response till the timer expires and then moves its search to the next channel. In the meantime, if a response is received from an SU, then negotiations with the SU will take place while the CBS continues its search for other potential channels. After all channels

are searched, the CBS will restart from the lowest frequency again. During the process, SUs listen to channels and wait for beacons coming from the CBS. The wait period is denoted by T_w, which is equal to NT_s. If a beacon is received, SUs respond by sending a beacon signal.

In RAN protocol, the CBS follows the same mechanism, except that the channels are searched in a random manner. SUs shift from a channel to another if they do not hear beacon signals. The wait time in this case is chosen as $T_w = W_s T_s$, where $W_s \leq N$.

* Multihop distributed architecture

In multihop distributed architecture, there is no central CBS. The EX protocol is modified such that an SU will wait for the beacons and see if the network is already initiated. Otherwise it will initiate the network by sending beacons. To know whether the network is initiated or not, an SU has to make sure that it is not the first user in the network. Thus, it will wait for a beacon in every channel for T_w seconds (T_w is randomly chosen from a set of predefined values). If it receives a beacon, it acknowledges it (and starts sharing the control data). If it does not receive a beacon, then it will consider itself as the first SU in the network and it starts sending beacons (as the CBS did in the centralized architecture).

In RAN protocol, SUs will follow the same rules as they did in a centralized scenario, and additionally, they will send beacons for every T_s seconds for a total wait period of T_w seconds ($T_w = W_s T_s$ where W_s is chosen at each cycle from a predefined set). Upon successful reception of a beacon from an SU, an SU acknowledges it and exchanges the channel information with the sender.

In SEQ protocol, the SU also starts from a random channel. The next channel is chosen in the increasing order of frequencies. After the last channel is reached, the next channel is chosen in decreasing order of frequencies (and not from the lowest). If the chosen channel is not available, the SU moves its search to the next channel. If the channel is available, it stays for a period of T_s in that channel and sends a beacon during that period. If it receives an acknowledgment from a neighboring SU (confirming that it has received the beacon), then the two SUs exchange their control data. See also [10] for other multihop protocols.

The protocols have been evaluated through computer simulations in [16]. In Figure 7.9, the average search time for each protocol has been displayed as a function of the number of channels. The search time is defined as the time taken for an SU to receive a beacon from the CBS, that is, the time taken before an SU connects to the CBS. Note that the total network setup time is directly proportional to the search time.

The following parameters are taken in the simulation [16]: Duration of a time slot, $T_s = 1$ s; beacon time duration $= 100$ ms; Time taken to move to a channel and check its availability $= 100$ ms. It can be seen that the EX protocol takes the least search time and the SEQ protocol takes the largest search time. Therefore, EX protocol is efficient when the total number of channels is known. However, if the number of channels is unknown, then RAN protocol is a good choice compared to SEQ protocol.

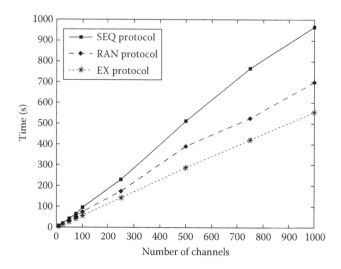

FIGURE 7.9 Average search time performance [16], the duration of a time slot, $T_s = 1$ s. (From Kondareddy, Y. et al., Cognitive radio network setup without a common control channel, in *Proceedings of IEEE Military Communications Conference (MILCOM)*, San Diego, CA, 2008.)

7.2.4 FAIR ALLOCATION OF CCC

The authors in [12] propose a decentralized nonglobal MAC protocol (DNG-MAC) based on time division multiplexing mechanism and using a fair allocation of control channels to all candidate SUs.

Basic access procedure of DNG-MAC [12]: This protocol works as follows. An SU node initiates the operation of the protocol by selecting one of the best channels as the CCC as illustrated in Figure 7.10. The selection criterion for the

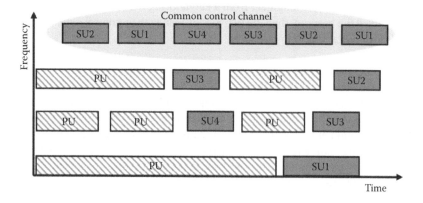

FIGURE 7.10 Allocation of CCC in the DNG-MAC protocol.

best channel in this case is arbitrary. The control channel is divided into time slots of fixed length. Each time slot has a listen period and a transceiving period. All CR nodes in the network are synchronized in the listening period of each time slot. Control information is exchanged in the transceiving period between the SU communication pair (SUCP). The durations of the time slots are carefully selected by calculating the average time required for each secondary pair to complete negotiation on the available CCC.

This mechanism has a shorter overall waiting time for the SUs to access CCC, compared to the waiting time in other MAC protocols. Here, every SU is queued to access the CCC. The vacant channels on the spectrum can be simultaneously used by more than one communicating pair. This is expected to improve the overall throughput of the CRN. However, the limitation of the protocol is that it assumes synchronization among the CR nodes.

We report here the simulation work done by Shah et al. [12] where an ad hoc–based scenario with 10 CR nodes has been considered. All devices used 1 Mbps maximum transmission rate, 5 mW transmission power, differential phase-shift keying (DPSK) modulation, and time slots of 10 ms. The size of a control frame is 20 bytes. Figure 7.11 illustrates the actual data exchange rate between two CR nodes. Due to the burst nature of the traffic, uneven curves have been obtained. Since a CR node cannot transmit until a negotiation has taken place and the opportunity to transmit has to be seized, the traffic sent remains less than 50 kbps in the first 50 s and gradually increases till it reaches about 100 kbps after 200 s.

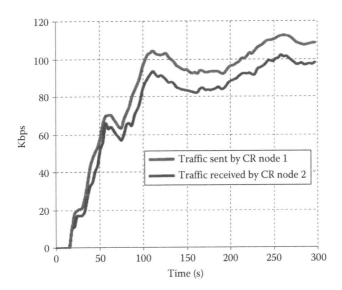

FIGURE 7.11 Traffic sent and received by two CR nodes in DNG-MAC protocol. (From Shah, M. et al., An analysis on decentralized adaptive MAC protocols for cognitive radio networks, in *Proceedings of the 18th International Conference on Automation & Computing*, Leicestershire, U.K., September 2012.)

7.3   QOS IN CR MAC

SUs may carry multiple types of traffic with different QoS requirements. On the other hand, the resources that are available to SUs are highly dynamic. This is because the reserved spectrum bands experience random use by PUs. The access of SUs is determined by several factors such as the presence of PUs, presence of other SUs, interference levels, channel quality, and so on. Therefore, it is very challenging to design efficient spectrum-aware MAC protocols with QoS provisioning to SUs.

QoS-aware MAC protocols' requirements include real-time (RT) and heterogeneous traffic (such as video, voice, data, and so on) to ensure fairness among competing nodes, minimize access delay, maximize channel utilization and throughput, etc.

In the following, we discuss two protocols that illustrate the challenges of QoS provisioning in CRNs and how these challenges have been addressed.

7.3.1   Distributed QoS-Aware Cognitive MAC (QC-MAC) Protocol

This protocol has been proposed in [9] to assure QoS for delay-sensitive multimedia applications for CR users. A priority spectrum access scheme has been applied to achieve multiple QoS levels for heterogeneous SUs. In addition, an analytical model has been developed to study the delay performance of the proposed MAC, considering the activities of both PUs and SUs.

In the system model, SUs are assumed to be in a single-hop mode, where they can directly communicate with each other in a distributed manner. SUs opportunistically access the data channels and can only sense one channel at a time. It is assumed that an SU can accurately determine the channel status after a basic sensing period, for example, 1 ms. An SU senses the first channel and starts data transmission if the channel is sensed idle for a sensing interval. To reduce possible collisions among SUs, each SU will sense the channel for an arbitrary sensing period (ASP). The ASP consists of the basic sensing period (that assures satisfactory sensing accuracy) plus some random time slots selected from a window [0, SWi]. If the channel is sensed busy, the SU switches to the second channel. For transceiver synchronization, the sender will initiate a handshake with its receiver over the control channel at the beginning of the channel sensing period. If a PU appears during an SU's data transmission, the SU will switch to the next available channel. The SUs can determine the channel sensing sequence according to the two different policies:

1. *Greedy policy*: The SUs simply sort channels in a descending order and always use channels with the highest success probability (calculated according to an analytical model) for achieving a low delay and high throughput. However, the channel with less PU activity is more likely to be selected by SUs, which may cause a high contention level among SUs sharing the same radio resources.

2. *QoS-based policy*: An SU selects a set of channels that can satisfy its QoS requirements and starts sensing these channels. For example, if delay is a QoS requirement, then the channels with the potential least delay will be sensed in an ascending order.

QoS provisioning of QC–MAC protocol is further enhanced by introducing service differentiation in the ASPs of different traffic flows. The idea is to apply a smaller sensing window (SW) for higher priority real-time applications so that they have a higher chance to access data channels, for example, $SW_{voice} < SW_{video} < SW_{data}$. Therefore, through a careful choice of the SWs, multiple levels of QoS provisioning can be achieved for different types of traffic.

7.3.2 QoS-Aware MAC (QA-MAC) Protocol

QA-MAC has been proposed in [13] where SUs with RT services have priority over SUs with non-real-time (NRT) traffic.

The model considers $N + 1$ channels available for use in the same bandwidth. These channels include one CCC, which is divided into two periods: reporting period and contention period. The reporting period is further slotted into N minislots, each of them corresponding to one of the N available channels (Figure 7.12).

At the beginning of each time slot, SUs sense the channels, and then report the channel state in the corresponding minislots. During the contention period, the SU includes its spectrum information (i.e., the result of the SU's sensing procedure) and service type (SST) into the RTS/CTS packets. When the SU exchanges the RTS/CTS packets, the neighboring SUs overhear these ongoing RTS/CTS packets, and then know whether they have sensed the same channels. If there are neighboring SUs that have sensed the same channels as the sender in the tth time slot, then each of them will sense another different licensed channel in the $(t + 1)$th time slot. After RTS/CTS exchange, each SU will get partial information of the leftover licensed channels. Based on the service type of each SU, contention-free and contention-based algorithms will be employed accordingly. This means that SUs with RT traffic will reserve specific channels, whereas SUs with NRT traffic will contend for the remaining leftover channels.

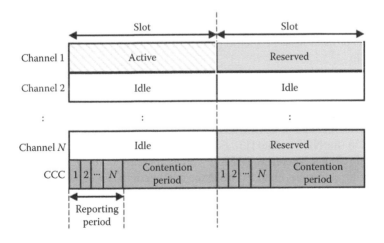

FIGURE 7.12 Channel description in the context of QA-MAC.

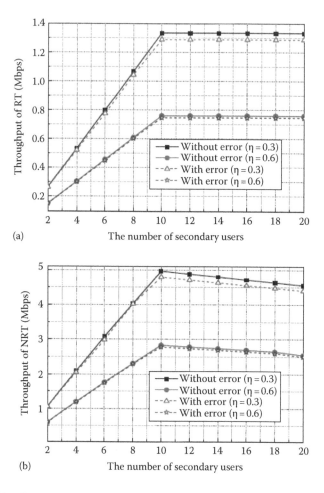

FIGURE 7.13 Throughput for (a) RT services and (b) NRT services in QA-MAC, where η is the channel utilization with respect to PUs and $N = 10$. (From Jiang, F. and Liu, X., A QoS aware MAC or multichannel cognitive radio networks, in *Proceedings of ICCTA*, Beijing, China, 2011.)

7.3.2.1 Performance Analysis

The numerical results of QA-MAC are shown in Figure 7.13a (for RT traffic) and Figure 7.13b (for NRT traffic) [13]. For RT traffic, when sensing error is not considered, the throughput increases linearly with an increase in the number of SUs. The maximum throughput is obtained when the number of SUs is equal to the total number of channels (here, the number of channels $N = 10$). It becomes constant when the number of SUs is larger than the number of available channels. This is due to the fact that RT SUs will reserve channels, and therefore, their overall traffic will not decrease. Note that the throughput decreases in the presence of sensing error. Results for NRT traffic are shown in Figure 7.13b. Note that NRT traffic decreases after the number of SUs reaches the total number of channels. This is because RT traffic is given priority over NRT traffic.

7.4 MOBILITY MANAGEMENT

Mobility support is a very important concern in realistic CRNs. It is important for MAC protocols to address the mobility issue while ensuring the QoS requirements. This section describes two protocols that can support mobility patterns of PUs and SUs. The first protocol requires a CCC, while the second one is completely distributed and does not require a CCC.

7.4.1 PRIMARY USER EXPERIENCE AND MOBILITY SUPPORT WITH CCC

To tackle the problem of mobility, several models have been proposed. For example, the authors in [7] consider that the PU transmitter, located at the center of the network, communicates with PU receivers within a disc called the primary exclusive region (PER). Inside the PER, no SU may transmit. This will ensure an outage probability for the PUs within a predefined interval. Outside the PER, SUs may transmit provided that they are at a certain protected radius from a primary receiver as shown in Figure 7.14.

The PER presents several challenges. For example, when a CR moves into a PER without stopping its transmission, both CR and PU will experience interference. This situation may result from sensing or geolocation errors. Moreover, the MAC protocol should be PER aware, so that it can manage the SU's access especially when the SU and/or the PU is moving.

A cognitive MAC protocol with mobility support (CM-MAC) has been proposed in [8] based on a CSMA/CA technique, where CM-MAC protocol can respond to the CRs' vicinity to PERs. An out-of-band CCC is used for exchanging control

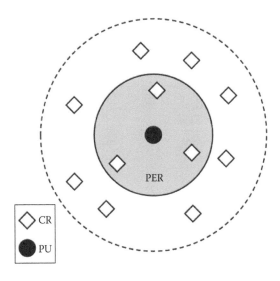

FIGURE 7.14 Concept of PER in CRN.

information. To overcome the challenge of CR knowledge on the PER vicinity, a radio signal strength indicator (RSSI) is adopted at the PHY layer. This protocol is discussed in Chapter 9.

7.4.2 MOBILITY SUPPORT WITHOUT CCC

A distributed channel selection strategy is adopted in [11] to dynamically select a channel with a minimal interference (Figure 7.15).

This protocol uses multichannel carrier sensing principle where an SU sequentially scans all channels in the pool. The transmitting SU ensures that the transmission in the selected channel lasts for long enough duration that the asynchronous receiving SUs detect it when scanning that particular channel. Upon detecting a packet transmission activity, the receiving SUs do not scan subsequent channels and keep on listening to the channel until a data packet is received. In order to engage the channel, the SU transmitter repeats the data packet. The total number of packet repetitions N_{pkt} governs an upper bound,

$$N_{pkt} \geq \left(T_{CS} + T_{Switch}\right) N_{ch} / T_{pkt}$$

where

T_{CS} is the carrier sensing duration
T_{Switch} is the channel switching duration
N_{ch} is the number of channels
T_{pkt} is the time required to send a packet

As described in [11], the SU transmitter first scans all channels in the pool to ensure that there is no other ongoing packet transmission before attempting to send a packet. Since an activity in the medium can also be due to an interferer or a PU, the protocol uses a time-out scheme in order to characterize an interferer. If a channel activity is detected and no valid packet is received within a preset

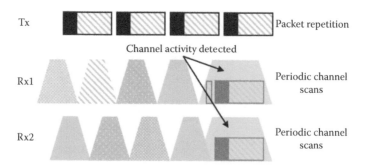

FIGURE 7.15 Multichannel sensing and packet transmission. (Adapted from Ansari, J. et al., A decentralized MAC for opportunistic spectrum access in cognitive wireless networks, in *Proceedings of CoRoNet'10*, Chicago, IL, September 2010.)

interval (e.g., two maximum-size packet transmissions), the channel is declared as an interfering channel.

Performance analysis

The experimental results reported in [11] are plotted in Figure 7.16 using the parameters given in Table 7.1. Figure 7.16 shows the successful packet delivery ratio of this protocol across different sensing durations by varying number of channels. It can be observed that the overall packet delivery ratio is not highly affected by the channel

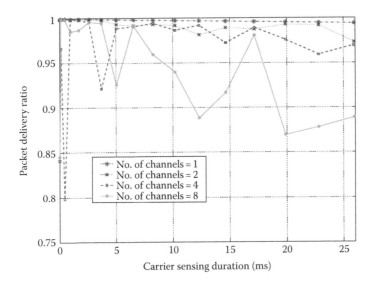

FIGURE 7.16 Performance in terms of successful packet delivery ratio versus the sensing duration for different number of channels. (From Ansari, J. et al., A decentralized MAC for opportunistic spectrum access in cognitive wireless networks, in *Proceedings of CoRoNet'10*, Chicago, IL, September 2010.)

TABLE 7.1
PHY/MAC Parameter Values

Parameter	Value
Max. packet size used ($L_{max-pck}$)	1000 bytes
Max. packet transmission time ($T_{max-pck}$)	1.48 ms
Channel switching interval ($T_{ch-switch}$)	35 μs
Interferer time-out interval ($T_{int-time-out}$)	2.96 ms

Source: From Ansari, J. et al., A decentralized MAC for opportunistic spectrum access in cognitive wireless networks, in *Proceedings of CoRoNet'10*, Chicago, IL, September 2010.

sensing durations. For example, with eight channels and a sensing duration of 25 ms, the packet delivery ratio is close to 90%.

7.5   CONCLUSION

The chapter discusses the different issues related to MAC protocol design. This includes network setup, absence of a CCC, distributed architectures, mobility of PUs and SUs, exchange of control data between SUs, and QoS support. The chapter focuses on a number of representative protocols and explains how these protocols address the targeted objectives. For example, JMAC, ICSMAC, MMAC, and SYN-MAC have been designed for networks that do not have a CCC; EX, RAN, and SEQ protocols address the problem of initial network setup in both distributed and centralized scenarios; DNG-MAC protocol considers fair allocation of channels; QC-MAC and QA-MAC are distributed protocols that ensure QoS provisioning; and CM-MAC provides mobility support in dynamic CRNs.

REFERENCES

1. C. Cormio and K. Chowdhury, A survey on MAC protocols for cognitive radio networks, *Ad Hoc Networks*, 7, 1315–1329, 2009.
2. K. Kosek-Szott, A survey of MAC layer solutions to the hidden node problem in ad-hoc networks, *Ad Hoc Networks*, 10, 635–660, 2009.
3. I. Akyildiz, W. Lee, M. Vuran, and S. Mohanty, NeXt generation/dynamic spectrum access/cognitive radio wireless networks: A survey, *Computer Networks*, 50 (13), 2127–2159, September 2006.
4. S. Kumar, V. Raghavan, and J. Deng, Medium access control protocols for ad hoc wireless networks: A survey, *Ad Hoc Networks*, 4 (3), 326–358, 2006.
5. I. Akyildiz, W. Lee, and K. Chowdhury, CRAHNs: Cognitive radio ad hoc networks, *Ad Hoc Networks*, 7, 810–836, 2009.
6. S. Jha, M. Rashid, and V. Bhargava, Medium access control in distributed cognitive radio networks, *IEEE Wireless Communications*, 18 (4), 41–51, August 2011.
7. M. Vu, N. Devroye, and V. Tarokh, On the primary exclusive region of cognitive networks, *IEEE Transactions on Wireless Communications*, 8 (7), 3380–3385, July 2008.
8. P. Hu and M. Ibnkahla, CM-MAC: A cognitive MAC protocol with mobility support in cognitive radio ad hoc networks, in *IEEE International Conference on Communications (ICC)*, Ottawa, Ontario, Canada, June 2012, pp. 430–434.
9. L. Cai, Y. Liu, X. Shen, J. Mark, and D. Zhao, Distributed QoS-aware MAC for multimedia over cognitive radio networks, in *Proceedings of IEEE GLOBECOM*, Washington, DC, 2010.
10. M. Zeeshan, M. Fahad, and J. Qadir, Backup channel and cooperative channel switching on-demand routing protocol for multi-hop cognitive radio ad hoc networks (BCCCS), in *Proceedings of the Sixth International Conference on Emerging Technologies (ICET)*, Islamabad, Pakistan, 2010.
11. J. Ansari, X. Zhang, and P. Mähönen, A decentralized MAC for opportunistic spectrum access in cognitive wireless networks, in *Proceedings of CoRoNet'10*, Chicago, IL, September 2010.
12. M. Shah, S. Zhang, and C. Maple, An analysis on decentralized adaptive MAC protocols for cognitive radio networks, in *Proceedings of the 18th International Conference on Automation & Computing*, Leicestershire, U.K., September 2012.

13. F. Jiang and X. Liu, A QoS aware MAC or multichannel cognitive radio networks, in *Proceedings of ICCTA*, Beijing, China, 2011.

14. K. Cheng and R. Prasad, *Cognitive Radio Networks*, John Wiley & Sons Ltd., Chichester, U.K., 2009.

15. A. Tanenbaum and D. Wetherall, *Computer Networks*, 5th ed., Prentice Hall, Upper Saddle River, NJ, October 2010.

16. Y. Kondareddy, P. Agrawal, and K. Sivalingam, Cognitive radio network setup without a common control channel, in *Proceedings of IEEE Military Communications Conference (MILCOM)*, San Diego, CA, 2008.

17. I. F. Akyildiz, W. Y. Lee, M. C. Vuran, and S. Mohatny, A survey on spectrum management in cognitive radio networks, *IEEE Communications Magazine*, 46 (4), 40–48, April 2008.

18. Y. Liang, K. Chen, G. Li, and P. Mahonen, Cognitive radio networking and communications: An overview, *IEEE Transactions on Vehicular Technology*, 60 (7), 3386–3407, September 2011.

19. K. Kosek-Szott, A survey of MAC layer solutions to the hidden node problem in ad-hoc networks, *Ad Hoc Networks*, 10 (3), 635–660, May 2012.

8 Cognitive Radio Ad Hoc and Sensor Networks
Network Models and Local Control Schemes

8.1 INTRODUCTION

As a result of the development of cognitive radio technology, the concept of cognitive radio ad hoc networks (CRAHN) has recently been proposed in the literature [1], which involves more challenges than classical cognitive radio networks (CRNs). These challenges are due to variable radio environments caused by spectrum-dependent communication links, hop-by-hop transmission, changing topology, and node mobility.

A CRAHN is a network composed of cognitive radio (CR) nodes and primary users (PUs) in an ad hoc manner in a changing radio environment induced by the time and location and PU activities. In order to ensure successful data transmissions, accessing the spectrum resource needs to be coordinated to prevent collisions. As such, with a spectrum sharing module, a CR is able to share spectrum resources with other CRs [1]. As an example of a CRAHN shown in Figure 8.1a, the CRs are colocated with PUs, where PUs and CRs are able to move. In order to make CRs aware of the available spectrum bands, the spectrum sharing module is required to ensure that spectrum resources in a region are fairly shared among CRs [2]. Similarly, the cognitive radio sensor network (CRSN) [3] requires that the spectrum sharing module ensures that spectrum resources are available to sensor nodes (SNs) as shown in Figure 8.1b.

This chapter is based on the findings of [4,71–73] and is organized as follows: Section 8.2 presents the main differences between CRAHN and regular cognitive radio networks. Section 8.3 addresses the spectrum sharing problem in CRAHN. Section 8.4 gives a brief survey of MAC protocols in CRAHN. Scaling laws and CRAHN models are presented in Sections 8.5 and 8.6, respectively. Local control schemes and fairness protocols for spectrum sharing are introduced in Section 8.7.

8.2 COGNITIVE RADIO NETWORKS VERSUS CRAHN

Cognitive radio networks are defined as wireless networks that consist of primary users and secondary users (SUs) called also CR users [5]. Traditional CRNs are often modeled as small networks in licensed bands with one PU and multiple SUs as seen in current IEEE 802.22 networks. However, the CRN paradigm can be extended to the unlicensed industrial, scientific, and medical (ISM) radio bands and therefore

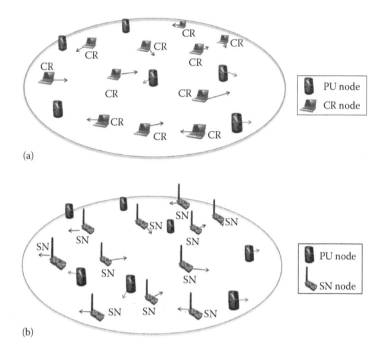

FIGURE 8.1   An example of (a) CRAHN and (b) CRSN.

can be used in the context of ad hoc networks and wireless sensor networks. Some current research topics of CRNs can be found in the recent studies [6,7].

The CRAHN has different specific research focus compared to CRNs. Inherited from the features of traditional ad hoc networks, nodes in a CRAHN can communicate with each other without a fixed infrastructure [8]. The ad hoc topology and data transmissions of ad hoc networks as well as the cognitive capabilities of CRNs bring new features and new challenges to CRAHNs.

8.3   SPECTRUM SHARING IN CRAHN

Spectrum sharing is an important function of spectrum management in CRAHNs. In [1], spectrum sharing is defined to provide the capability of sharing the spectrum resource opportunistically among multiple CRs while avoiding interference caused to the primary network. Basically, spectrum sharing involves spectrum access, spectrum allocation, and spectrum sensing with cross-layer information. In this sense, from the protocol architecture point of view, the spectrum sharing function needs to collaborate with physical (PHY), medium access control (MAC), and network (NWK) layers.

In order to ensure data communication, CRs need to maximize their own share of spectrum resources for data transmission sessions. Also, CRs need to perform channel selection and power allocation while choosing the best channel. Cooperation among neighbors can help to enhance the performance of spectrum sharing.

However, with the local observation of radio environment, CRs have limited radio information from their neighbors. This constraint may affect the performance of the network in terms of throughput and spectrum utilization.

Several distributed schemes or algorithms have been proposed in the literature to solve the spectrum sharing problems. A single-channel asynchronous distributed pricing scheme for spectrum selection and power control is proposed in [9], where each CR determines the transmit power by maximizing the received utility minus the total cost of the associated interference. A graph coloring–based scheme is proposed in [10], which is essentially a global optimization algorithm. This global optimization algorithm is centralized in nature and is required to be recomputed whenever there is a change in CRAHNs. Compared to a centralized scheme, a distributed scheme is more suitable for the CRAHN due to its robustness in varying radio environments (e.g., topology and spectrum availability and so on). A distributed spectrum allocation scheme, referred to as local bargaining, is proposed in [11], where CRs can self-organize and form a local group to improve system utility. Results in [11] show that the communication overhead using local bargaining can be significantly reduced compared to a greedy coloring algorithm. A device-centric spectrum access approach for spectrum allocation problem is introduced in [12], where five different rules are applied to individual CRs. Although these rules have a slightly worse performance than local bargaining [12], they have lower computational complexity and communication overhead. Furthermore, learning algorithms like reinforcement learning [13,14] can be involved in the spectrum sharing problem, but they may need much more information and collaboration efforts across layers and hops.

Another type of algorithms, known as swarm intelligence, has been proposed in the literature to solve spectrum sharing problems. In [15], the spectrum sharing problem is solved by an insect colony–based algorithm. In [16], an algorithm based on the schooling mechanism of fish is studied. Consensus protocols have been used for data fusion problems in sensor networks, robotic control, and multiagent systems (MASs). Recently, Li et al. [17] have applied consensus protocols to spectrum sensing in order to control the fusion of sensing data. Authors in [18] have proposed a distributed and scalable scheme for spectrum sensing based on consensus algorithms.

The aforementioned references have given hints of how to use consensus protocols in CRNs, but they hardly address spectrum sharing fairness in CRAHNs. This chapter shows the applicability of consensus protocols in spectrum sharing for CRAHNs and CRSNs. Moreover, this chapter discusses how to use the consensus protocol for spectrum sharing fairness.

8.4 MAC PROTOCOLS IN CRAHN

The objectives of CRAHN MAC protocols not only include the improvement of channel utilization and throughput without degrading PU communications, but also include the control of spectrum management modules such as spectrum access and spectrum sharing functions to determine the timing for data transmissions [1].

The use of multiple channels for throughput improvement has been addressed in several MAC protocols. A feasible solution for throughput improvement is to find a set of good quality channels. A dual-channel MAC protocol (DUCHA) is proposed in [19], which can improve the one-hop throughput up to 1.2 times and multihop throughput up to 5 times compared to the IEEE 802.11 MAC protocol. An opportunistic multiradio MAC (OMMAC) is proposed in [20], where a multichannel-based packet scheduling algorithm is employed and packets are sent on a channel having best spectral efficiency (i.e., the channel with the highest bit rate). A carrier sense medium access with collision avoidance (CSMA-/CA)-based multichannel cognitive radio–medium access control (MCR-MAC) protocol is proposed in [21].

In a CRN, the spectrum utilization can be improved if we choose the appropriate set of channels that meet the transmission rate requirement. A MAC protocol based on statistical channel allocation (SCA) is proposed in [22] which uses a channel aggregation approach to improve the throughput and dynamic operating range to reduce the computational complexity. Results of [22] show that SCA-MAC can use spectrum holes effectively to improve spectrum efficiency while keeping the performance of coexisting PUs. In order to meet data rate requirement for data transmissions, a MAC with the so-called multichannel parallel transmission protocol is proposed in [23], where the minimum number of channels was selected to meet a certain data rate. The results of [23] show that the proposed MAC protocol has better spectrum utilization and system throughput than the protocol presented in [24], which only selects the channels by the best signal-to-interference-plus-noise ratio (SINR) value. In [25], an opportunistic autorate MAC protocol is used to maximize the utilization on individual channels.

Spectrum sharing and spectrum access functions are explicitly addressed in [26], where spectrum access and spectrum allocation schemes are introduced into the proposed cognitive radio MAC (COMAC) protocol. Specifically, the spectrum utilization is improved by providing enough channels instead of assigning all the possible channels to a CR node, such that the other available channels could be reserved for other CR transmissions. In [27], the authors employ a distance-dependent channel assignment scheme in a distance-dependent MAC (DDMAC) protocol.

In fact, the aforementioned works do not comprehensively consider several important factors. First, although the spectrum sensing can be simultaneously performed in one shot [28], the sensing time cannot be ignored, as it may be relatively large and lead to throughput degradation [29]. Second, with the existence of the primary exclusive region (PER) where CR communications will interfere with PU communications, the CR should keep silent when moving into this region if maintaining PU communication is a priority.

As CRAHN MAC protocols favor distributed solutions, a distributed function like distributed coordination function (DCF) is a good option for protocol design. In fact, most of the aforementioned MAC protocols [20,21,23–29] are DCF based with request-to-send (RTS)/clear-to-send (CTS) handshaking procedures, which intrinsically deal with the hidden terminal problem. Other non-CSMA-/CA-based MAC protocols like multichannel MAC (MMAC) [30] and cognitive MAC (C-MAC) [31] can also solve the hidden terminal problem, but they need a periodic synchronization, which can hardly be applied to large-scale CRAHNs.

As indicated in Chapter 7, CSMA-/CA-based MAC protocols have the advantage of dealing with hidden terminal problems and having distributed operations (e.g., distributed coordination function in IEEE 802.11 MAC). Thus, some state-of-the-art

MAC protocols [21–27,32] for CRNs have been proposed. However, PER, PU activity, and CR mobility have not been comprehensively addressed in the literature.

8.5 SCALING LAWS OF CRAHN

The scaling law analysis for wireless networks can give hints to the theoretical bounds of throughput performance. Guptar and Kumar [33] give the throughput bounds for a general wireless network. They show that the throughput will decrease with an increased number of nodes. However, the bounds given by Gupta and Kumar [33] are loose for the CR network in CRAHNs, because, in CRAHNs, communications between CR nodes can be affected by the PU activities. By utilizing the multiple spectrum bands for data communications, system capacity, multipath diversity, and data rate can be improved [34]. However, how to comprehensively address the design parameters across different layers in the randomly deployed CRAHN is a challenge. Vu et al. [35] have analyzed the throughput for cognitive networks, where the authors merely discussed the network model with one PU transmitter. This analysis is suitable for some cognitive networks, such as the cognitive network with one TV tower and multiple CRs. However, the analysis is not suitable for CRAHNs, as more than one PU transmitter can be present with CRs. Moreover, considering the possible flexible deployment of CRAHNs, the scaling law of throughput should be analyzed in different transmission scenarios.

Some research work has been done regarding the throughput scaling law for CRAHNs. Shi et al. [36] have recently given lower and upper bounds for the throughput in a randomly distributed CRAHN by using two auxiliary networks. However, PU activities and multihop transmission scenarios have not been considered in the discussion.

The PER concept is addressed in [37], where interference and outage probability are derived for bipolar and nearest-neighbor network models. Opportunistic multichannel MAC protocols for CRAHNs are analyzed in [38], where a Markov model was used to estimate the number of sensed channels. The relationship of delay, connectivity, and interference was analyzed in [39]. Besides, with new features brought to CRAHNs, different spectrum management schemes can result in new scaling laws in the CRAHN. Although some recently proposed physical-layer techniques, such as physical-layer network coding (PLNC) [40,41] or interference-based network, may help to derive new scaling laws in CRAHNs, there is still a need to explore the essential factors that affect the CRAHN throughput performance. Moreover, stochastic geometry has been employed as an analytical tool for fundamental limits of wireless networks [42], enabling the inclusion of several transmission scenarios [43] in the analysis.

8.6 CRAHN MODELS

8.6.1 Spectrum Availability Map

Spectrum availability varies from node to node and from link to link in CRAHNs. In the same radio environment, node spectrum availability and link spectrum availability can be converted to each other. It is known that spectrum availability in a CRN is usually modeled as conflict graph [11,44]. However, in this chapter, we model the spectrum availability in the perspective of PUs. In this sense, we can start from the introduction of spectrum availability map (SAM) in a CRAHN with grid topology.

SAM is defined against time, and it is the probability of having some available spectrum bands in a time slot Δt. Although in a time slot, a CR can perform the spectrum hopping from one spectrum band to another, here we start with considering the extreme case: in a time slot Δt, there are only two available spectrum bands for data communication. It is worth noting that the correlation between two SAMs is based on the previous time slot $\Delta(t-1)$ and the immediate next time slot Δt.

The knowledge of SAM known a priori can be considered as global information; the knowledge of local SAM known a priori is considered as local information.

For a CR in a CRAHN, the local SAM is enough and this local SAM can be constructed by

1. Sensing the available spectrum bands.
2. Capturing the available spectrum bands from different PUs and storing them into the internal memory.

8.6.1.1 Cell-Based Spectrum Availability Map

The cellular automaton (CA) is a discrete model that has been broadly studied in different disciplines including computer science [45]. A CA is composed by a regular grid of cells with a finite number of states in each cell.

The spectrum availability of a CRAHN can be molded as a map by the concept of CA and we name it cell-based spectrum availability map (C-SAM). Suppose each CR has different spectrums at a time t, we can explore the dynamics of the available spectrums in a large-scale CRAHN. With this model, the dynamics of the CRAHN's system behavior can be evaluated by this two-D CA model. Figure 8.2 shows an example of C-SAM in a CRAHN with three spectrum bands for each CR. Numbers in the figure represent the different spectrum indexes. Assumptions regarding this CA-based model are as follows:

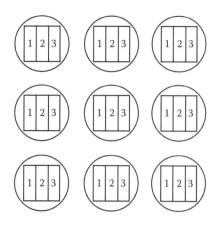

FIGURE 8.2 An example of C-SAM in a CRAHN with three spectrum bands for each CR.

1. Available spectrums at a time t are identical to all CRs.
2. Each CR can only communicate with the immediate neighbors. Their states decide the availability of the spectrums of CR i.

8.6.1.2 Radio Environment Map

Instead of obtaining the radio environment parameters at CR nodes, the radio environment map (REM) proposed in [46,48] can be used to store environmental and operational information. An REM can provide many kinds of radio environment information over a CRN, such as geographical features, available services, spectral regulations, location and radio activities, and experience. The REM can be classified as global REM and local REM [47]. These two classes of REM can be used by cognitive radio regional area networks (e.g., IEEE 802.22 networks) or cognitive radio local area networks (e.g., CRAHNs). According to the link-level and network-level analysis in [48], using REM can significantly improve network performance in terms of reduced adaptation time, average packet delay, and mitigation of the hidden terminal problem.

The REM is a practical solution when reliable information (e.g., a certain amount of local information and global information) regarding radio environment is needed in CRAHNs. As an example of the REM-based architecture, in Figure 8.3, the cluster head (CH) is responsible for exchanging information to the local REM server. The local REM server contains the information collected from CRs in each cluster. The data in local REMs will be sent to the global REM server.

8.6.2 SPECTRUM AVAILABILITY PROBABILITY

For the spectrum sharing protocols, it is natural to see the relationship between the SAM and the CRs. In fact, the two models can be converted from one to another. With the spectrum availability probability (SAP), we can divide a CRAHN into

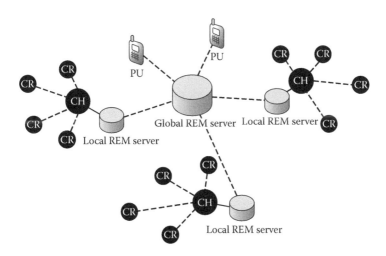

FIGURE 8.3 The architecture of a CRAHN with REM servers.

different subareas. In this sense, the data transmission scenario can be converted to the probability of a CR transmitter at the center of a subarea and the SAP of this transmitter at this location.

Definition 8.1. Spectrum availability probability $[SAP = \rho(\Delta t, k, s)]$ is defined as the probability of when a CR is able to access a spectrum band k in a time period Δt in an area s.

With a Poisson traffic flow of PUs deployed in an area S, we know that in an area $s \in S$, SAP can be determined by three parameters Δt, k, and s.

Note that if we consider that each data transmission flow needs different bandwidths, we have to improve the aforementioned SAP and SAM. With an application-specific Quality of Service (QoS) requirement, if the rate cannot be met by the available spectrum band, then the spectrum band is considered not available.

8.6.3 VARIABLE-SIZE SPECTRUM BANDS

We assume that the size of the spectrum bands is identical in terms of same traffic model. The problem is more complicated when we consider a more general case of spectrum bands having variable sizes. This means that a large chunk of spectrum can be split into two or more smaller chunks of spectrum, or a smaller chunks of spectrum can be combined into a larger chunk. We consider that this variation occurs only when the current available spectrum bands cannot meet the flow bandwidth requirement.

In fact, multiple available spectrum bands can be virtually combined to form one band using channel aggregation to boost the throughput, where, for example, a large packet can be split into two and transmitted in the two channels in a faster speed. With these assumptions, we can convert this case into a case similar to SAP in which spectrum bands have identical sizes. This will enable us to calculate the probability bound of the presence of variable spectrum bands.

8.6.4 MULTICHANNEL MULTIRADIO SUPPORT

The CRAHN can be considered a network paradigm with multichannel multiradio support. The network throughput performance can be boosted by multichannel multiradio capability in CRNs. The network layer schemes can take advantage of that capability in CRs, because the multiple routes brought by the CR capability can increase the data transmitted per unit time. To illustrate this, Figure 8.4 shows the throughput performance of a CRAHN based on different routing protocols with K spectrum bands and R multiple radios. We can see from the figure that when more channels and radios are available, the weighted cumulative expected transmission time (WCETT) protocol [63], which can take advantage of multichannel multiradio capability, has better performance than the ad hoc on-demand distance vector (AODV) routing protocol. More cognitive routing protocols have been discussed in [48–53,55].

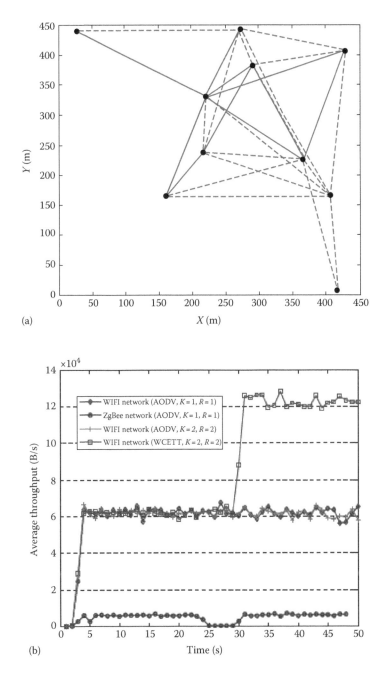

(a)

(b)

FIGURE 8.4 Throughput performance in a CRAHN with multichannel multiradio support in different settings. The network has 10 CRs and the communication range per node is 250 m in 2 GHz band: (a) network deployment and (b) performance results.

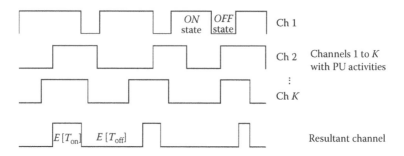

FIGURE 8.5   Example of the resultant channel model.

8.6.5   Resultant Channel Model

With the proposed concept of SAM, it is easy to visualize spectrum availability at a given time t. It will be more useful if the spectrum availability in different bands can be represented by one spectrum band at a given time t. This can be achieved by using the resultant channel model [54].

The resultant channel model can be seen in Figure 8.5, where for the ith PU, the time spent in *busy* and *idle* states is exponentially distributed, with mean α_i and β_i, respectively. In this model, the PU activity is determined by an ON–OFF model, where ON or "1" means PU is busy and is occupying a channel; OFF or "0" means PU is not transmitting and is not occupying a channel. It is worth noting that by using the resultant channel model, multiple PU transmitters can be modeled as one virtual PU transmitter [55].

8.6.6   Local and Global Information

Local information is the information that can be acquired through local observation (e.g., local sensing) or communications with neighbors. We can refer to the categorization for local information in IEEE 1900.4 standard [56], where information is categorized into terminal class and network class. The former can be used for classifying the local information, and the latter can be used for classifying the global information. Terminal class includes application information and device information. Application information contains information about measurements supported by applications, such as delay, packet loss, and bandwidth. Device information contains information about the current active links and channels. Information about links includes block error rate, power, signal to interference and noise ratio (SINR), etc., while information about channels includes channel ID, frequency range, etc.

When obtaining local information, we should consider the communication cost of obtaining the information. Due to the changing radio environments in CRAHNs, some cost values may be dynamic, while others are not. Moreover, the cost values can be considered in the metrics for distributed protocol design. As an example, some pieces of local information with cost values are shown in Table 8.1.

In Table 8.1, we can see the cost of obtaining the channel state and the cost of obtaining channel utilization are C1 and C3, respectively. The cost of obtaining the

TABLE 8.1

Local Information Associated with Cost Values

Local Information	Cost
Channel state (idle/busy)	C1
Number of neighbors	C2
Immediate neighbors' spectrum usage	C2
Spectrum utilization	C3
Overheard MAC information	0
Signal strength	0

number of neighbors and the cost of obtaining the neighboring spectrum usage are the same, that is, C2. This is true when some information such as the number of neighbors can be estimated from overheard incoming packets, which contain MAC address fields and data fields with spectrum utilization of neighbors. Therefore, we can assume that this MAC information needs no cost to obtain. For the signal strength that can be easily estimated by most receivers, we assume the cost of obtaining it is zero. If we initiate a particular communication process to obtain channel state and channel utilization, then the values of C1 or C3 would be larger than C2.

The global information refers to information over the network. For example, from the IEEE 1900.4 standard [56], the network information includes channel information, cell information, and base station information. Channel information is mainly about the frequency channel, including frequency channel ID, frequency range, etc. Cell information is the general information about a cell configuration, including cell ID, location, coverage area, etc. Base station information contains the general information about the current base station configuration, including transmission power, load, etc.

8.6.7 LOCAL CONTROL IN SPECTRUM MANAGEMENT

The local control can be considered as a distributed control of individual CRs in CRAHNs. Because of the lack of a central controller and changing radio conditions, a centralized control is not suitable. Moreover, the cooperation between CRs can help create and distribute radio environment information, which makes an individual node have a macroscopic view of the network status. It has been proven that cooperation between CRs can help to improve the spectrum sharing process. However, cooperation can lead to increasing communication overhead and underlying interference. As such, the approach of spectrum management based on global information would be costly.

The difference between local control schemes and spectrum etiquette is discussed [57]. The former may include a set of protocols, rules, or schemes, enabling system-level and protocol-level modeling and analysis for spectrum management problems, such as spectrum sharing, spectrum mobility, and spectrum decision. The latter may be considered as a mere set of rules, which regulate access to spectrum and its usage

[57] (i.e., a set of rules dictating when, where, and how devices may transmit [58]). Therefore, the two concepts may overlap to some extent, but, in fact, they focus on different problems.

8.6.8 GAME-THEORETIC APPROACH

Due to the features of the CRAHN, a noncooperative scheme is desirable for spectrum sharing and allocation as it can reduce the communication overhead and underlying interference. In game-theoretic approach, Nash equilibrium is an important tool to measure the outcome of a noncooperative game [59–64] in the spectrum management problems.

A game-theoretic approach for spectrum allocation is proposed in [65], where the CR nodes (i.e., players) make decision based on the utility function to select a channel without causing interference to other nodes. In [66], a spectrum sharing solution based on game-theoretic approach for the primary–secondary model is proposed, where an oligopoly market model is used to maximize the profit of all CRs based on the equilibrium adopted by all CRs.

To model a spectrum sharing problem in game-theoretic approach, the players have to make decisions sequentially, that is, a coordinator to control the playing order is required. To transform the game-theoretic scheme into a distributed version, a Bernoulli trial is used to make the sequential decision-making process happen at players by probability. In other words, at the beginning of every iteration, the decision-making process is performed at players who win a Bernoulli trial.

From the earlier discussion, the game-theoretic approach can model strategic interactions among agents using formalized incentive structures [22]. The general methodology in game-theoretic approach is to

1. Find a suitable game model for a problem
2. Formulate a utilization function
3. Prove the equilibrium condition

Due to the autonomous and learning properties of CRs, the game-theoretic approach may be a suitable way to solve problems in CRAHNs.

However, it should be noted that modeling a problem as a game cannot always get an optimal solution. For example, the authors of [54] show that when the nodes have complete information about the network, the steady-state topologies are suboptimal. In order to make a game convergence, the utility function also has to meet some conditions.

In [67], the authors indicate that Nash equilibrium assumes that the players are rational, meaning each player has a view of the world. They also argue that the Nash equilibrium has two practical limitations:

1. Best response strategy required to achieve Nash equilibrium does not always hold. For example, in a two-player game, if only one player adopts a nonequilibrium strategy, the optimal response of the other player is of a nonequilibrium kind too.

2. The description of a noncooperative game is essentially confined to an equilibrium condition, which is not enough to be used in cognitive radio with underlying dynamics.

In the state-of-the-art research work, although the game-theoretic approach is popular for decision making in spectrum allocation and spectrum sharing, the realization in this approach is dependent on a certain centralized flow control protocol in the MAC or NWK layer. A zero-player game may be included in a local control scheme to show the system-level characteristics.

8.6.9 GRAPH COLORING–BASED ALGORITHMS

Graph coloring–based algorithms can be directly used to solve the spectrum allocation problem. As soon as the available spectrum bands for each CR are transformed to the colors of a map, the objective of the graph-coloring algorithm for spectrum allocation is to minimize the use of colors.

Here, we show the classical graph-coloring algorithm proposed in [10]. In a undirected graph $G = (V, E)$, the number of users is $N = |V|$, and $E = e_{ij}$, where $e_{ij} = 1$ if there is an edge between vertices i and j and $e_{ij} = 0$ if i and j use the same spectrum bands. The availability of spectrum bands at vertices of G is represented by an $N \times K$ matrix $L = (l_{ik})$, referred to as a coloring matrix. For example, $l_{ik} = 1$ means a color (spectrum band) k is available at vertex i.

A channel assignment policy is denoted by an $N \times K$ matrix $S = (s_{ik})$, where $s_{ik} = 0, 1$. If $s_{ik} = 0$, channel k is assigned to node i and 0 otherwise. S is a feasible assignment if the assignments satisfy the interference constraint and the color availability constraint, which can be denoted by

$$s_{ik} s_{jk} e_{ij} = 0, \quad \text{for all } i, j = 1, \ldots, N, k = 1, \ldots, K \tag{8.1}$$

The earlier constraint means that two connected nodes cannot be assigned to the same colors (channels).

The objective of the resource allocation is to maximize the spectrum utilization.

The spectrum allocation problem can be represented by:

$$\text{Maximize:} \sum_{i=1}^{N} \sum_{k=1}^{K} s_{ik} \tag{8.2}$$

subject to: for all $\quad i, j = 1, \ldots, N, k = 1, \ldots, K$:

$$s_{ik} \leq l_{ik}$$

$$s_{ik} = 0, 1$$

$$s_{ik} s_{jk} e_{ij} = 0$$

If a time-slotted communication between the network nodes is considered, then at each time unit, the optimization problem in (8.2) needs to be recomputed.

It can be seen from (8.2) that, in the varying radio environment in CRAHNs, the optimization problem has to be executed many times, which may make the graph-coloring algorithm inefficient. Moreover, the graph-coloring algorithm is an innate centralized algorithm, so it is not suitable for the CRAHN. However, it can be used as a benchmark for comparison with distributed algorithms.

8.6.10 PARTIALLY OBSERVABLE MARKOV DECISION PROCESS

The partially observable Markov decision process (POMDP) is a generalization of a Markov decision process (MDP). A POMDP models a decision process of a CR where the system dynamics are determined by an MDP, but the CR cannot directly observe the underlying state of a channel. Therefore, the POMDP is more practical than an MDP model when solving spectrum access problems.

For example, if the channel is modeled as a Markov channel with two states *good* and *bad* and four transition probabilities given by p_{ij}, $i, j = 0, 1$, a transmitter can select one of the channels to sense based on its prior observations, and the selected channels obtain some fixed award if it is in the good state. This problem can be described as a POMDP, as the states of the Markov chains are not fully observable. In [68], the myopic policy (i.e., a policy that maximizes one-step reward) is studied. Results suggest that when $p_{11} \geq p_{01}$, it is optimal for any number of channels; when $p_{11} < p_{01}$, it is optimal when the number of channels $n = 3$.

We can see that a POMDP is suitable for modeling a channel access problem, as the channel states are not fully observable to a CR, and an advantage of using POMDP is that it runs in a distributed fashion. However, POMDP involves some limitations. For example, POMDP is often computationally intractable to be solved. Another problem is that a POMDP is suited to the single player with multiple states.

8.6.11 BIOINSPIRED SCHEMES

There are some swarm intelligence algorithms that have been proposed recently. Atakan and Akan [15] propose a spectrum sharing algorithm called biologically inspired spectrum sharing (BIOSS) based on the task allocation model of an insect colony. This algorithm does not need any coordination among the CRs compared to nonbioinspired ones. Another swarm intelligence algorithm is proposed by Doerr et al. [16], which is inspired by the emergent behavior of fish groups. In [16], CRs' behavior can be analogous to a school of fish, where CRs can sense the radio environment by local observation and react to the changing radio environment. Each CR has limited intelligence, but in the entire network, they have better overall intelligence than individual intelligence for a certain task.

8.7 LOCAL CONTROL SCHEMES FOR SPECTRUM SHARING

This section first introduces the concept of local control schemes by which a CR can locally perform a spectrum sharing process. Then it defines the spectrum sharing fairness issue and investigates the convergence condition when applying a consensus-based protocol to spectrum sharing. Based on the local observation and local control scheme using spectrum-related information, an individual cognitive node can effectively perform spectrum sharing.

8.7.1 How to Apply Local Control Schemes in CRAHN?

Compared to a classical ad hoc network, a CRAHN is able to deal with the problems caused by changing radio environment and to protect licensed users transmissions. Compared to classical CRNs, CRAHNs inherit some important features from ad hoc networks, such as node mobility, hop-by-hop spectrum availability, and uni-directional links. Other features in CRAHNs include spectrum-dependent links, topology control, multichannel transmission, and spectrum mobility, implying more challenges than those in either classical CRNs or ad hoc networks. Due to the lack of central network entities in CRAHNs [69], each CR node necessitates that all spectrum-related CR capabilities and distributed operations must be based mostly on local observations.

In CRSNs, each cognitive SN has cognitive capability and the network is usually intensively deployed with colocated PUs. Therefore, this type of network inherits the similar cognitive modules as those in CRAHNs. A CRSN can use similar local control schemes in the spectrum sharing module. A CRSN, which has limited coverage and power supply, can be considered as the extension of a CRAHN, so that the local control schemes can be applied to CRSNs.

Moreover, a local control scheme is suitable for another network paradigm called sensor networks for CRAHNs, where SNs are aided for cognitive actuation. With local observation and local knowledge, SNs perform the collective behavior for spectrum sharing, monitoring, and decision. The enabling technology for this network, called sensor network–aided cognitive radio, is discussed in [70]. As the local control scheme on SNs in this network is very similar to the CRs in a CRAHN, a detailed discussion for this network will not be provided in this section.

Based on the aforementioned discussion, we see that CRAHN is a more general network framework than CRSN, and a local control scheme for CRAHN is also applicable to CRSN. Therefore, this chapter will focus on how the local control scheme can be applied to CRAHN.

The radio environment in CRAHN is subject to change from time to time, which is the major problem for the spectrum sharing function. Typically, a change of radio environment can be caused by

- PU activities
- Interference during communications
- Spatial–temporal characteristics of radio signals

Here, we consider only the first two factors.

As an example, Figure 8.6 shows that CRs are deployed in an area with a changing radio environment. Each CR senses and observes the local radio environment.

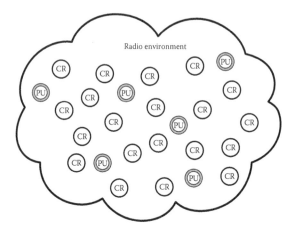

FIGURE 8.6   A CRAHN deployed in a radio environment.

When CRs request the spectrum bands occupied by PUs, they need to invoke local control to share spectrum resources. A natural question one may raise is how the local control for spectrum sharing can be performed by using local observation? In order to answer the question, we introduce a block diagram to present a local control scheme, in which each CR will run for a spectrum sharing process.

8.7.2   FRAMEWORK OF LOCAL CONTROL SCHEMES

In Figure 8.7, a local control scheme framework is represented in a block diagram. Here, when a CR receives a sensing input from sensors (e.g., a spectrum sensor or a global positioning system device and so on), together with feedback information, a CR will process the information and make a spectrum sharing decision. At the sensing input, due to the different sensing capabilities, a CR may have comprehensive, partial, or strictly limited sensing information. At the junction of sensing input and feedback, we can adopt arbitrary types of combinations, where we use the symbol "·" to represent any combination. In Figure 8.7, the feedback block is important for a decision-making process, which may contain a consensus feedback (i.e., feedback from consensus process of nodes in a CRAHN), or a partial consensus feedback (i.e., feedback partially from consensus), or no feedback. In the spectrum sharing process block, a dynamic or a static process may be involved. A dynamic process occurs

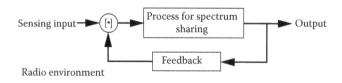

FIGURE 8.7   Framework of a local control scheme.

at an individual CR when the positions or spectrum availabilities of PUs and CRs in CRAHNs change. A static process occurs when the positions or spectrum availabilities of PUs and CRs do not change. At the decision output, we can have different kinds of solutions, such as optimal solution, suboptimal solution, or intermediate solution. If an optimal solution is not achievable, it may be feasible to find a suboptimal solution. The intermediate solution may be neither optimal nor suboptimal; however, this solution may achieve the optimal or suboptimal solution by iterations.

To give an example for the aforementioned framework, we can consider a local control scheme in each CR in a CRAHN where each CR only takes the local information as sensing inputs, such as the network-related information and spectrum information from neighbors. After running a process for spectrum sharing functions in the local control scheme, a CR will make a decision of what spectrum bands to use based on the available spectrum resources.

8.7.3 FAIRNESS IN SPECTRUM SHARING

Definition 8.2. *Fairness*: The spectrum resource allocation is fair to each CR at time t if the available spectrum resources at time t are evenly distributed among CRs.

We mainly consider the available spectrum bands as a spectrum resource requiring fairness. The fairness of spectrum sharing is important as

1. It can help to ensure equal communication opportunity for each CR.
2. It best responds to the changing radio environment in terms of available spectrum bands.

In order to achieve fairness by local observation, each CR tries to achieve fairness by considering the number of available spectrum bands of surrounding neighbors. In order to achieve this goal, a consensus feedback is used in the local control scheme, which can be mathematically formulated as follows.

Suppose the CRAHN can be represented by a graph $G = (V(t), E(t))$, where $V(t)$ is the set of vertices at time t and $E(t)$ is the set of communication edges at time t. We can analyze the system performance using a local control scheme executed by each CR node. In an ideal condition (without any time delay), the consensus feedback is defined in the form of variation as follows:

$$\dot{x}_i(t) = \sum_{j \in N_i(t)} a_{ij}\left(x_j(t) - x_i(t)\right) \tag{8.3}$$

where
the dot operator in $\dot{x}_i(t)$ is used to show the variation of the values of $x(t)$ between CR i and neighboring nodes
$x_i(t)$ indicates the number of spectrum bands available to a node i at time t
$N_i(t)$ is the set of neighbors of node i at time t
a_{ij} is the 0–1 element in adjacency matrix of the network G

Equation 8.3 shows that the spectrum allocation decision is made based on the feedback information $\dot{x}_i(t)$ calculated from $x_i(t)$ and the neighbors' spectrum information $x_j(t)$. With the aforementioned notations, in order to measure fairness, the following expression is used:

$$\sigma_F = \sqrt{\frac{\sum_{i=1}^{M}\left(x_i(t)-m\right)^2}{M}} \tag{8.4}$$

where
    m is the fairness goal (e.g., m equals to the desired number of spectrum bands of a CR)
    M is the total number of CR nodes

From (8.3), the fairness can be ensured if we can make sure that the number of spectrum bands is evenly distributed among CRs. However, since the CRAHN performs hop-by-hop communication, a one-hop time delay τ is inevitable when receiving the information of spectrum availability from immediate neighbors. Then (8.3) can be transformed as

$$\dot{x}_i(t) = \sum_{j\in N_i(t-\tau)} a_{ij}\left(x_j(t-\tau)-x_i(t-\tau)\right) \tag{8.5}$$

Note that (8.3) and (8.5) are inherited from the Vicsek model [71].

For some cases where not all the CRs have the same need of spectrum resources, that is, different groups of nodes have different degrees of fairness, the degree of fairness needs to be defined.

Definition 8.3. *Degree of fairness*: We refer the value of consensus feedback as a degree of fairness for a node, which is defined as

$$DF(i) = \min E\left(X_{i,j} - X_{i,j+1}\right), \quad j \in N_i \tag{8.6}$$

where X_{ij} is the number of spectrum bands of the jth node in the ith group of nodes. We denote by $DF_j(i)$ the degree of fairness for a node j in the ith group of nodes.

Definition 8.4. *Fairness group (FG)*: A set of CRs with the same degree of fairness is called a fairness group (FG), that is, group i and group j are in the same FG, if $DF(i)=DF(j)=p$, where p is a constant. The notation FG can be used to denote the number of FGs in CRAHNs.

The concept of FG is useful when describing heterogeneous nodes that have different spectrum band requirements. Moreover, the concept can be used to virtually divide a large-scale network into different groups with different degrees of fairness.

8.7.4 Protocol Design and Illustrations

The general system model for computer simulations is based on Figures 8.6 and 8.7, where each CR runs a local control scheme in a CRAHN. CRs will use a common control channel to communicate with each other for the information of spectrum availability. Moreover, the focus here will be on the spectrum allocation and convergence performance of the local control scheme.

In order to evaluate an open-loop local control scheme (i.e., the local control scheme without a consensus feedback), a grid topology is adopted for the CRAHN. Let M denote the number of CR nodes. Each CR is denoted by its row and column coordinates in the grid, that is, (i, j).

In the open-loop local control scheme, the sensing input is the spectrum bands chosen by the neighboring CRs. The initial spectrum bands are randomly allocated to each CR. The local information used here is the spectrum bands selected by a CR's immediate neighbors. The process in this local control scheme is to randomly select the available bands of neighboring CRs, that is, the local information is the available spectrum bands chosen by eight immediate neighboring nodes (where in this case the average number of neighbors N_i equals 8). To make the local control scheme configurable, a control parameter λ is set in the process to represent the frequency parameter with which a CR randomly selects a portion of spectrum bands from a neighbor. This parameter can be considered as the feedback information shown in Figure 8.7.

Figure 8.8 shows the results of the aforementioned open-loop local control scheme (which can also be considered as a zero-player game) under different scenarios, where the spectrum utilization results can reflect the convergence performance of the scheme (the results are smoothed every 20 iterations). The spectrum utilization is defined as the ratio of already allocated spectrum bands to a CR and the total available spectrum bands to a CR. From Figure 8.8, it can be seen that although the spectrum bands are randomly selected based on the neighbor's spectrum availability and the parameter λ, the spectrum utilization can show a certain pattern. By changing the value of λ from 1.0 to $1 + \varepsilon$, where ε is a small positive number, a phase transition occurs when $\lambda > 1.0$, the spectrum utilizations are fluctuating among the available spectrum bands; however, when $\lambda = 1.0$, the spectrum utilization over the network is bifurcated into two groups—one is increasing and the other is decreasing. To show whether the phase transition is applicable to the case with more spectrum bands (i.e., $|K| > 3$), we plot Figure 8.8d, where the phase transition still occurs when $|K| = 8$. It is found that when $|K| > 1$, the results are similar. In fact, we found the phase transition is only dependent on the control parameter λ.

From this example, it can be concluded that the overall performance in terms of spectrum utilization is to some extent controllable by using the limited local information. However, as the convergence cannot be achieved, this controllability may not be sufficient to some applications as more variables should be considered. Furthermore, we can see the possible structure of a local control scheme with local information, where the local control scheme described earlier is an open-loop local control scheme without any feedback. Therefore, the local information may be helpful to spectrum sharing if we employ it in a closed-loop local control with a feedback.

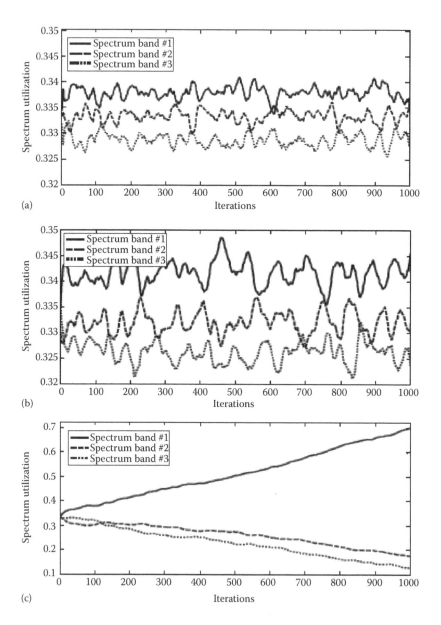

FIGURE 8.8 Results of the open-loop local control scheme for spectrum allocation in a CRAHN. (a) $\lambda = 1.2$, $M = 300$, $|K| = 3$. (b) $\lambda = 1.1$, $M = 300$, $|K| = 3$. (c) $\lambda = 1.0$, $M = 300$, $|K| = 3$. (*Continued*)

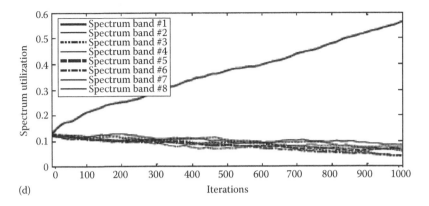

(d)

FIGURE 8.8 (**Continued**)   Results of the open-loop local control scheme for spectrum allocation in a CRAHN. (d) $\lambda = 1.0$, $M = 300$, $|K| = 8$.

8.7.4.1   Protocol

The protocol is briefly described in Figure 8.9, where Step 1 aims to process the proposed consensus-based feedback from neighboring nodes, while Step 2 performs standard data communications in the RTS/CTS MAC protocol.

For example, after a CR receives the handshaking frames with spectrum information from neighboring CRs, it will update its local cache with available spectrum band indexes, and the FG it belongs to from the value p. Then, a CR can know the available spectrum bands from the neighbors' feedbacks and then inform the other CRs in a similar way.

In addition, in order to determine which spectrum bands the neighboring CRs are using, it is assumed that a CR node can acquire this information by overhearing the neighboring CRs' communications. A feasible and economic way of overhearing that information is by encapsulating a data field containing that information and piggybacking it in a frame sent by a neighboring CR. For example, the consensus

```
FOR EACH CR node i at time slot t

IF a spectrum change is detected do spectrum sensing
    do overhear incoming RTS/CTS frames from neighboring CRs
    (1) IF the frame piggybacked spectrum information
    do parse the information from frames, x_i(t) = k, DF_i(j) = p
    do perform the local control scheme based on the consensus feedback of the
    spectrum information in the same FG
    ELSE
    (2) Do normal communication with other CRs
    END IF
    ELSE
    do normal communication with other CRs
    END IF
END FOR
```

FIGURE 8.9   Pseudocode of consensus-based spectrum allocation protocol.

feedback is derived from the spectrum information piggybacked in protocol-specific frames or packets, such as RTS or CTS frames in an IEEE 802.11–based MAC protocol. The slotted time characteristic in the 802.11-like MAC protocol can also meet the requirements of the proposed protocol. Therefore, the proposed protocol can be readily integrated in the IEEE 802.11–based CRAHN. More importantly, the proposed consensus-based protocol will not result in extra communication efforts or cause delays affecting throughput.

Now, we analyze the complexity of the proposed protocol. We denote by M the number of CRs and d the average degree of a CR. The spectrum sensing takes s_1 time units; Step 1 and Step 2 take s_2 and s_3 time units, respectively. Step 1 will repeat at maximum dM times, so the average time spent on each CR is $dM(cs_2+(1-c)s_3)+s_1$, where c is a constant denoting the fraction of time that the protocol will go to Step 1. As CRs in different FGs can individually perform the protocol, and in each group, we have approximately (M/FG) nodes, the total number is therefore $(M/FG)(dM(s_2+s_3)+s_1)$. We can obtain the time complexity as $O((M/FG)^2)$, and therefore, more FGs result in less complexity.

8.7.4.2 Simulation/Performance Analysis and Comparison with Other Protocols

The protocol is compared to a classical scheme called device-centric Rule-A [12], where the so-called property line measure calculated from the available spectrum bands of neighboring CRs is used for spectrum sharing. The reason we make comparison with Rule-A is that Rule-A is the most similar scheme to our proposed protocol with basic local information (i.e., connectivity and spectrum availability of neighboring CRs) without extra communication efforts, while some other schemes like the local bargaining scheme or graph-coloring scheme are centralized and require extra information through extra communication efforts. Moreover, the max–min fairness-based schemes, which have a different spectrum sharing objective from the proposed Definition 8.2, will not be considered in the comparisons.

Suppose that the number of spectrum bands at a CR is randomly allocated initially. Each CR performs the proposed consensus-based protocol to ensure that the number of spectrum bands is decided by the consensus feedback from immediate neighbors. The consensus-based communication protocol will be executed once every iteration, and therefore, a successful run needs several iterations. Considering PU activities, the spectrum band availability varies at the beginning of each run. Moreover, the total number of available spectrum bands in the network is kept equal to 1900 for the following simulations.

Figure 8.10 shows a dense CRAHN with 350 CR nodes (i.e., $M=350$). Each link has a negligible data transmission delay. The darkness of the node color indicates a higher number of available spectrum bands. The darker the CR node color, the more the number of available spectrums.

The convergence performance is compared using the metric of unallocated spectrum bands after running the proposed consensus-based protocol and device-centric Rule-A. The results are shown in Figures 8.11 and 8.12.

The convergence performance of Rule-A and the proposed consensus-based protocol are shown in Figure 8.11, where we can see that the proposed consensus protocol

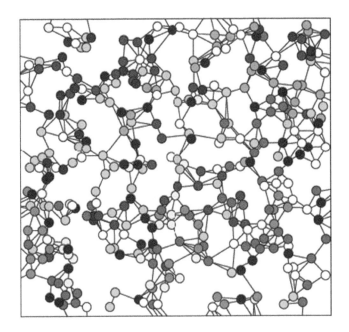

FIGURE 8.10   A randomly distributed CRAHN with 350 CRs and initially allocated spectrum bands.

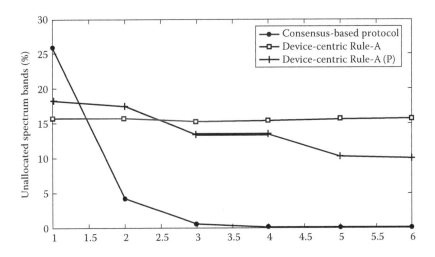

FIGURE 8.11   Convergence performance of the proposed consensus-based protocol, Rule-A, and Rule-A (P).

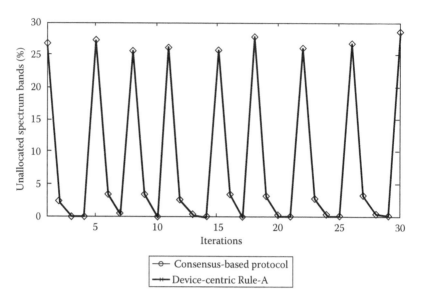

FIGURE 8.12 Convergence performance of the consensus-based protocol and Rule-A in multiple iterations.

converges very quickly, whereas Rule-A and Rule-A (P) have stable convergence performance in one iteration. Rule-A (P) in Figure 8.11 is the improved version of Rule-A, where we increase the accuracy of the poverty line as the feedback; it has better performance than the original Rule-A. At the end of the iteration, all the CRs are allocated with spectrum bands. The reason that the consensus protocol is better is that the consensus spectrum availability information is accurate during the spectrum sharing process, while the device-centric Rule-A and Rule-A (P) use the feedback based on poverty line, which does not accurately estimate the spectrum availability of CR nodes.

In Figure 8.12, we can see how the consensus-based protocol converges over multiple iterations, where at the end of each iteration (i.e., each run of the protocol), all the nodes can be successfully assigned with desired spectrum bands. We use a randomly generated topology for the CRAHN in each iteration epoch, during which the consensus protocol will converge as expected, that is, it can make all the spectrum bands be shared among all the CRs. Moreover, the convergence time is quite stable even if we change the network topology before each iteration. In addition, the spectrum information in this simulation mentioned in Step 1 of Figure 8.9 is the spectrum band.

This scheme outperforms Rule-A, because the inaccurate so-called poverty line is used as a feedback in Rule-A. This inaccurate feedback is considered as a low-quality feedback in the view of local control framework.

8.7.4.3 Fairness Performance in Various Network Sizes

The fairness performance is shown in Figure 8.13 for different network sizes. It can be seen that the fairness measure of Rule-A is larger than the fairness measure of the

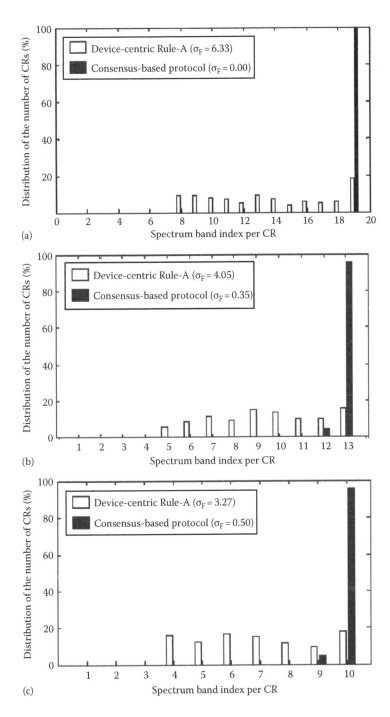

FIGURE 8.13 Fairness performance versus different network sizes when (a) $M = 100$, (b) $M = 150$, (c) $M = 200$.   (*Continued*)

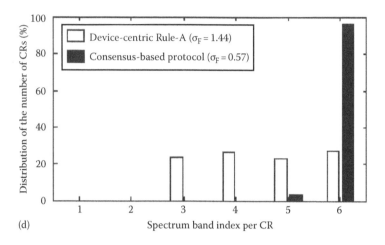

(d)

FIGURE 8.13 (*Continued*) Fairness performance versus different network sizes when (d) $M = 350$.

consensus-based protocol when network size varies. This means that the consensus protocol has better fairness performance than Rule-A. Furthermore, it can be seen how the spectrum sharing goal is achieved by these two algorithms. Figure 8.13a through d shows that the proposed consensus-based protocol can fairly distribute and meet the spectrum sharing goal in a better way than Rule-A.

The FG is discussed in the following. First, Figure 8.14 shows the intermediate results of using the consensus protocol for spectrum allocation. Here, the nodes pointed by arrows indicate different leading nodes in FGs, and, in the initial stage, only the leading nodes have been allocated with spectrum bands. From Figure 8.14a, it can be seen that the CRs running the consensus protocol can adjust the spectrum availability based on a leading node, which gives response to the changing spectrum bands and thus causes the spectrum reallocation to neighboring CRs. The neighboring CRs will run the consensus-based protocol to spontaneously change their spectrum bands. In other words, the leading node can share the spectrum bands to the rest of CRs in this case. Similarly, in Figure 8.14b and c, the leading nodes can share the spectrum bands to the other CRs. Therefore, it can be seen that the nodes following the spectrum information of the leading node belong to the same FG. Furthermore, if the leading nodes are considered as cluster heads reflecting the accurate changing radio environment, all the CRs in a cluster can instantly be informed of the spectrum change accordingly. If the extreme case is considered where the number of FGs equals the number of CRs, this case can actually be converted to the CRAHN similar to Figure 8.10.

In order to see the convergence performance of the consensus-based protocol versus different FGs, the convergence performance is evaluated for three networks with one FG, two FGs, and three FGs, respectively. The experimental results are shown in Figure 8.15, where all nodes can share spectrum bands at the end of each run. Moreover, it is shown that the convergence time per iteration when FG = 1 is in general longer than the convergence time when FG = 2 or FG = 3. This makes sense

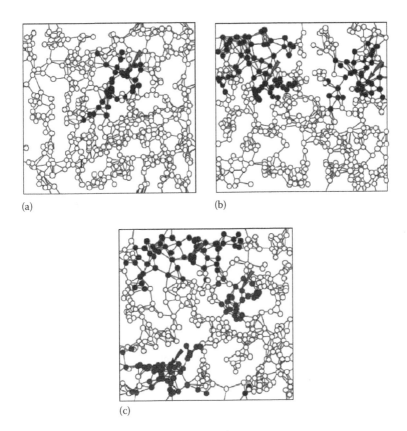

(a)   (b)

(c)

FIGURE 8.14   Intermediate spectrum sharing results in CRAHN when (a) FG = 1, (b) FG = 2, and (c) FG = 3.

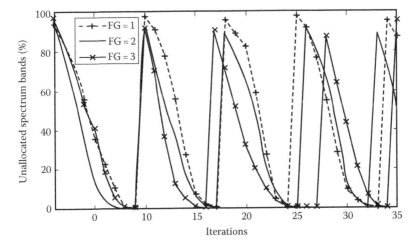

FIGURE 8.15   Convergence performance with different number of FGs.

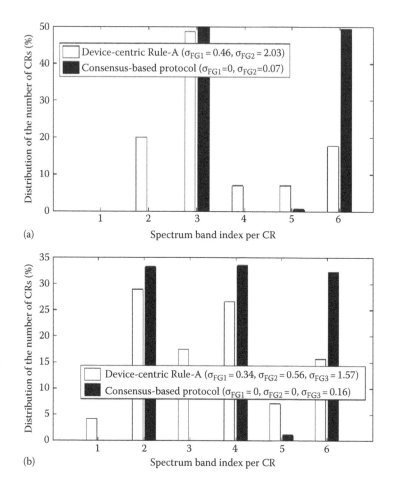

FIGURE 8.16 Fairness performance in a network when (a) FG = 2 and (b) FG = 3.

because the more FGs in a network, the fewer nodes in each FG, and therefore, the quicker decision can be made by a consensus protocol.

Next, the fairness performance is evaluated when FG = 2 and FG = 3 (Figure 8.16). In Figure 8.16a, the network has been evenly divided into two FGs with two separate spectrum sharing goals, where one group wants to get three spectrum bands ($m = 3$) and the other group wants to get six spectrum bands ($m = 6$). From the results shown in Figure 8.16a, it can be seen that after running the two schemes, the consensus-based protocol can fairly distribute the spectrum bands and meet the desired spectrum goal. Similarly, in Figure 8.16b, the network is divided into three FGs and the spectrum goals are $m = 2$, $m = 4$, and $m = 6$, respectively. From Figure 8.16b, the consensus-based protocol can meet the spectrum sharing goal while obtaining much better fairness performance than Rule-A. In conclusion, the consensus-based protocol can meet the spectrum sharing goals in different FGs compared to Rule-A.

8.8 CONCLUSION

This chapter provided an overview of CRAHN models based on the SAM and the resultant channel model. MAC protocols have been discussed in the context of CRAHN where the fairness criteria play an important role. This chapter also explored the effectiveness of using a consensus-based protocol to solve the fairness problem in spectrum sharing. Consensus-based protocols can provide lightweight and efficient solutions for CRAHNs; however, the theoretical grounds need to be investigated for spectrum sharing fairness [4,71–73].

REFERENCES

1. I.F. Akyildiz, W.Y. Lee, and K.R. Chowdhury, CRAHNs: Cognitive radio ad hoc networks, *Ad Hoc Networks*, 7 (5), 810–836, 2009.
2. J.M. Peha, Sharing spectrum through spectrum policy reform and cognitive radio, *Proceedings of the IEEE*, 97 (4), 708–719, 2009.
3. O. Akan, O. Karli, and O. Ergul, Cognitive radio sensor networks, *IEEE Network*, 23, 34–40, 2009.
4. P. Hu, Cognitive radio ad hoc networks: A local control approach, PhD dissertation, Queen's University, Kingston, Ontario, Canada, 2013.
5. N. Devroye, M. Vu, and V. Tarokh, Cognitive radio networks, *IEEE Signal Processing Magazine*, 25 (6), 12–23, November 2008.
6. B. Wang and K.J.R. Liu, Advances in cognitive radio networks: A survey, *IEEE Journal of Selected Topics in Signal Processing*, 5 (1), 5–23, February 2011.
7. P. Lassila and A. Penttinen, Survey on performance analysis of cognitive radio networks, Technical report, Helsinki University of Technology, 2008.
8. L. Zhou and Z.J. Haas, Securing ad hoc networks, *IEEE Network*, 13 (6), 24–30, 1999.
9. J. Huang, R.A. Berry, and M.L. Honig, Spectrum sharing with distributed interference compensation, in *Proceedings of the First IEEE International Symposium on New Frontiers in Dynamic Spectrum Access Networks (DySPAN)*, November 2005, pp. 88–93.
10. C. Peng, H. Zheng, and B. Zhao, Utilization and fairness in spectrum assignment for opportunistic spectrum access, *Mobile Networks and Applications*, 11 (4), 555–576, 2006.
11. L. Cao and H. Zheng, Distributed spectrum allocation via local bargaining, in *Proceedings of the Second Annual IEEE Communications Society Conference on Sensor and Ad Hoc Communications and Networks*, 2005, pp. 475–486.
12. L. Cao and H. Zheng, Distributed rule-regulated spectrum sharing, *IEEE Journal on Selected Areas in Communications*, 26 (1), 130–145, 2008.
13. C. Wu, K. Chowdhury, M. Di Felice, and W. Meleis, Spectrum management of cognitive radio using multi-agent reinforcement learning, in *Proceedings of the Ninth International Conference on Autonomous Agents and Multiagent Systems: Industry track (AAMAS'10)*, Richland, SC, 2010, pp. 1705–1712.
14. T. Jiang, D. Grace, and P.D. Mitchell, Efficient exploration in reinforcement learning-based cognitive radio spectrum sharing, *IET Communications*, 5 (10), 1309–1317, 2011.
15. B. Atakan and O.B. Akan, Biologically-inspired spectrum sharing in cognitive radio networks, in *Proceedings of the IEEE Wireless Communications and Networking Conference (WCNC)*, 2007, pp. 43–48.
16. C. Doerr, D. Grunwald, and D. Sicker, Local control of cognitive radio networks, *Annals of Telecommunications*, 64 (7), 503–534, 2009.
17. Z. Li, F. Yu, and M Huang, A cooperative spectrum sensing consensus scheme in cognitive radios, in *Proceedings of the IEEE International Conference on Computer Communications (INFOCOM)*, 2009, pp. 2546–2550.

18. F. Richard Yu (ed.), *Cognitive Radio Mobile Ad Hoc Networks*, Springer, New York, 2011.
19. H. Zhai, J. Wang, and Y. Fang, DUCHA: A new dual-channel MAC protocol for multihop ad hoc networks, *IEEE Transactions on Wireless Communications*, 5 (11), 3224–3233, 2006.
20. F. Chen, H. Zhai, and Y. Fang, An opportunistic multiradio MAC protocol in multirate wireless ad hoc networks, *IEEE Transactions on Wireless Communications*, 8 (5), 2642–2651, 2009.
21. R. Hasan and M. Murshed, A novel multichannel cognitive radio network with throughput analysis at saturation load, in *Proceedings of the 10th IEEE International Symposium on Network Computing and Applications (NCA)*, August 2011, pp. 211–218.
22. A. Chia-Chun Hsu, D.S.L. Weit, and C.C.J. Kuo, A cognitive MAC protocol using statistical channel allocation for wireless ad-hoc networks, in *Proceedings of the IEEE Wireless Communications and Networking Conference (WCNC)*, 2007, pp. 105–110.
23. T. Shu, S. Cui, and M. Krunz, Medium access control for multi-channel parallel transmission in cognitive radio networks, in *Proceedings of the IEEE Global Telecommunications Conference (GLOBECOM)*, 2006, pp. 1–5.
24. N. Jain, S.R. Das, and A. Nasipuri, A multichannel CSMA MAC protocol with receiver-based channel selection for multihop wireless networks, in *Proceedings of the International Conference on Computer Communication Networks*, 2001, pp. 432–439.
25. B. Sadeghi, V. Kanodia, A. Sabharwal, and E. Knightly, OAR: An opportunistic auto-rate media access protocol for ad hoc networks, *Wireless Networks*, 11 (1–2), 39–53, 2005.
26. H.B. Salameh, M. Krunz, and O. Younis, MAC protocol for opportunistic cognitive radio networks with soft guarantees, *IEEE Transactions on Mobile Computing*, 8 (10), 1339–1352, 2009.
27. H.B. Salameh, M. Krunz, and O. Younis, Distance- and traffic-aware channel assignment in cognitive radio networks, in *Proceedings of the IEEE International Conference on Sensing, Communication, and Networking (SECON)*, 2008, pp. 10–18.
28. D. Cabric and R.W. Brodersen, Physical layer design issues unique to cognitive radio systems, in *Proceedings of the International Symposium on Personal, Indoor and Mobile Radio Communications (PIMRC)*, vol. 2, 2005, pp. 759–763.
29. K.R. Chowdhury, M. Di Felice, and I.F. Akyildiz, TP-CRAHN: A transport protocol for cognitive radio ad-hoc networks, in *Proceedings of the IEEE International Conference on Computer Communications (INFOCOM)*, Vols. 1–5, 2009, pp. 2482–2490.
30. J. So and N.H. Vaidya, Multi-channel MAC for ad hoc networks: Handling multichannel hidden terminals using a single transceiver, *Proc. MOBIHOC'04*, May 2004, Roppongi, Japan.
31. C. Cordeiro and K. Challapali, C-MAC: A cognitive MAC protocol for multi-channel wireless networks, in *Proceedings of the IEEE International Symposium on New Frontiers in Dynamic Spectrum Access Networks (DySPAN)*, 2007, pp. 147–157.
32. H. Su and X. Zhang, Cream-MAC: An efficient cognitive radio-enabled multi-channel MAC protocol for wireless networks, in *Proceedings of the 2008 International Symposium on a World of Wireless, Mobile and Multimedia Networks (WoWMoM)*, 2008, pp. 1–8.
33. P. Gupta and P.R. Kumar, The capacity of wireless networks, *IEEE Transactions on Information Theory*, 46 (2), 388–404, 2000.
34. C. Cordeiro, K. Challapali, and M. Ghosh, Cognitive PHY and MAC layers for dynamic spectrum access and sharing of TV bands, in *Proceedings of the First International Workshop on Technology and Policy for Accessing Spectrum (TAPAS'06)*. ACM, New York, 2006.
35. M. Vu, N. Devroye, and V. Tarokh, On the primary exclusive region of cognitive networks, *IEEE Transactions on Wireless Communications*, 8 (7), 3380–3385, 2009.

36. Y. Shi, C. Jiang, Y.T. Hou, and S. Kompella, On capacity scaling law of cognitive radio ad hoc networks, in *Proceedings of the IEEE International Conference on Computer Communication Networks (ICCCN)*, 2011 Maui, HI, pp. 1–8.

37. C. Lee and M. Haenggi, Interference and outage in doubly Poisson cognitive networks, in *Proceedings of the 19th International Conference on Computer Communications and Networks (ICCCN)*, August 2010, pp. 1–6.

38. H. Su and X. Zhang, Cross-layer based opportunistic MAC protocols for QoS provisioning over cognitive radio wireless networks, *IEEE Journal on Selected Areas in Communications*, 26 (1), 118–129, 2008.

39. W.C. Ao, S.M. Cheng, and K.C. Chen, Phase transition diagram for underlay heterogeneous cognitive radio networks, in *Proceedings of the IEEE Global Telecommunications Conference (GLOBECOM)*, December 2010, pp. 1–6.

40. C. Chen and X. Haige, The throughput order of ad hoc networks with physical-layer network coding and analog network coding, in *Proceedings of the 18th IEEE International Conference on Communications (ICC)*, 2008, pp. 2146–2152.

41. P. Hu and M. Ibnkahla, A survey of physical-layer network coding in wireless networks, in *Proceedings of the 25th Biennial Symposium on Communications (QBSC)*, May 2010, pp. 311–314.

42. M. Haenggi, J.G. Andrews, F. Baccelli, O. Dousse, and M. Franceschetti, Stochastic geometry and random graphs for the analysis and design of wireless networks, *IEEE Journal on Selected Areas in Communications*, 27 (7), 1029–1046, September 2009.

43. P.H.J. Nardelli, M. Kaynia, and M. Latva-aho, Efficiency of the ALOHA protocol in multi-hop networks, in *Proceedings of the 11th IEEE International Workshop on Signal Processing Advances in Wireless Communications (SPAWC)*, June 2010, pp. 1–5.

44. M.C. Golumbic, *Algorithmic Graph Theory and Perfect Graphs*, 2nd ed., Elsevier, February 2004.

45. S. Wolfram, *A New Kind of Science*, Wolfram Media, 2002.

46. Y. Zhao, L. Morales, J. Gaeddert, K.K. Bae, J.-S. Um, and J.H. Reed, Applying radio environment maps to cognitive wireless regional area networks, in *Proceedings of the Second IEEE International Symposium on New Frontiers in Dynamic Spectrum Access Networks (DySPAN)*, April 2007, pp. 115–118.

47. A.B. MacKenzie, J.H. Reed, P. Athanas, C.W. Bostian, R.M. Buehrer, L.A. DaSilva, S.W. Ellingson et al., Cognitive radio and networking research at Virginia Tech, *Proceedings of the IEEE*, 97 (4), 660–688, April 2009.

48. Y. Zhao, J. Gaeddert, K.K. Bae, and J.H. Reed, Radio environment map enabled situation-aware cognitive radio learning algorithms, in *Proceedings of the Software Defined Radio Forum (SDRF) Technical Conference*, Orlando, FL, 2006.

49. R. Draves, J. Padhye, and B. Zill, Routing in multi-radio, multi-hop wireless mesh networks, in *Proceedings of the 10th Annual International Conference on Mobile Computing and Networking (MobiCom'04)*, ACM, New York, 2004, pp. 114–128.

50. J. Li, Y. Zhou, and L. Lamont, Routing schemes for cognitive radio mobile ad hoc networks, in F. Richard Yu (ed.), *Cognitive Radio Mobile Ad Hoc Networks*, Springer, New York, 2011, pp. 227–248.

51. I. Pefkianakis, S.H.Y. Wong, and S. Lu, SAMER: Spectrum aware mesh routing in cognitive radio networks, in *Proceedings of the Third IEEE International Symposium on New Frontiers in Dynamic Spectrum Access Networks (DySPAN)*, October 2008, pp. 1–5.

52. G. Cheng, W. Liu, Y. Li, and W. Cheng, Spectrum aware on-demand routing in cognitive radio networks, in *Proceedings of the Second IEEE International Symposium on New Frontiers in Dynamic Spectrum Access Networks (DySPAN)*, April 2007, pp. 571–574.

53. M. Cesana, F. Cuomo, and E. Ekici, Routing in cognitive radio networks: Challenges and solutions, *Ad Hoc Networks*, 9 (3), 228–248, 2011.

54. R. Hasan and M. Murshed, Provisioning delay sensitive services in cognitive radio networks with multiple radio interfaces, in *Proceedings of the IEEE Wireless Communications and Networking Conference (WCNC)*, March 2011, pp. 162–167.

55. R. Olfati-Saber, J.A. Fax, and R.M. Murray, Consensus and cooperation in networked multi-agent systems, *Proceedings of the IEEE*, 95 (1), 215–233, January 2007.

56. S. Filin, H. Harada, H. Murakami, K. Ishizu, and G. Miyamoto, IEEE 1900.4 architecture and enablers for optimized radio & spectrum resource usage, in *Proceedings of the International Conference on Ultra Modern Telecommunications and Workshops (ICUMT)*, 2009, pp. 1–8.

57. D.P. Satapathy and J.M. Peha, Etiquette modification for unlicensed spectrum: Approach and impact, in *Proceedings of the 48th IEEE Vehicular Technology Conference (VTC)*, vol. 1, 1998, pp. 272–276.

58. D.P. Satapathy and J.M. Peha, Performance of unlicensed devices with a spectrum etiquette, in *Proceedings of the IEEE Global Telecommunications Conference (GLOBECOM)*, vol. 1, 1997, pp. 414–418.

59. Z. Ji and K.J.R. Liu, Cognitive radios for dynamic spectrum access—Dynamic spectrum sharing: A game theoretical overview, *IEEE Communications Magazine*, 45 (5), 88–94, May 2007.

60. M.M. Halldórsson, J.Y. Halpern, L. Li, and V.S. Mirrokni, On spectrum sharing games, in *Proceedings of the 23rd Annual ACM Symposium on Principles of Distributed Computing (PODC'04)*, ACM, New York, 2004, pp. 107–114.

61. R. Etkin, A. Parekh, and D. Tse, Spectrum sharing for unlicensed bands, *IEEE Journal on Selected Areas in Communications*, 25 (3), 517–528, April 2007.

62. I.F. Akyildiz, W.-Y. Lee, M.C. Vuran, and S. Mohanty, Next generation/dynamic spectrum access/cognitive radio wireless networks: A survey, *Computer Networks*, 50 (13), 2127–2159, September 2006.

63. C. Zou, T. Jin, C. Chigan, and Z. Tian, QoS-aware distributed spectrum sharing for heterogeneous wireless cognitive networks, *Computer Networks*, 52 (4), 864–878, March 2008.

64. R.S. Komali, A.B. MacKenzie, and R.P. Gilles, Effect of selfish node behavior on efficient topology design, *IEEE Transactions on Mobile Computing*, 7 (9), 1057–1070, September 2008.

65. N. Nie and C. Comaniciu, Adaptive channel allocation spectrum etiquette for cognitive radio networks, in *Proceedings of the First IEEE International Symposium on New Frontiers in Dynamic Spectrum Access Networks (DySPAN)*, November 2005, pp. 269–278.

66. D. Niyato and E. Hossain, A game-theoretic approach to competitive spectrum sharing in cognitive radio networks, in *Proceedings of the IEEE Wireless Communications and Networking Conference (WCNC)*, March 2007, pp. 16–20.

67. S. Haykin, Cognitive radio: Brain-empowered wireless communications, *IEEE Journal on Selected Areas in Communications*, 23 (2), 201–220, 2005.

68. S. Ahmad, M. Liu, T. Javidi, Q. Zhao, and B. Krishnamachari, Optimality of myopic sensing in multichannel opportunistic access, *IEEE Transactions on Information Theory*, 55 (9), 4040–4050, September 2009.

69. I.F. Akyildiz, W.Y. Lee, and K.R. Chowdhury, Spectrum management in cognitive radio ad hoc networks, *IEEE Network*, 23 (4), 6–12, 2009.

70. B. Mercier, V. Fodor, R. Thobaben, M. Skoglund, V. Koivunen, S. Lindfors, J. Ryynanen et al., Sensor networks for cognitive radio: Theory and system design, *Proceedings of the ICT Mobile and Wireless Communications Summit (ICT)*, 2008.

71. P. Hu and M. Ibnkahla, Consensus-based local control schemes for spectrum sharing in cognitive radio sensor networks, *Proceedings of the 26th Biennial Symposium on Communications (QBSC)*, May 2012, pp. 115–118.

72. P. Hu and M. Ibnkahla, Fairness and consensus protocol for cognitive radio networks, *Proceedings of the IEEE International Conference on Communications (ICC)*, June 2012, pp. 93–97.

73. P. Hu and M. Ibnkahla, A consensus-based protocol for spectrum sharing fairness in cognitive radio ad hoc and sensor networks, *International Journal of Distributed Sensor Networks*, 2012.

9 Medium Access in Cognitive Radio Ad Hoc Networks

9.1 INTRODUCTION

Medium access control (MAC) protocols used in ad hoc networks are different from MAC protocols for cognitive radio ad hoc networks (CRAHNs). In CRAHNs [1], MAC protocols have to address the spectrum sharing function [2] while improving the overall throughput and spectral efficiency. Furthermore, primary exclusive regions (PERs) [3] in CRAHNs can make a significant impact on secondary user (SU) and primary user (PU) communications; therefore, MAC protocols should address PERs, especially when PUs and/or SUs are mobile.

Classical carrier sense medium access with collision avoidance (CSMA-/CA)-based MAC protocols have the advantage of solving hidden terminal problems and having distributed operations (e.g., distributed coordination function in IEEE 802.11 MAC); thus, state-of-the-art MAC protocols [4–9] for CRNs have been proposed.

This chapter presents a CSMA-/CA-based MAC protocol called CM-MAC for CRAHNs. This protocol has been initially presented in [32,33]. Its goal is to improve the performance of the network. The protocol is supported with a mobility support algorithm (MSA) to specifically address PER issues [32,33]. The choice of the MAC layer to solve PER issues is motivated by the following: (1) the MAC layer is the right place to address the PER issues. If an upper-layer scheme like a routing scheme is employed, a MAC-layer mechanism is still needed to address the issues caused by PER-like spectrum sharing, mobility, and PU detection, and (2) solving the PER issues in MAC layer is lightweight compared to solving it at the network layer (as either the path formation or scheduling in routing is costly).

A survey of MAC protocols for CRAHNs has been provided in Chapter 8. Here, we give a summary of this survey. Dual channel MAC for ad hoc networks (DUCHA) protocol is proposed in [10], which can improve the single-hop throughput up to 1.2 times and multihop throughput up to five times compared to the IEEE 802.11 MAC protocol. Opportunistic multiradio MAC (OMMAC) protocol is proposed in [11], where a multichannel-based packet scheduling algorithm is employed and packets were sent on a channel having the highest bit rate. A MAC protocol based on statistical channel allocation (SCA) is proposed in [4], which uses a channel aggregation approach to improve the throughput and dynamic operating range to reduce the computational complexity. Results of [4] show that SCA can use spectrum holes effectively to improve spectrum efficiency while keeping the performance of coexisting PUs. In order to meet data rate requirement for data transmissions, a MAC with the so-called multichannel parallel transmission protocol is proposed in [7], where the minimum number of channels is selected to meet a certain data rate. The results of [7] outperform the protocol proposed in [8] in which the channel is selected according to the best signal to interference noise ratio (SINR) value. In [9], an opportunistic autorate MAC protocol is used to maximize the utilization on individual channels. Spectrum sharing and spectrum access functions are explicitly addressed in [6], where spectrum access and spectrum allocation schemes are introduced into the proposed cognitive opportunistic MAC (COMAC) protocol. In [5], the authors propose distance-dependent and traffic-aware MAC (DDMAC) protocol employing a distance-dependent channel assignment scheme. The aforementioned works do not comprehensively consider several important factors. For example, although spectrum sensing can be simultaneously performed in one shot [12], the sensing time cannot be ignored, as it may be relatively large, which may lead to throughput degradation [13]. Note that most protocols in the literature are based on CSMA/CA procedure. However, some non-CSMA-/CA-based MAC protocols have been proposed as well (such as MMAC [14] and cognitive multichannel MAC (C-MAC) [15]) and have been shown to solve the hidden terminal problem while requiring periodic synchronization.

The chapter is based on the results of [32,33] and is organized as follows. Section 9.2 presents the network model and its requirements. Section 9.3 describes the CM-MAC protocol with the mobility support scheme. Sections 9.4 is devoted to an in-depth analytical analysis of the protocol's performance. Some illustrative examples based on computer simulations are presented in Section 9.5. In particular, the simulations show the impact of the network parameters (such as PU and SU traffic and mobility patterns) on the protocol's performance.

9.2 NETWORK MODEL AND REQUIREMENTS

9.2.1 System Model

Before further discussion, we describe the system model used in this chapter. A CRAHN is deployed in a plane containing N_p PUs and N_{CR} cognitive radio (CR) nodes [32,33]. In a certain time period, a set of channels, denoted by $\mathbf{K}_i(t)$, are available to a CR node I, and thus, the total number of channels available to CR node i is $|\mathbf{K}_i(t)|$. The set of channels on a transmission link between ith CR and $(i+1)$th CR is $\mathbf{K}_{i,i+1}(t)$. There are K spectrum bands in total available to CRs and PUs, while the typically used $(K+1)$th out-of-band common control channel (CCC) [16] is employed to exchange control information. When the jth PU is active for a transmission, its traffic flow takes one channel C_k, that is, $\mathbf{K}_j^P(t) = \{C_k\}$. For simplicity, we use the following notation: $\mathbf{K}_j^P(t) = \{k\}$. When the jth PU is active, $\mathbf{K}_j^P(t) = \{k \mid k > 1 \text{ and } k \leq K\}$. In the chapter, the PU traffic flow is assumed to follow a Poisson process with parameter λ [17]. Note that in this chapter, a PU that occupies multiple channels is equivalent to multiple PUs that occupy different channels.

We consider a CRAHN with a PER as shown in Figure 9.1. The PER is located at the center of the network and the primary receivers are within the PER bounded by the circle of radius $R_0+\varepsilon$ and ε is a small guard band around the PER. A PER operating on channel k is denoted by $\mathbf{S}_{PER}(k)$. $\mathbf{S}_{PER}(k)$ (i.e., the shaded area shown

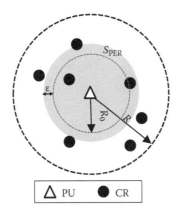

FIGURE 9.1   A CRAHN with a PER and multiple CRs.

in Figure 9.1) has a radius of $R_0 + \varepsilon$ and the radio coverage of a PU (i.e., the area circled by dotted line in Figure 9.1) has a radius of R, where $R_0 < R$. In the $S_{PER}(k)$, CR communications will severely affect the PU communication and vice versa.

9.2.2   REQUIREMENTS

A MAC protocol needs to decide the availability of spectrum bands for current and future data transmissions. These spectrum bands can facilitate the upper-layer protocol (e.g., routing protocol) in order to obtain an optimized path for data transmissions. Moreover, in order to keep a desirable throughput, a MAC protocol is expected to perform local observation without extra communication efforts. As such, the information exchange based on built-in handshaking procedures in a CSMA/CA MAC protocol is a preferred solution.

Traditionally, the MAC sublayer is at the link layer, where the link layer is in charge of the communication between adjacent nodes. Therefore, based on the layered perspective for the CRAHN protocol stack, maintaining the communications with adjacent nodes while sharing the spectrum resources among nodes is a major challenge.

In order to determine why a CRAHN MAC is important for data transmissions, consider the example of a typical data transmission scenario shown in Figure 9.2. SU S tries to transmit data frames to SU E through the path from node A to D. In the previous time slots, data were transmitted on channel 3, which was not occupied by either PU1 or PU2. However, channel 3 is now occupied by PU2. The links from node C to D and D to E are now broken. Therefore, the rest of nodes need to be informed of the spectrum change of nodes D and E. Due to the nature of changing PU's activity, it is efficient to perform updates of spectrum changes before each data transmission. Since the PU spectrum availability is dependent on the PER region induced by a PU, PER regions should be considered by the MAC protocol.

To see how PER region can affect CR and PU throughput, consider the network shown in Figure 9.1, where CRs are distributed outside PER when CRs and

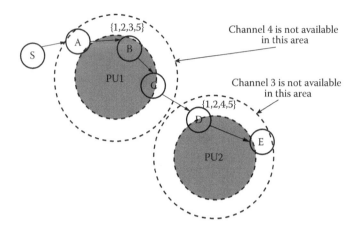

FIGURE 9.2 Necessity of a CRAHN MAC: The available spectrum bands for the nodes covered by a PU are shown in brackets; the links are broken (shown in dashed arrows) when the data transmission from S to E is operated on channel 3.

PUs are operating on channel k. In this sense, Mai et al. [3] derive the worst-case interference power that the PU transmitter experienced from all CR nodes where a PU transmitter communicates with a PU receiver at a distance R_0. This worst-case interference power is

$$E[I_0]_{\alpha=4} = d\pi P\left[\frac{R^2}{(R^2-R_0^2)^2} + \frac{(R+\varepsilon)^2}{\varepsilon^2(2R_0+\varepsilon)^2}\right] \tag{9.1}$$

where
   α is the path loss exponent
   R_0 is the radius of $\mathbf{S}_{\mathrm{PER}}(k)$
   R is the coverage radius of the PU
   ε is the guard band radius, which ensures that the interference caused by CRs will
        not affect PU communications

CR nodes are distributed in the outer circular area of PER with density θ.
   Because a CSMA/CA MAC protocol is based on time frames, we consider a time frame with an interval $[0, T]$. If the PU transmitter/receiver pair is active for ν time units, while CR nodes are active for the entire time frame, CRs can interfere with PU communications in the ν time units. Based on the data rate equation in [3], the data rate of PU, D_{PU}, and data rate of CR, D_{CR}, can be expressed as

$$D_{\mathrm{PU}} \leq \frac{\nu}{T}\log\left(1+\frac{P_{\mathrm{PU}}}{R_0^2(N_0+E[I_0])}\right) \tag{9.2}$$

$$D_{\mathrm{CR}} \le \frac{v}{T} \log\left(1 + \frac{P_{\mathrm{CR}}}{(R_0 + \varepsilon)^2 (N_0 + P_{\mathrm{PU}})}\right) + \left(1 - \frac{v}{T}\right) \log\left(1 + \frac{P_{\mathrm{CR}}}{(R_0 + \varepsilon)^2 N_0}\right) \quad (9.3)$$

where

N_0 is the noise power spectral density

P_{PU} and P_{CR} are the transmit power of a PU and a CR, respectively

In reality, the PU activities may not be continuous and v is not a constant, but a random variable. As the bursty traffic is an interrupted Poisson process with on–off periods, the PU activity is assumed to follow a Poisson process with parameter λ [18], and the mean of the probability of interarrival time is $1/\lambda$. In this sense, on average, CR nodes can be considered to have $v = T(1 - x/\lambda)$ time units without interfering with PUs during $[0, T]$, where x is the number of PU flows. Additionally, the value of v can be adjusted to reflect the spectrum sharing technique in (9.2) and (9.3). For example, let v be T in (9.3), then (9.2) and (9.3) can represent a spectrum sharing model where PUs and CRs can access the spectrum resources at the same time and can avoid interference without explicit signaling.

Based on (9.2) and (9.3), Figure 9.3 shows two cases in which adjusting the CR transmit power P_{CR} or the radius of PER can change the data rates for both CRs and PUs. These two cases exemplify the impact of PER region on throughput. Besides, in order to get an optimal D_{PU}, we should choose a proper value of R/R_0, which is 1.33 in the example shown in Figure 9.3. In Figure 9.3a, as P_{CR} increases, D_{CR} increases more quickly than the declining D_{PU} does. This is the reason that throughput always rises when P_{CR} increases. In Figure 9.3b, D_{CR} increases more slowly than the declining D_{PU} does. In fact, in order to choose a proper value of P_{CR}, we have to consider a feasible range of P_{CR}. A reasonable value of P_{CR} can be chosen based on the maximum transmit power for P_{PU} and P_{CR} regulated in the current wireless network standards, such as Global System for Mobile Communications (GSM) (where P_{PU} is about 1–2 W), IEEE 802.22 (where P_{PU} is less than 4 W [19]), IEEE 802.11 (where P_{CR} is less than 100 mW), and IEEE 802.15.4 (where P_{CR} is less than 100 mW).

With the aforementioned discussion, it can be seen that the PER region has significant impact on the throughput for both PUs and CRs. Furthermore, given a certain data rate for a PU, C_0, and a certain CR output power, P_{CR}, an optimal value of R/R_0 can be chosen.

9.3 CM-MAC: CSMA-/CA-BASED MAC PROTOCOL FOR CRAHNs

9.3.1 PROTOCOL DESCRIPTION

RSS: receiver spectrum sensing	CW: contention window
TSS: transmitter spectrum sensing	RTS: request to send
SIFS: short interframe space	CTS: clear to send
MPDU: MAC protocol data unit	ACK: acknowledgment
ACTS: acknowledgment CTS	

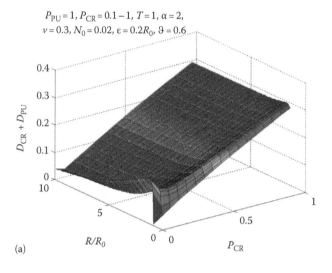

(a)

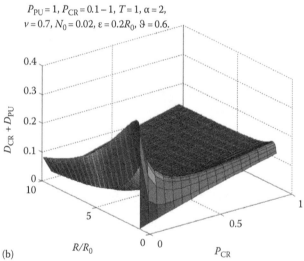

(b)

FIGURE 9.3   Normalized throughput of a PU and CRs versus P_{CR} and R/R_0, when (a) $\nu=0.3$ and (b) $\nu=0.7$.

As discussed in Section 9.2, in order to meet the requirements of a CRAHN MAC protocol, the traditional CSMA-/CA-based MAC protocol (Figure 9.4a) needs to be improved. In Figure 9.4b, a common control channel (CCC) is used in order to exchange the control frames such as request to send (RTS) frames, clear to send (CTS) frames, and acknowledgment (ACK) frames. Following the MAC protocol data unit (MPDU) transmission, a node will wait for a short interframe space (SIFS) period and then transmit the ACK frame. Before sending an RTS frame, the

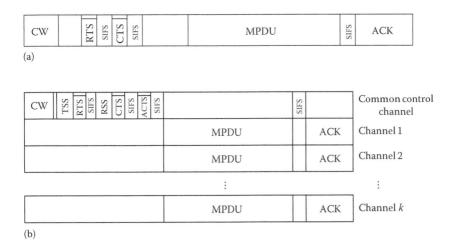

FIGURE 9.4 Frame structures of (a) the traditional CSMA-/CA-based MAC protocol and (b) CM-MAC protocol.

spectrum sensing process will be initiated by a CR to make sure that there is a data transmission link on a certain channel k.

There are two advantages to using a CCC in CRAHNs. First, possible collisions of control frames and data frames can be avoided. Second, when spectrum availability changes, assigning a CCC can alleviate the communication efforts required to consult other CRs in a new spectrum band for exchanging control information.

The transmitter spectrum sensing (TSS) and receiver spectrum sensing (RSS) procedures are employed, which are dedicated to ensuring the spectrum availability of links for upcoming data transmissions. Checking the spectrum availability before transmission can avoid transmission failures. TSS is done by a CR transmitter which combines the spectrum information into the immediate RTS frame field. Meanwhile, RSS is completed on a receiver, which combines the spectrum information into the CTS frame. After the broadcasting stage of RTS/CTS frames with piggybacked spectrum information, the neighboring CRs of the transmitter and receiver obtain the local knowledge of one-hop spectrum availability.

It is worth noting that integrating the spectrum information into the RTS/CTS routine, the update frequency of spectrum information on neighboring CRs is dependent on the RTS/CTS request frequency (i.e., the data transmission load). It is expected that in the saturated mode of a CRAHN (i.e., a CR always has data payload to send), the spectrum information can be frequently updated. For CRAHNs with less data load, the spectrum information may be updated less frequently, subject to the possible failures of data transmissions caused by inaccurate spectrum information on the links.

Another solution to inform CRs about spectrum availability is to use a periodic updating mechanism that maintains broadcast frames containing spectrum availability information. Since this solution may cause collisions with the routine control frames and may result in significant delays, it will not be further discussed in this chapter.

9.3.2 CHANNEL AGGREGATION

The separation of CCC and data channel cannot significantly improve the throughput. This is because the data transmission channel cannot be utilized at all before a successful RTS/CTS process.

A feasible way to further improve the throughput is to decrease the transmission time of a data frame. The channel aggregation will be considered in this chapter in a similar manner to the method suggested in [4,20]. An example of channel aggregation is shown in Figure 9.5a, where, compared with that in Figure 9.4, the transmission time for a data payload in an MPDU is reduced as the MPDU is split into three segments and transmitted on three channels simultaneously. In each segment, the sequence number is added for each split data payload.

It should be noted that the channels used in this technique are dependent on the spectrum sharing scheme. In Figure 9.5b, it can be seen that the actual channels for transmission are obtained after the negotiation stage. This negotiation stage can be either an RSS/TSS procedure or the SPEC_CHANGE notification procedure in

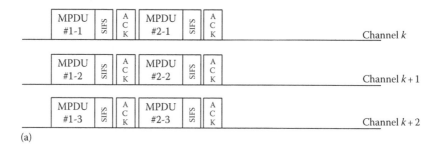

(a)

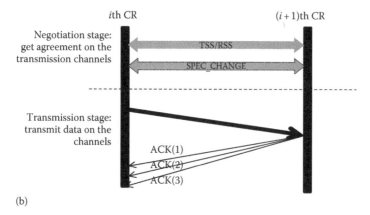

(b)

FIGURE 9.5 Example of channel aggregation in the view of (a) the MAC frame and (b) the sequence diagram.

MSA, which will be discussed later. Since channel aggregation is employed, the sender is expected to receive three ACKs for three split MPDUs.

9.3.3 SPECTRUM ACCESS AND SHARING

In the CM-MAC protocol, a simplified spectrum access scheme is employed where a CR can access the minimum available channels that can meet a certain rate, D_{CR}, on a link between ith CR and $(i+1)$th CR. Therefore, the set of channels accessed by a CR on this link is $\mathbf{K}_{i,i+1}(t)=\mathbf{K}_i(t)\cap\mathbf{K}_{i+1}(t)$ and $D_{CR} = \sum_{k=1}^{|\mathbf{K}_{i,i+1}(t)|} r(k)$, where $r(k)$ is the rate supported on channel k. Besides, if we consider the case where a CR uses all available channels to meet the rate D_{CR}, the link will have $|\mathbf{K}_{i,i+1}(t)|$ available channels for data transmissions.

For spectrum sharing, instead of using the central coordination in IEEE 802.22 standard [21,22], distributed spectrum information exchange will be employed. The main goal in the proposed CM-MAC is to ensure successful next one-hop transmission. Therefore, it is necessary to show the convergence of spectrum information exchange of the TSS/RSS procedure.

For example, Figure 9.6 illustrates that after the TSS procedure, CR nodes 2, 4, and 6 have the updated spectrum information of CR node 1, and after the RSS procedure, CR nodes 1, 3, 5, and 7 can receive updated spectrum information from node 6. Although CR nodes 2 and 4 cannot get the updated spectrum information of CR node 6, it is not a problem as CR nodes 2 and 4 are not on the next transmission link of node 6. The candidate CR nodes for the next transmission are CR nodes 1, 3, 5, and 7. As such, it can be seen that the TSS/RSS procedure integrated in RTS/CTS/ACTS handshaking is sufficient enough and no significant communication overhead is required. All neighboring CRs can receive the spectrum information, which assures successful next one-hop transmission.

The time spent on spectrum sensing is not negligible in the proposed CM-MAC protocol because a spectrum sensing process usually can take as long as 20 μs [11], which is similar to a typical short interframe space (SIFS) duration. Although it is

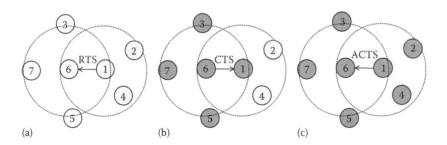

FIGURE 9.6 Example of intermediate results of the spectrum sharing procedure after (a) an RTS transmission, (b) a CTS transmission, and (c) an ACTS transmission. The dotted lines are transmission ranges of CR node 1 and CR node 6.

more desirable to reduce the number of times of spectrum sensing processes when PUs are inactive, the TSS/RSS procedure with RTS/CTS is flexible as PU activity may not be known in advance. Furthermore, the effect of sensing errors on the TSS/RSS procedure is discussed. This chapter assumes that there are no sensing errors during the TSS/RSS procedure. However, the occurrence of sensing errors caused by a spectrum sensing function may exist during the transmissions of TSS-/RSS-related frames. If there is a spectrum sensing error caused by the spectrum sensing function, the data rate will be affected. From [23], the sensing error may affect the data rate if a transmission fails when a CR falsely identifies an idle channel and the delay arisen by the re-detection efforts. However, the sensing errors can be mitigated by physical-layer (PHY) techniques [24,25], sensing scheduling protocols in an upper layer [26], and multiple access protocol design [27]. For example, in the TSS/RSS procedure, if a spectrum sensing error occurs, it can be mitigated by using the PHY techniques or rescheduling schemes of the spectrum sensing during RTS/CTS/ACTS transmissions in the link layer.

9.3.4 Mobility Support

Since CRs in a CRAHN are able to move in the network and cause significant interference to PU traffic, it is important to consider the case that CRs may move into a PER. The negative effects when a CR moves into a PER are as follows: (1) PU communications experience interference from CRs and (2) CR communications experience interference from PU communications. Both effects result from the case that a PU is not aware of the spectrum band being used by CRs. As such, an efficient algorithm is required to solve these issues.

One issue here is how a CR can readily know its vicinity to a PER region. The radio signal strength indicator (RSSI) at the PHY can be used to solve this problem. The RSSI value is proportional to the radio signal strength, with the beacons received from PU transmissions, an SU can know its vicinity to the PU. Although an RSSI itself usually cannot give accurate distance estimations and it may cause some *false alarms*. However, we considered the use of RSSI as a low-cost method to obtain the proximity of a node to a PER and assume the sufficient accuracy of the RSSI method, which can be improved by triangulation or more sophisticated schemes. As the signal is received by an ith CR on the CCC, $RSSI(i, j)$ is inversely proportional to the distance (d) between the ith CR and the jth PU, and the RSSI value can readily indicate the vicinity to a PU if $RSSI(i, j)$ is close to a constant threshold $RSSI_{thres}$. If we assume all the PUs have the same transmit power, the $RSSI_{thres}$ value is sufficient for all CRs. The $(i+1)$th CR node is assumed to communicate with the ith CR node. The MSA scheme can be described as follows [32,33]:

MOBILITY SUPPORT ALGORITHM

Input: $RSSI(i, j)$, State(i), $\mathbf{K}_{i,i+1}(t)$, $\mathbf{K}_j(t)$
for each CR
   if $RSSI(i, j) > RSSI_{thres}$ **AND** State(i) == MAC_OPER
     $\mathbf{K}_j(t) \leftarrow$ func_SS(j)

if $\mathbf{K}_j(t) \in \mathbf{K}_{i,i+1}(t)$
  if State(i) == MAC_TRANSMIT
    if $|\mathbf{K}_{i,i+1}(t)| == 1$
      send a STOP frame over CCC to the (i+1)th node on channel
      the (i+1)th node will stop the data transmission
    else if $|\mathbf{K}_{i,i+1}(t)| > 1$
    $\mathbf{K}_{i,i+1}(t) \leftarrow \{k | k \in \mathbf{K}_{i,i+1}(t), k \notin \mathbf{K}_j(t)\}$
      send the SPEC_CHANGE frame with $\mathbf{K}_{i,i+1}(t)$ to the ($i+1$)th node
    end if
  end if
  if State(i) == MAC_IN_TRANSMIT
    send a STOP frame over CCC to the ($i+1$)th node
    the ($i+1$)th node will record the data frames/segments already transmitted
    the ($i+1$)th node will reinitiate the transmission for the remaining frames
  end if
  if State (i) == MAC_CTS/ACTS
  $\mathbf{K}_{i,i+1}(t) \leftarrow \{k | k \in \mathbf{K}_{i,i+1}(t), k \notin \mathbf{K}_j(t)\}$
    send a CTS or ACTS frame piggybacking $\mathbf{K}_{i,i+1}(t)$ to the transmitter over CCC
  end if
  State(i) $\leftarrow$ MAC_PER
  end if
  end if
if RSSI(i, j) ≤ RSSI$_{thres}$ **AND** State(i) == MAC_PER
  State(i) $\leftarrow$ MAC_OPER
end if
end for

State(i) records the current MAC state on a CR. If the ith CR is in a PER region, then State(i)=MAC_PER; if the CR is out of a PER, then State(i)=MAC_OPER; if the CR is in a CTS/ACTS procedure with a transmitter, then State(i)=MAC_CTS/ACTS; if a CR is transmitting data frames, then State(i)=MAC_TRANSMIT; State(i)=MAC_IN_TRANSMIT means that some frame segments have been transmitted with the channel aggregation technique; func_SS(j) is the spectrum sensing routine to sense which channel is occupied by the jth PU. The STOP frame contains short control information on the current channel, while the SPEC_CHANGE frame contains the available channels for transmission. When the ($i+1$)th CR receives a SPEC_CHANGE frame, it then uses the available channels to send the data frames. If a CR is in the process of sending CTS/ACTS frames, then an updated CTS/ACTS will be resent to the transmitter/receiver with the updated channel information.

From the MSA, it can be seen that once the CR is in the PER region, the data transmission should immediately stop and may cause retransmissions of frames. When the state is MAC_IN_TANSMIT, meaning the frames or frame segments have been in the process of transmission, the CR in a PER region should notify the transmitter immediately in order to resume the transmission of remaining frames or frame segments on the other CRs.

9.4 ANALYTICAL ANALYSIS

This section presents the throughput analysis of CM-MAC protocol as it is investigated in [32,33]. The PU topology can affect the performance in terms of throughput. The distance between the centers of two PERs will be assumed to be at least $2R$, as shown in Figure 9.7.

9.4.1 MOBILITY EFFECT

The mobility should be considered in the throughput analysis as it affects the time spent on spectrum sensing and MSA. We let the coverage of a CR be S_{CR}, where $\|S_{PER}\| > \|S_{CR}\|$, and it is assumed that all CRs have identical coverage disk in the CRAHN. The CR nodes are deployed in the disk area S_{PER} (with radius R), following a homogeneous Poisson process with density θ per unit area.

When a CR node is moving into the PER, it will run the MSA. We are interested in the number of moving nodes in an annulus area with radius $[R_0 - r_0, R_0 + \varepsilon + r_0]$ shown in Figure 9.7. The number of neighbors of a node (i.e., the average degree) [28] between the CRs inside the PER and the CRs outside the PER at a distance r_0 can be obtained, that is,

$$E[\text{Deg}] = 2\pi\theta \int_{R_0-r_0}^{R_0+\varepsilon+r_0} P(\Lambda(i,i+1) \mid s(i,i+1))s \, ds \qquad (9.4)$$

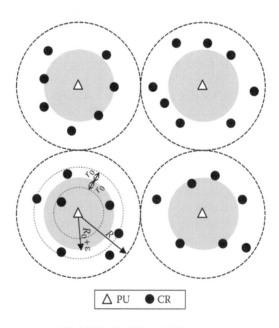

FIGURE 9.7   Example of a CRAHN with PUs and CRs.

where $\Lambda(i, i+1)$ means the event that ith CR and $(i+1)$th CR have a radio link while $s(i, i+1)$ is the distance between them. From [28], assuming $s(i,i+1)=r_0$, then

$$P(\Lambda(i,i+1)\,|\,s(i,i+1)) = \frac{1}{2} - \frac{1}{2}\operatorname{erf}\left(\frac{10\alpha}{\sqrt{2\vartheta}}\log\frac{r_0}{10^{\frac{\beta_{th}}{\alpha 10\,dB}}}\right) \qquad (9.5)$$

where

β_{th} is the threshold value of the received power to maintain the radio link
ϑ is the shadow fading variance
α is the path loss exponent

When mobility of CRs occurs, there is a probability of retaining connectivity among CRs outside the PER, and we know that probability $P(\mathrm{Deg}>0)=1-e^{-E[\mathrm{Deg}]}$. By algebraically manipulating (9.4) and (9.5), the value of r_0 can be determined.

We assume all CRs are moving. If at time t, a CR in the area with radius $[R_0 - r_0, R_0+\varepsilon+r_0]$ just moved in the PER, then the average time spent regarding MSA is

$$E[T_{\mathrm{MSA}}] = P_0(T_{\mathrm{SS}} + (P_{11}T_{\mathrm{STOP}} + P_{12}T_{\mathrm{S\_CHANGE}})P_1 + T_{\mathrm{STOP}}P_2 + T_{\mathrm{CTS}}P_3) \qquad (9.6)$$

where

$P_1 = P(\text{state} = \text{MAC\_TRANSMIT})$
$P_2 = P(\text{state} = \text{MAC\_IN\_TRANSMIT})$
$P_3 = P(\text{state} = \text{MAC\_CTS/ACTS})$
P_{11} is the probability of sending STOP frame
P_{12} is the probability of sensing S_CHANGE frame

The average time spent on one shot of spectrum sensing is T_{SS}; T_{STOP} and $T_{\mathrm{S\_CHAGE}}$ are the time spent on transmitting the STOP and SPEC_CHANGE frames, respectively. P_0 is the probability that the ith CR just moves in a PER region, we can obtain $P_0 = P(\text{state}(t)=\text{MAC\_PER}|\text{state}(t-1)=\text{MAC\_OPER})$.

Because it is difficult to give an exact value of P_1, P_2, or P_3 due to the fact that they are application specific, we use an estimated value for each of them.

For P_1, we can take $\dfrac{T_{\mathrm{data}}}{T_{\mathrm{data}}+T_{\mathrm{CTS}}+\omega}$ as its value, where ω is the delay and empty slot time to consider.

For P_3, we can take $\dfrac{T_{\mathrm{CTS}}}{T_{\mathrm{data}}+T_{\mathrm{CTS}}+\omega}$ as its value. P_2 is difficult to know, because when there is a bulk of data to send, P_2 is large; when there is a small bulk of data to send, P_2 is small. However, we can know the maximum value of P_2, that is,

$\dfrac{\omega}{T_{\mathrm{data}}+T_{\mathrm{CTS}}+\omega}$.

P_0 is usually dependent on the mobility pattern of the CRs. If the CR motion is assumed to follow a 1-D-correlated random walk model (with bounds $[0, 2R]$ with equal probability of moving in two opposite directions), the steady-state probability at any location is $1/4R$ [29] (assuming the speed is one-unit length per time unit). This can be used to estimate the value of P_0.

9.4.2 THROUGHPUT

The link throughput performance for a CR is considered in this chapter, and the normalized throughput defined in [30] is adopted:

$$\eta = \frac{E[\text{Payload transmitted in a slot time}]}{E[\text{length of a slot time}]} \tag{9.7}$$

If a CR has a successful transmission as shown in Figure 9.8, it should be noted that the spectrum availability may change because of the PU activity. Let the time spent during the TSS procedure and data transmission be T_{ct}, and the time spent during the RSS procedure and data transmission be T_{cr}.

It can be obtained that

$$T_{ct} = T_{\text{CTS+RTS}} + SIFS + D + SIFS\_T_{\text{RSS}} \tag{9.8}$$

$$T_{cr} = T_{\text{CTS}} + SIFS + D \tag{9.9}$$

where D is the propagation delay.

The PU activity is assumed to follow a Poisson process ($\{N(t), t \geq 0\}$ with rate parameter λ) during the time interval $[0, T_s]$. Thus, we denote by $P_{re}(k)$ the probability that an available channel will be taken by a PU activity on data channel k during T_s. If the number of PU flow at a time frame is larger than zero, the data channel k will be taken by PUs, and in this case, we have

$$P_{re}(k) = P(N(t+T_s) - N(t) > 0) = 1 - e^{-\lambda T_s} \tag{9.10}$$

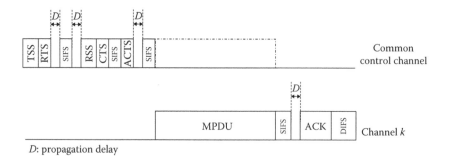

D: propagation delay

FIGURE 9.8 Description of a successful data transmission.

Apart from $P_{re}(k)$, the probabilities that affect the length of a slot time include

1. The probability that no CR is transmitting, $1 - P_{tr}$, where P_{tr} is the probability that there is at least one transmission in the considered slot time
2. The probability that a data payload is successfully transmitted, $P_{tr}P_s$, where P_s is the probability of a transmission occurring on the channel is successful
3. The probability that a data payload is not successfully transmitted because of a collision $P_{tr}(1 - P_s)$

Based on the aforementioned discussion, the throughput in (9.11) can be determined from the definition in (9.7):

$$\eta = \frac{P_s P_{tr} E[P]}{(1 - P_0)\left((1 - P_{tr})\sigma + P_{tr}P_s T_s + P_{tr}(1 - P_s)T_c + \dfrac{P_{re}(k)T_s}{1 - P_{re}(k)} \right) + P_0 T_{MSA}} \tag{9.11}$$

where

T_s is the time length of a successful data transmission
T_c is the time length that a channel is busy because of a collision
σ is the empty slot time
P_0 is the probability that CRs are moving into PER regions ($P_0 < 1$)
T_{MSA} is the average time length spent on the MSA when mobility occurs and the data transmission time after running MSA
P is the data frame length (i.e., the length of MPDU)

To derive (9.11), from (9.7) the numerator can be easily deducted as it is the length of payload; the denominator is determined by the frame structure and aforementioned probabilities associated with the estimated length of each chunk of the frame as well as the possibility of transmitting a packet, which is determined by P_0.

Moreover, from Figure 9.8, although we have a CCC and the other channel for data transmission, we can combine the factors on these two channels together so that

$$T_s = DIFS + T_{RTS+CTS+ACTS} + 4D + 4SIFS + T_{RSS}$$
$$+ T_{TSS} + T_{data} + T_{ACK} \tag{9.12}$$

where $T_{RTS+CTS+ACTS} = T_{RTS} + T_{CTS} + T_{ACTS}$, while T_{TSS} and T_{RSS} are spectrum sensing time of transmitter and receiver, respectively; T_c is related to the RTS frame collision on CCC as

$$T_c = SIFS + D + DIFS \tag{9.13}$$

At this point, from the assumption that the data frames have the same length (i.e., $E[P] = P$), and the expressions of functions P_{tr} and P_s of p, where p is the stationary probability of a frame transmission by a CR, as well as Equations 9.8 through 9.13, we can get the average throughput result as

$$\eta = \frac{P\zeta}{(1-P_0)\left(\sigma+(T_c-\sigma)\zeta'+(T_s-T_c)\zeta+\frac{CT_s}{1-C}\right)+P_0 T_{MSA}} \qquad (9.14)$$

where

$$\zeta = np(1-p)^{n-1}$$

$$\zeta' = (1-p)^n$$

$$C = \frac{1-e^{-\lambda T_s}}{e^{-\lambda T_s}}$$

n is the number of CRs for transmission

If we assume that the payload in MPDU will be transmitted on the available channels on the link between the ith CR and $(i+1)$ CR, the time spent on each available channel can be $\mathbf{K}_{i,i+1}(t)$ times less at most (when each available channel has the same bandwidth). However, in order to assemble the split data frames on the receiver side, we keep the same MPDU header for each available channel. Therefore, the average time length on a data frame transmission is

$$T'_{data} = T_{data}\left(\varphi+\frac{1-\varphi}{|\mathbf{K}_{i,i+1}(t)|}\right) \qquad (9.15)$$

where φ is the ratio of header length to payload length in MPDU, which is usually less than 1.

The new throughput can be derived by substituting the T_{data} with T'_{data} in (9.12). η can be expressed as $\eta\left(n,p,\lambda,|\mathbf{K}_{i,i+1}(t)|,P_0\right)$.

9.4.3   CASE STUDY [32,33]

If CR traffic model follows a Poisson process $\{N'(t), t \geq 0\}$ with mean arrival parameter λ, a similar analysis can be conducted in order to estimate the CR link throughput. The preceding analysis considered the saturated-mode case (i.e., a CR always has a data payload to send). The special case of a nonsaturated mode (i.e., a CR does not always have a data payload to send) is discussed now. The introduction of Poisson's traffic model imposes two changes to the aforementioned analytical model. One is the stationary transmission probability p of a data frame; the other is retransmission times of a data frame.

We denote by p' as the new stationary transmission probability of a data frame. From [31], by assuming each CR has a frame buffer and the probability of a frame arrival is q, the nonsaturated mode of CRs will finally affect the value of transmission probability. Moreover, for the Poisson traffic model, $q = P\{N'(t) = 1\} = 1 - e^{-\lambda'T}$. The value of p' can be calculated by the collision probability p_c and the total number of stages ρ, as well as q.

The retransmission probability on a channel k, $P'_{re}(k)$, can be obtained as follows:

$$P'_{re}(k) = P\{N(t) > 0 \mid N'(t) > 0\} = 1 - e^{-\lambda Ts} \tag{9.16}$$

As such, the estimated link throughput results can be derived in the case of Poisson's traffic model for both PUs and CRs.

If there is no MSA and TSS/RSS procedure, the retransmission probability, $P_{re}(k)$, can be calculated by the channel availability and spectrum hole sufficiency [4]. If the channel aggregation factor is taken as one, $P_{re}(k)$ can be obtained as [4]

$$P_{re}(k) = 1 - \left(1 - \frac{U_{CR} \cdot n}{(1 - U_{CR+PU})r}\right)\frac{1}{m+1} \tag{9.17}$$

where

r is the dynamic operating range (i.e., the number of channels a PU is operating on)
U_{CR} is the channel utilization of a CR
U_{CR+PU} is the channel utilization by the PU and CR
m is the aggregation factor on the channel k

Therefore, we can substitute the variable C in (9.14) with (9.17). In this sense, we can compare the proposed CM-MAC protocol with SCA-MAC protocol, because SCA-MAC belongs to the CSMA-/CA RTS-/CTS-based MAC, which has the same nature of the proposed MAC.

9.5 NUMERICAL RESULTS

This section shows some numerical results based on the aforementioned analysis, which were initially presented in [32,33]. The parameters are listed in Table 9.1. Besides, all the switching intervals from transmitting to receiving are set to zero. We assume that the number of CRs, N, is identical to n in (9.14), and these CRs are transmitters in a CRAHN and can interfere with each other. Moreover, the channel aggregation is not considered in SCA-MAC and CM-MAC in order to let the three protocols be compared in the same condition. The essential parameters represented in Table 9.1, for example, are, that is, $\varphi = 0.03$, $P = 8584$ bits, $T_{RTS+CTS+ACTS} = 768$ μs, $T_c = 141$ μs, and $T_{ACK} = 240$ μs.

With the parameter values, we can get $T_s = 1151.03 + \dfrac{7938.48}{\left|\mathbf{K}_{i,i+1}(t)\right|}$. Then, we can derive (9.14) as

TABLE 9.1

Parameter Values for Evaluation

Parameter	Value	Parameter	Value
MAC data payload	8184 bits	Propagation delay (D)	1 μs
MAC header	272 bits	No. of spectrum bands (K)	6
PHY header	128 bits	PHY max transmit power	100 mW
RTS payload	160 bits + PHY header	PHY sensitivity	−100 dBm
CTS payload	112 bits + PHY header	Rx spectrum sensing time (T_{RSS})	20 μs
ACTS payload	112 bits + PHY header	Tx spectrum sensing time (T_{TSS})	20 μs
SIFS	20 μs	Empty slot time (σ)	50 μs
DIFS	120 μs	Receiving threshold power (β_{th})	50 dB
Slot time	50 μs	Path loss exponent (α)	4
Channel bit rate	1 Mbps	Dynamic operating range (r)	1000
ACK length	112 bits + PHY header	Stationary probability of a data transmission by a CR (p) (saturated mode)	0.02

$$\eta = \frac{8584}{\left(\dfrac{91\zeta'}{\zeta} + \dfrac{257.6P_0 + 50 + \left(1151.03 + 7938.48\dfrac{1}{K - K_p} \right)C/(1-C)}{\zeta} \right)}$$

$$+ \left(1010.03 + \frac{7938.48}{K - K_p} \right) \tag{9.18}$$

where

K is the total number of available spectrum bands

ζ and C are defined in (9.14)

$K_p = K - E\left[\left| \mathbf{K}_{i,i+1}(t) \right| \right]$, that is, the average number of channels occupied by a PU

For the nonsaturated mode throughput, note that the variable C and p will change as mentioned in Section 9.5. Moreover, it is expected that the larger the value of λ, the more frequent the PU traffic occupies the available spectrum bands. Note that the throughput η is defined as the probability of successful transmitted frames per frame time.

Figure 9.9a shows the throughput performance versus the number of CRs (assumed to be in the saturated mode) when we take the value $K_p = 1$ for all links. Using $K_p = 1$ for all nodes is to make the three protocols be compared in a fair condition. For example, the CSMA/CA RTS/CTS MAC is the baseline, which can support the data communication in one spectrum band. The CR throughput decreases when the value of λ increases. This is because the PU traffic with increasing λ has

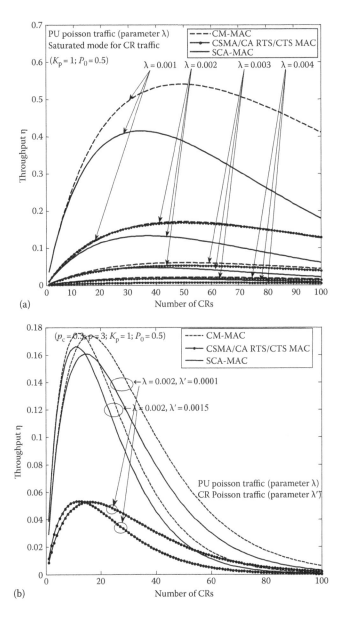

FIGURE 9.9 Theoretical results of CR link throughput in the (a) saturated mode and (b) nonsaturated mode.

a high possibility of affecting TSS/RSS procedures. In this case, the PU becomes more active, which reduces CR access. Moreover, as shown in Figure 9.9a, with any given value of λ and N, the throughput performance of CM-MAC outperforms that of SCA-MAC. The reason for this result is that the delay caused by MSA and TSS/RSS procedures in CM-MAC is smaller than that of SCA-MAC.

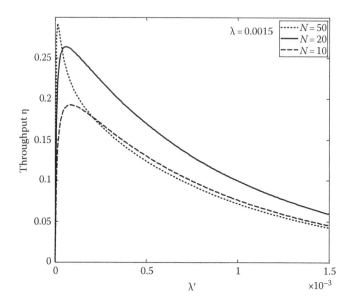

FIGURE 9.10   CR link throughput versus N and λ'.

Furthermore, under heavy PU traffic, CM-MAC can successfully reduce the effect of the existence of PER regions. Besides, if the nonsaturated mode is considered for CR traffic (Figure 9.9b), it be can seen that the CM-MAC still outperforms CSMA/CA MAC and SCA-MAC protocols.

Figure 9.10 shows the variation of CR throughput versus N and λ'. It can be seen that when the CR traffic intensity increases, the throughput curves rise, reach a maximum, and then decline. Moreover, in Figure 9.10, when N increases from 10 to 20, the CR throughput increases as well. However, when $N=50$, the throughput sharply increases when λ' is small and decreases faster than the other throughput curves. The reason for this phenomenon is that having more CRs will increase the traffic, which increases the chances of more data transmission conflicts. This results in decreased CR throughput.

Figure 9.11 shows how the CR throughput changes versus N and PU traffic parameter, λ. It can be seen that the throughput curves decline with the increasing intensity of PU traffic. In Figure 9.11, when N increases from 10 to 20, the CR throughput increases correspondingly. However, when $N=50$, the overall throughput is slightly less than the throughput when $N=20$. This is expected, because the increasing number of CR nodes results in increasing conflicts in the data transmission, which affects the throughput performance.

Figure 9.12 shows how the throughput performance can be affected by the different values of K_p (i.e., the number of exclusive channels occupied by PUs). Figure 9.12a shows the analytical results when CR traffic is in the saturated mode. The results for Poisson's CR traffic (i.e., nonsaturated mode) are presented in Figure 9.12b and c. We can see that when we consider different intensities of CR Poisson's traffic (reflected

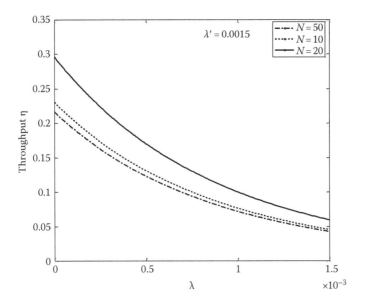

FIGURE 9.11   CR link throughput versus N and λ.

by λ'), the throughput performance will change accordingly. Furthermore, it can be seen from Figure 9.12a through c that when the number of available spectrum bands to CRs increases (i.e., K_p decreases), the throughput performance improves correspondingly. These results are expected because with additional channels available to CRs, the overall throughput will increase.

Figure 9.13 displays how the CR mobility factor, P_0, affects the CM-MAC throughput performance. Figure 9.13a mainly shows the MSA when the CR traffic is in the saturated mode; Figure 9.13b and c shows the throughput performance results for CR Poisson's traffic. It is clear that P_0 only slightly decreases the throughput performance because the MSA can deal with the mobility that affects the disruption of the data transmissions. Therefore, it can be concluded that the proposed CM-MAC is robust in the CR mobility case.

9.6   CONCLUSIONS

This chapter presents the CM-MAC protocol, which is a CRAHN MAC protocol that mainly considers the mobility of CRs and PU PER regions [32,33]. The protocol involves spectrum sensing in the handshaking procedure, and thus, the spectrum information updates on CRs are highly dependent on the PU traffic and the CR data traffic. Moreover, the effectiveness of CM-MAC is demonstrated through the analytical link throughput performance. The performance is mainly determined by the number of CRs, stationary probability of a frame transmission by a CR, probability of CR mobility, PU traffic, CR traffic, and the set of available channels [32,33]. The results show that the CM-MAC protocol outperforms the IEEE 802.11 MAC and

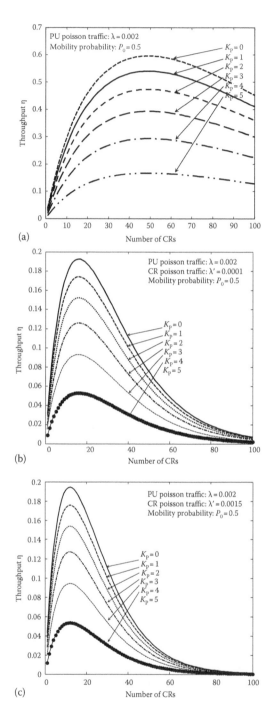

FIGURE 9.12 CR link throughput performance with different values of K_p in the (a) saturated mode, and (b and c) nonsaturated mode.

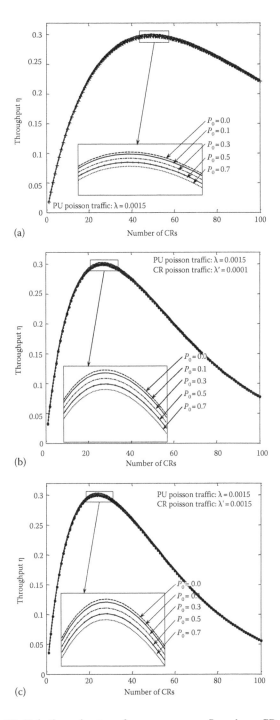

FIGURE 9.13   CR link throughput performance versus P_0, where CR traffic is in the (a) saturated mode and (b and c) nonsaturated mode with PU traffic.

SCA-MAC protocols in terms of throughput performance. The results also show that the CM-MAC protocol is robust with respect to CR mobility. The properties of the protocol that have been investigated in this chapter make it a suitable candidate for real-world CRAHN deployments.

REFERENCES

1. I. F. Akyildiz, W.-Y. Lee, and K. R. Chowdhury, CRAHNs: Cognitive radio ad hoc networks, *Ad Hoc Networks*, 7, 810–836, 2009.
2. I. F. Akyildiz, W. Y. Lee, M. C. Vuran, and S. Mohanty, A survey on spectrum management in cognitive radio networks, *IEEE Communications Magazine*, 46, 40–48, 2008.
3. V. Mai, N. Devroye, and V. Tarokh, On the primary exclusive region of cognitive networks, *IEEE Transactions on Wireless Communications*, 8, 3380–3385, 2009.
4. A. Chia-Chun Hsu, D. S. L. Weit, and C. C. J. Kuo, A cognitive MAC protocol using statistical channel allocation for wireless ad-hoc networks, in *Proceedings of WCNC'07*, Hong Kong, China, 2007, pp. 105–110.
5. H. B. Salameh, M. Krunz, and O. Younis, Distance- and traffic-aware channel assignment in cognitive radio networks, in *Proceedings of SECON'08*, San Francisco, CA, 2008, pp. 10–18.
6. H. B. Salameh, M. Krunz, and O. Younis, MAC protocol for opportunistic cognitive radio networks with soft guarantees, *IEEE Transactions on Mobile Computing*, 8, 1339–1352, 2009.
7. S. Tao, C. Shuguang, and M. Krunz, Medium access control for multi-channel parallel transmission in cognitive radio networks, in *Proceedings of GLOBECOM'06*, San Francisco, CA, 2006, pp. 1–5.
8. N. Jain, S. R. Das, and A. Nasipuri, A multichannel CSMA MAC protocol with receiver-based channel selection for multihop wireless networks, in *Proceedings of ICCCN'01*, Scottsdale, AZ, 2001, pp. 432–439.
9. B. Sadeghi, V. Kanodia, A. Sabharwal, and E. Knightly, OAR: An opportunistic auto-rate media access protocol for ad hoc networks, *Wireless Networks*, 11, 39–53, 2005.
10. Z. Hongqiang, W. Jianfeng, and F. Yuguang, DUCHA: A new dual-channel MAC protocol for multihop ad hoc networks, *IEEE Transactions on Wireless Communications*, 5, 3224–3233, 2006.
11. C. Feng, Z. Hongqiang, and F. Yuguang, An opportunistic multiradio MAC protocol in multirate wireless ad hoc networks, *IEEE Transactions on Wireless Communications*, 8, 2642–2651, 2009.
12. D. Cabric and R. W. Brodersen, Physical layer design issues unique to cognitive radio systems, in *Proceedings of PIMRC'05*, Berlin, Germany, vol. 752, 2005, pp. 759–763.
13. K. R. Chowdhury, M. Di Felice, and I. F. Akyildiz, TP-CRAHN: A transport protocol for cognitive radio ad-hoc networks, in *Proceedings of Infocom'09*, Washington, DC, 2009, pp. 2482–2490.
14. J. So and N. H. Vaidya, Multi-channel MAC for ad hoc networks: Handling multi-channel hidden terminals using a single transceiver, in *Proceedings of MobiHoc'04*, Tokyo, Japan, ACM, 2004, pp. 222–233.
15. C. Cordeiro and K. Challapali, C-MAC: A cognitive MAC protocol for multi-channel wireless networks, in *Proceedings of DySPAN'07*, Dublin, Ireland, 2007, pp. 147–157.
16. P. Pawelczak, R. Venkatesha Prasad, L. Xia, and I. G. M. M. Niemegeers, Cognitive radio emergency networks—Requirements and design, in *Proceedings of DySPAN'05*, Baltimore, MD, 2005, pp. 601–606.
17. V. S. Frost and B. Melamed, Traffic modeling for telecommunications networks, *IEEE Communications Magazine*, 32, 70–81, 1994.

18. P. Tran-Gia, D. Staehle, and K. Leibnitz, Source traffic modeling of wireless applications, *AEU—International Journal of Electronics and Communications*, 55, 27–36, 2001.

19. C. Corderio, K. Challapali, D. Birru, and S. Shankar, IEEE 802.22: An introduction to the first wireless standard based on cognitive radios, *Journal of Communications*, 1, 38–47, 2006.

20. D. Skordoulis, N. Qiang, C. Hsiao-Hwa, A. P. Stephens, L. Changwen, and A. Jamalipour, IEEE 802.11n MAC frame aggregation mechanisms for next-generation high-throughput WLANs, *IEEE Wireless Communications*, 15, 40–47, 2008.

21. J. M. Peha, Sharing spectrum through spectrum policy reform and cognitive radio, *Proceedings of the IEEE*, 97, 708–719, 2009.

22. C. Stevenson, G. Chouinard, L. Zhongding, H. Wendong, S. Shellhammer, and W. Caldwell, IEEE 802.22: The first cognitive radio wireless regional area network standard, *IEEE Communications Magazine*, 47, 130–138, 2009.

23. T. Shu and M. Krunz, Throughput-efficient sequential channel sensing and probing in cognitive radio networks under sensing errors, in *Proceedings of the 15th Annual International Conference on Mobile Computing and Networking*, Beijing, China, ACM, 2009, pp. 37–48.

24. D. Cabric, S. M. Mishra, and R. W. Brodersen, Implementation issues in spectrum sensing for cognitive radios, in *Signals, Systems and Computers, 2004 Conference Record of the 38th Asilomar Conference on*, vol.771, Pacific Grove, CA, 2004, pp. 772–776.

25. A. Ghasemi and E. S. Sousa, Collaborative spectrum sensing for opportunistic access in fading environments, in *Proceedings of DySPAN'05*, Baltimore, MD, 2005, pp. 131–136.

26. I. F. Akyildiz, B. F. Lo, and R. Balakrishnan, Cooperative spectrum sensing in cognitive radio networks: A survey, *Physics Communications*, 4, 40–62, 2011.

27. A. A. El-Sherif and K. J. R. Liu, Joint design of spectrum sensing and channel access in cognitive radio networks, *IEEE Transactions on Wireless Communications*, 10, 1743–1753, 2011.

28. C. Bettstetter and C. Hartmann, Connectivity of wireless multihop networks in a shadow fading environment, *Wireless Networks*, 11, 571–579, 2005.

29. S. Bandyopadhyay, E. J. Coyle, and T. Falck, Stochastic properties of mobility models in mobile ad hoc networks, *IEEE Transactions on Mobile Computing*, 6, 1218–1229, 2007.

30. G. Bianchi, Performance analysis of the IEEE 802.11 distributed coordination function, *IEEE Journal on Selected Areas in Communications*, 18, 535–547, 2000.

31. K. Duffy, D. Malone, and D. J. Leith, Modeling the 802.11 distributed coordination function in non-saturated conditions, *IEEE Communications Letters*, 9, 715–717, 2005.

32. P. Hu, Cognitive radio ad hoc networks: A local control approach, PhD dissertation, Queen's University, Kingston, Ontario, Canada, 2013.

33. P. Hu and M. Ibnkahla, A MAC protocol with mobility support in cognitive radio ad hoc networks: Protocol design and performance analysis, *Elsevier Ad Hoc Networks Journal*, 17, 114–128, June 2014.

10 Routing in Multihop Cognitive Radio Networks

10.1 INTRODUCTION

The main task of a routing protocol is to select a path in the network and move the packets from source to destination. Since all users will not necessarily be in range of each other, intermediate nodes must be used as relays to forward the packets [14].

Routing in multihop cognitive radio networks (CRNs) must consider the creation and maintenance of wireless multihop paths among secondary users (SUs). For each hop along the path, the SU must decide which relay nodes to use and the spectrum channels they will transmit on. These considerations are similar to routing in traditional multichannel multihop ad hoc networks, however, with the added complexity of dealing with primary users (PUs) and switching between channels to utilize the spectrum effectively.

This chapter covers the main issues related to routing in multihop CRNs, including mobility, spectrum awareness, network topology changes, and scalability. A number of protocols will be discussed in detail based on the study in [14]. This will highlight the design challenges of routing protocols in CRNs.

The remainder of this chapter is organized as follows. Section 10.2 describes the routing problems in cognitive ratio networks, and some classifications of networks are described in Section 10.3. In Section 10.4, an overview is provided on some basic centralized and distributed solutions and an in-depth look at a layered graph protocol. Section 10.5 describes a geographical algorithm SEARCH (spectrum-aware routing for cognitive ad hoc network), which is a distributed protocol. Section 10.6 provides an extension of the *SEARCH* protocol that allows it to operate on highly dynamic networks. Section 10.7 describes opportunistic cognitive radio multihop (OCR) protocol which has been designed for CRNs with high PU mobility. Section 10.8 presents a summary of the protocols, and finally, Section 10.9 provides some conclusions on the presented material and discusses future work.

10.2 ROUTING PROBLEMS IN COGNITIVE RADIO NETWORKS

Routing protocols in traditional multihop networks must overcome a large number of challenges to provide the best results for the network. Protocol designers must keep the following challenges in mind when designing their protocols [1].

Energy efficiency: Many multihop networks do not have a central fixed infrastructure to provide unlimited power to the nodes. Thus, the protocol must keep in mind its energy consumption to ensure the algorithm does not deplete the user's power rapidly.

Network topology changes: If a node fails, for example, due to power depletion or technical difficulties, the routing protocol must be able to detect and adapt to this failure and determine a new route. During this process, it must consider the latency introduced and try to reduce it.

Scalability: The algorithm's performance should not be affected if the size of the network increases. This is particularly important in networks of varying topology, such as mobile ad hoc networks.

Mobility: Some nodes in a network may be constantly moving; thus, the routing protocol must be able to adapt to these moving nodes and detect broken routes and provide replacement paths.

In CRNs, the aforementioned challenges must be considered as well as the following issues, which are introduced by the presence of PUs [3]:

Spectrum awareness: Due to restrictions created by PUs, cognitive radio (CR) nodes must always be aware of their local spectrum to ensure that they are not affecting a PU. Due to the multichannel aspect of cognitive radio networks, nodes may have different views of the spectrum and its availability even if they are neighbors. Thus, some form of communication must be implemented to ensure that when two nodes communicate, they will choose a channel that is appropriate for both of them [4,5,6]. To ensure awareness of the entire network, the following three scenarios may be implemented:

1. The information on the spectrum availability is provided to the nodes by a central all-knowing entity that monitors the entire network and the available spectrum.
2. The spectrum availability is gathered locally by the nodes and distributed to all nearby nodes that require this information.
3. A combination of 1 and 2.

Quality routes: Route quality is a major consideration in traditional multihop networks, but in CRNs, it has to be redefined with additional variables. The topology of the network is highly influenced by the PU's behavior; thus, the traditional way of measuring the quality of a route (bandwidth, throughput, delay, energy efficiency, and fairness) must be coupled with new measures like path stability, spectrum availability/efficiency, and PU presence.

Route maintenance: In CRNs, PUs leave and join quite often. Therefore, there is a high probability that during a transmission, the route will break. Thus, the protocol must be able to detect quickly a break in the route, stop transmission, and provide a replacement path. Several protocols deal with this challenge by storing a set of backup routes that they are able to switch quickly to another route in case of existing route failure.

Complexity: Unlike traditional multihop networks, routing must be coupled with spectrum sensing; this increases the algorithm complexity. Higher complexity could affect some of the other challenges such as power efficiency or cause a long convergence time.

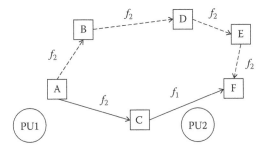

FIGURE 10.1 Routing and spectrum cycle example.

Consider the small network shown in Figure 10.1.

In this network, there are two potentially available channels f_1 and f_2. If SU A wants to transmit to SU F, it has two choices, either $A \to C \to F$ or $A \to B \to D \to E \to F$. The times it takes for a packet to travel over each path are as follows:

$$A \to C \to F : 10 \text{ ms}$$

$$A \to B \to D \to E \to F : 14 \text{ ms}$$

Moreover, it is assumed that it takes γ seconds to change channels.

If the routing process is separate from the spectrum cycle, then the routing protocol will choose $A \to C \to F$ since it has the lower number of hops. However, this does not consider that the channel has to be changed twice to get to F (i.e., switching to channel f_2 at A and to channel f_1 at C). Consider $\gamma = 6$ ms. In this case, the total time for the $A \to C \to F$ path will actually be 22 ms. However, the $A \to B \to D \to E \to F$ path involves only one channel change; that is, it has to switch to channel f_2 at A, and the other nodes will use the same channel f_2. Thus, the total cost for this path is only 20 ms. Therefore, if the routing process is coupled with the spectrum cycle, then the $A \to B \to D \to E \to F$ path will be chosen.

10.3 CLASSIFICATION OF COGNITIVE RADIO NETWORKS

There are a number of requirements that need to be fulfilled when designing a routing protocol for CRNs. In the following, we propose a classification of CR networks based on the design requirements and assumptions.

10.3.1 SPECTRUM KNOWLEDGE

Spectrum knowledge is the amount of information about channel availability and quality that each SU must have when it makes its routing decisions. There are two main classifications in this area, centralized and distributed. Figure 10.2 shows a flowchart of the main types of spectrum knowledge classifications and some examples of the protocols that are based on them.

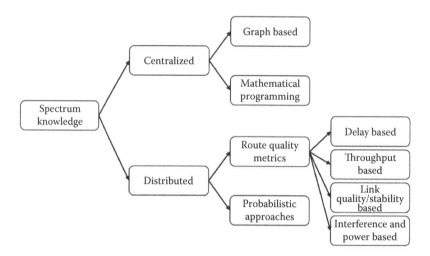

FIGURE 10.2   Routing protocols classification based on spectrum knowledge.

10.3.1.1   Centralized Network

A centralized network, also called a full-spectrum knowledge network, assumes that each SU has full-spectrum knowledge of the entire network. This is accomplished by having an all-knowing central devise with a purpose to monitor the available spectrum and relay that information to the users in the network.

10.3.1.2   Distributed Network

Distributed networks, or local spectrum knowledge networks, assume that the spectrum knowledge is determined by each SU locally and then forwarded to their neighbors. This requires the routing protocol to be able to determine the appropriate routing paths while measuring the spectrum and identifying potential bands that the user is able to transmit on.

The choice of the local spectrum can be accomplished with either a more deterministic approach or in a probabilistic way. In a deterministic approach, the nodes measure the spectrum and make their decisions on the current spectrum band situation. In a probabilistic approach, SUs periodically measure the bands and generate statistics. Based on these statistics, SUs can make better decisions based on the past nature of a particular band. For example, if a particular band has shown that the PU's on/off time is changing quickly, then it may not be the best band to choose for a long transmission.

10.3.2   PU Activity

PUs have a large impact on routing in CRNs. In [8], the authors have classified routing protocols based on the PU's activity as static, dynamic, and opportunistic protocols. Figure 10.3 shows how the networks are classified with respect to the average PU's idle time.

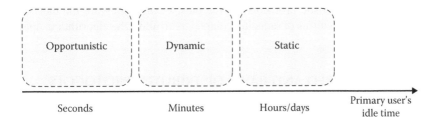

FIGURE 10.3 PU activity classification.

10.3.2.1 Static Cognitive Multihop Networks

In this classification, PUs' bands are available for a duration that, on average, exceeds the communication time of SUs. The SUs usually make the assumption that an available frequency band is a permanent resource. Here, the main difference between a traditional multihop network and a static cognitive network is that the static cognitive network uses different spectrum bands and has the physical capability of transmitting over several frequency bands.

In most cases, a traditional multihop protocol can be modified and used.

10.3.2.2 Dynamic Cognitive Multihop Networks

Here, PUs are assumed to be idle for minutes at a time. In these networks, the frequency band cannot be assumed as a permanent resource. Therefore, traditional multihop protocols cannot be used. The protocol must incorporate route stability, exchanging control information, and channel synchronization.

Path stability can be improved by incorporating the routing protocol with the spectrum cycle. This can be further extended by using past channel statistics and favoring stable spectrum bands (less dynamic) over unstable ones.

Fast computation is required for this type of networks to make the protocol adaptive to the dynamic changes in the spectrum.

10.3.2.3 Opportunistic Cognitive Multihop Networks

In opportunistic networks, the time in which the spectrum is available is smaller than the average communication duration. Thus, an end-to-end path will not work since each packet may experience different network properties due to the highly dynamic nature. Thus, it is usually better to use an opportunistic solution, in which every packet is sent and forwarded over opportunistically available channels. This means that each packet makes its own decision on which channel and relay to choose for each hop. Having the algorithm deal with each packet, or a group of packets, can reduce the complexity of establishing end-to-end routes and increase the efficiency of the proposed solutions.

The choice of the channel to forward data is very important in these types of networks since a packet may operate on more than one channel during its trip from source to destination. Channel history can be a good measure to determine which channel to choose at each hop. For example, if a band has been unreliable for a long period of time, it should be avoided even though the PU is currently not there.

Although opportunistic protocols are the most flexible, they are the most challenging, especially in terms of reducing complexity to allow the algorithm to operate fast enough.

10.4 CENTRALIZED AND BASIC DISTRIBUTED PROTOCOLS

10.4.1 CENTRALIZED PROTOCOLS

Centralized protocols assume that each SU has spectrum knowledge of the entire network. This knowledge is usually determined through a central device. We will first give an overview of three basic centralized solutions and then provide a detailed description of a more complex protocol.

10.4.1.1 Color Graph Routing

The coloring graph method [11] uses graph theory to model the network and all the channels available. This method can be split up into two parts:

1. Graph creation
2. Route calculation

When creating the graph, the SUs are represented by the vertices of the graph and if two users want to communicate, they can choose one of M colored edges, where M is the number of available channels. The color of the edge corresponds to the carrier frequency that will be used. The network can be represented using (10.1)

$$G_C = \left(N_C, E_C\right) \tag{10.1}$$

where
  N_C is the vertex set
  E_C is the edge set

Each of the vertices can be connected by a number of different edges as seen in Figure 10.4. After the graph has been created, it begins to generate routes. The goal is to generate all the routes needed in the network while trying to minimize the number of channels used in the network and ensuring the interference is at a minimum.

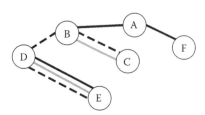

FIGURE 10.4   Color graph protocol.

This algorithm is very basic and only works on static networks with PUs that have low mobility. This algorithm is not scalable. The convergence time for searching the best route has a high probability of taking so long that the graph will change by the time it completes.

Route maintenance is not addressed here; thus, if a route breaks or the network changes, it needs to generate another graph and run the routing portion of the algorithm again. This causes a large amount of overhead when the network conditions change.

10.4.1.2   Conflict Graph Routing

The conflict graph method [12] is similar to the color graph protocol. However, it decouples choosing channels and routing into two separate parts. The graph aspect is used for selecting the channel, and the routing is done with a basic routing algorithm. The protocol considers all available routes between a source and a destination, and for each route, it considers all the available channels for each relay. The best combination is chosen for the route.

An edge is drawn between two vertices if the two users cannot be active at the same time. The graph can be represented by (10.2)

$$G_F = \left(N_F, E_F\right) \tag{10.2}$$

where

N_F represents the set of vertices
E_F represents the set of edges between the vertices as seen in Figure 10.5.

The graph is used to determine the channels that will provide no conflict between nearby SUs. Once the channels are determined, the routing can be done with any basic routing protocol. This is NP-hard, which means that it either requires a large amount of computational power or it will take a long time to converge.

This algorithm is similar to the color graph. However, it will have a shorter computation time since there will be fewer edges on the map. However, since the route is exhaustive, the convergence time may still be a problem.

10.4.1.3   Optimization Approach

Hou et al. modeled the routing problem as a mixed integer optimization problem [7]. The objective of the protocol is to maximize the spectrum reuse factor throughout the network. A number of constraints are imposed on the system:

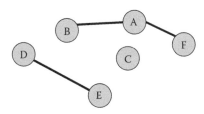

FIGURE 10.5   Conflict graph protocol.

1. *Link Capacity*: This restriction forces the total traffic flow in each link not to exceed the capacity of the link. Shannon law is used to define the link capacity, as shown in the following equation for link (i, j)

$$c_{ij}^{L} = W^{L} \log_2\left(1 + \frac{g_{ij}Q}{\eta}\right) \qquad (10.3)$$

where
   W^{L} is the bandwidth of subband L
   $g_{i,j}$ is the propagation gain of link (i, j)
   Q is the transmission power spectral density
   η is the noise power density

2. *Range*: The transmission range, R_T, is defined in the following equation,

$$R_T = \left(\frac{Q}{Q_T}\right)^{1/\alpha} \qquad (10.4)$$

where Q_T is the threshold power spectral density guaranteeing correct reception and α is the path loss exponent. Thus, if two users fall in the range of each other, they cannot transmit on the same channel.

3. *Routing*: Routing is managed by using flow-balancing constraints for each node that restricts the incoming flow to be equal to the outgoing flow. This allows the creation of split routing paths, which adds robustness to the network.

10.4.1.4   Layered Graph Routing

This layered graph protocol [13] is an extension of a basic weight graph that is used in networks that operate on a single channel.

The goal of the protocol is to maximize the network capacity while minimizing the interference from neighboring nodes. To accomplish this, the protocol generates a layered graph, and then computes routing paths to maximize the network connectivity and provide diverse channel selection to prevent interference from two adjacent hops in a path.

10.4.1.4.1   Creation of Layered Graph

Each layer of the graph represents a corresponding frequency band that can be used by the nodes. Let M denote the number of frequency bands (layer) available in the network and N the number of nodes in the network. Figure 10.6 shows a simple network with $M = 2$ and $N = 4$. In this example, each node in the network is represented by two subnodes on different layers (i.e., node 3 has subnodes 3_1 and 3_2). A subnode can be either active or inactive. If the node is operating on frequency 1, then the subnode on frequency 1 is an active subnode, and all the other subnodes would be inactive.

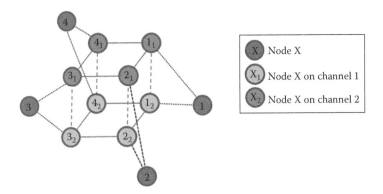

FIGURE 10.6 Example of layered graph, $M = 2$, $N = 4$.

An edge in the graph is represented by a line connecting two subnodes. There are three classifications of edges in the network:

1. *Horizontal edges*: Horizontal edges connect subnodes that are on the same layer. A horizontal edge represents two nodes that can transmit to each other on a particular frequency.
2. *Vertical edges*: Vertical edges connect nodes to their own subnodes on different layers. A vertical edge is created between a node on layer Y and its subnodes on layer X if the nodes can operate on either channel Y or channel X.
3. *Internal edges*: Internal edges connect a subnode to its auxiliary node. Auxiliary nodes are used to allow nodes to be connected to layer/channel that are not directly above or below them.

10.4.1.4.2 Routing and Interface Assignment

The basic idea behind this routing protocol is to traverse the layered graph finding the route with the smallest total sum of the edge weights. Thus, the edge weights are very important in ensuring optimality of the search.

The vertical costs are suggested to be negative to have as many channel changes as possible in order to reduce interference in the network. The horizontal edges are positive and are relevant to the *effective* capacity of the edges. Thus, their weights depend on the number of nodes using a given channel and the traffic volumes currently being transmitted through the node.

The routing algorithm operates as follows:

1. The graph is generated using the current network situation.
2. Choose a path to reduce the cost of trip.
3. Determine which channel to operate each connection between nodes to reduce interference.

Route maintenance has not been addressed by the protocol (other than generating a whole new graph). This results in a potential waste of resources, since there may only be a few changes in the network, but a whole new graph has to be created. Therefore, a possible extension to this algorithm would be to have the nodes notify the central node or base station when their situation has changed so that it can update only a smaller section of the graph.

10.4.1.4.3 Simulations

Simulations are completed using MATLAB® with two networks, one with 15 nodes and another with 30 nodes [13]. The nodes are placed on a Cartesian grid and their X and Y coordinates are uniformly distributed between [0, 100]. It is assumed that there are six available frequency bands each with a bandwidth of 10 Mbps. The number of available channels for each node is randomly selected between 1 and 3. The randomness in the number of channels for each node is supposed to represent PUs acting in a particular area. Carrier sense medium access with collision avoidance (CSMA/CA) is used at the medium access control (MAC) layer. The cost of edges is shown in Table 10.1.

Whenever a path is completed, the horizontal edge weights in that path increase by one. The internal edge is set reasonably large to prevent a path from leaving its layer and going back to the same layer.

The throughput of the 15-node network is shown in Figure 10.7. The four graphs represent different transmission ranges and the protocol is compared to a sequential interference assignment (SeqAssign) method. SeqAssign allocates channels to nodes in descending order of the number of neighbors that the channel can reach.

The throughput results show that the proposed algorithm always works better than the SeqAssign method, due to its optimization in the path computation and interference management. As the load increases, the throughput gap increases between the two algorithms, this is due to the fact that the proposed algorithm selects different channels on adjacent hops to effectively avoid interference, this results in higher channel utilization. SeqAssign does not have any mechanism to select different channels on adjacent hops; thus, its throughput is very poor at high load.

TABLE 10.1
Edge Weights

Type of Edge	Weight	Extra Information
Horizontal	10	Only at start of network
Vertical	−10	
Internal	40	

Source: Xin, C. et al., A novel layered graph model for topology formation and routing in dynamic spectrum access networks, *in Proc. First IEEE International Symposium on New Frontiers in Dynamic Spectrum Access Networks (DYSPAN)*, Baltimore, MD, 2005, pp. 308–317.

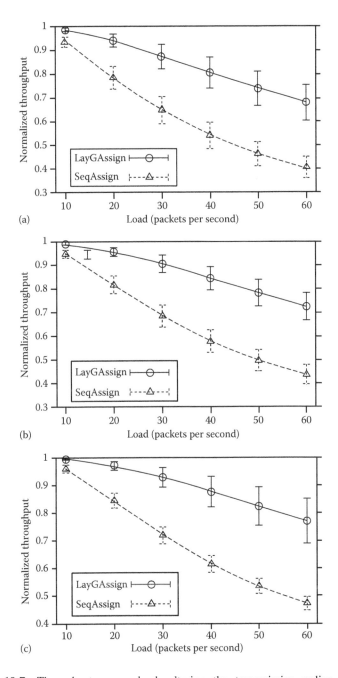

FIGURE 10.7 Throughput versus load, altering the transmission radius, 15 users. (From Xin, C. et al., A novel layered graph model for topology formation and routing in dynamic spectrum access networks, *in Proc. First IEEE International Symposium on New Frontiers in Dynamic Spectrum Access Networks (DYSPAN)*, Baltimore, MD, 2005, pp. 308–317.) Transmission radius range = [0.1, 0.4] (a), [0.2, 0.4] (b), [0.1, 0.5] (c).   (*Continued*)

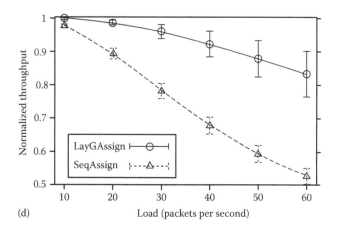

(d)

FIGURE 10.7 (*Continued*) Throughput versus load, altering the transmission radius, 15 users. (From Xin, C. et al., A novel layered graph model for topology formation and routing in dynamic spectrum access networks, *in Proc. First IEEE International Symposium on New Frontiers in Dynamic Spectrum Access Networks (DYSPAN)*, Baltimore, MD, 2005, pp. 308–317.) Transmission radius range = [0.2, 0.5] (d).

Figure 10.8 shows a 30-node network simulation with a transmission range of [0.1, 0.3]. This graph shows the limit of this protocol in terms of scalability. An increase from 15 to 30 users greatly decreased its throughput (~42%), especially with a high load.

10.5 DISTRIBUTED PROTOCOLS

Unlike centralized protocols, distributed protocols do not make the assumption that all SUs know all information about the spectrum. Thus, each SU must sense the spectrum, make decisions about the spectrum availability, and broadcast the results across the network. We discus here the problems that arise when working with distributed networks.

10.5.1 Control Information

The major difference between distributed and centralized networks is that users can only measure their local spectrum bands. If SUs make their routing decisions based only on their sensed spectrum information, their choices may negatively affect the network as a whole. Thus, they must gather global information to make better decisions, which are usually done through a control channel. When SUs broadcast on the control channel, they have to ensure that they do not interrupt PUs. Therefore, the protocols must find intelligent techniques to convey the control information without affecting the primary nodes' traffic.

10.5.2 Source- or Destination-Based Routing

Routing protocols must choose whether to use destination-based or source-based routing approaches. Destination-based routing requires periodic routing tables and

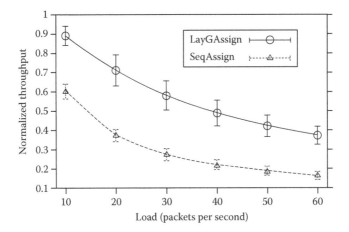

FIGURE 10.8   Throughput versus load, transmission radius range = [0.1, 0.3], 30 users. (From Xin, C. et al., A novel layered graph model for topology fomation and routing in dynamic spectrum access networks, *in Proc. First IEEE International Symposium on New Frontiers in Dynamic Spectrum Access Networks (DYSPAN)*, Baltimore, MD, 2005, pp. 308–317.)

potential broadcasting route requests and route replies. This type of routing is used in traditional ad hoc networks. In CRNs, there are usually multiple channels; thus, the messages will have to be broadcasted across all channels, which will increase overhead. An alternative is to transmit on the control channel. However, this may result in a large influx of messages on the control channel, thus causing it to lose its purpose of informing nodes about the network status.

For source-based routing protocols, every node, before starting communication, locally computes the path to the destination using control messages sent by the users. The advantage of source-based routing is that no routing tables are used (thus deceasing the overhead from updating them when the network changes). The packet is forwarded based on the information found in the header.

10.6   GEOGRAPHICAL PROTOCOL FOR DYNAMIC NETWORKS

This section describes a more complex distributed protocol that is designed for dynamic networks and incorporates geographic location. We will be discussing SEARCH protocol proposed in [4]. SEARCH is a completely distributed routing solution that accounts for PUs, the mobility of cognitive users, and jointly considers the path and channel choices to decrease path latency. It is developed to operate on networks with PUs that have a dynamic nature.

The algorithm provides the node with two choices if a PU is found:

1. Switch channels
2. Go around the PU

If the node chooses to go around the PU, it will usually result in more hops, thus increasing the latency seen by the packet. Switching channels may also create latency if the new channel is very busy. There is also a potential that switching to a nearby channel on the spectrum may not solve the problem since the PU may have power leaking into that channel.

Here, the authors make the following assumption for the network [4]:

- CCC is used.
- The sender knows the location of its destination.
- Each node knows its own location.
- There are M channels available with known bandwidths.

The algorithm can be split up into two parts, the initial route set up and the route enhancement/repair [4].

10.6.1 INITIAL ROUTE SETUP

A route request (RREQ) is transmitted by the source on each channel that is not affected by a nearby PU. This RREQ will be forwarded by intermediate users till it reaches the destination. Throughout the trip from the source to the destination, each intermediate node attaches its

1. ID
2. Current location
3. Time stamp
4. A flag representing, which mode it is operating in

Each node checks its focus region for any hops that comply with the greedy forwarding requirements. If none exist, it means that there may be a PU nearby. Therefore, the node flips its operating flag, changing its mode to PU avoidance. It will go around the PU and then switch back to greedy forwarding mode. Once all the routes have been received by the destination, it chooses the best route with a joint channel-path optimization.

10.6.2 GREEDY FORWARDING

The algorithm by default operates in greedy forwarding mode. At the beginning of the protocol, beacon messages are used to inform the nodes which users are in their transmission range. Using this information, the protocol can determine which of the candidate forwarders should be chosen as the next hop to minimize the distance to the destination. To be an eligible relay, the node must satisfy the following three conditions:

1. The next hop must be on the same channel.
2. The next hop must not interfere with a PU.
3. The user must be in the current node focus region.

The third condition is what allows the algorithm to determine the general locations of PUs. The focus region is generated by having a line, with a magnitude of

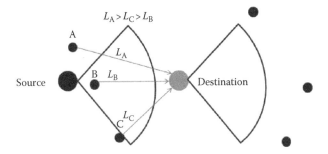

FIGURE 10.9 Focus region, example 1.

the transmission range, from the source to the destination, and then drawing a pie shape with the angle of θ_{max} above and below the line. A focus region example is shown in Figure 10.9. In the example, the source node only has one option which is to transmit its packets through B. Even if C is closer to destination, it is not in A's focus region; thus, it does not meet condition 3. Therefore, it is not a possible relay node.

Figure 10.10 shows a source node and a destination node, it also shows four possible relay nodes A, B, C, and E. If the PU is turned off, then the greedy algorithm will choose the path, source → A → E → destination. This is the shortest route and will provide the least delay. However, if the PU turns on, node A has no nodes in its focus region since condition 2 is not satisfied for node E. In this case, it goes into PU avoidance mode and goes around the PU.

If there are no nodes in a focus region, the node is known as decision point. The decision points are used to give the destination a general idea about the locations of PUs and occupied channels.

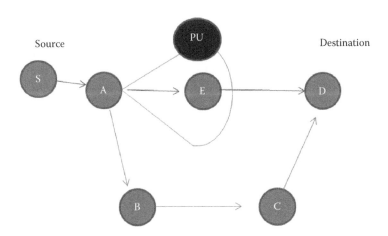

FIGURE 10.10 Focus region, example 2.

10.6.3 PU Avoidance

When a node determines that it is a decision node, it sets the PU avoidance flag in the RREQ before sending it to the next hop. It relaxes the constraints to just 1 and 2 and chooses the next relay hop. At the next hop, if there is a node in its focus region, it goes back to greedy operation and continues to the destination. In example 2, node A will choose B if the PU is operating. B will check its focus region and sees C; thus, it will go back to the greedy operation mode and continue to D.

10.6.4 Joint Channel-Path Optimization

If the licensed band is free of PUs, then there will be no regions that have to be avoided, and the RREQs sent on different channels should have similar paths.

If there are PUs in the network, then the destination will receive RREQs with decision nodes that represent where the PUs are. The destination will choose the route with the least latency. To improve this route even more, the destination will look at the RREQs on different channels and determine if it is faster to switch channels or to go around that PU. After it decides upon the optimal channels, it sends a route reply RREP back to the source along the optimal route.

10.6.5 Simulations and Illustrations

We present here the results of [4] where the PU's activity is modeled using an on–off model. Four hundred CR nodes are used in a 1000×1000 m area with 2–10 PUs. The coverage of a PU in its occupied channel is 300 m, and the transmission range of a cognitive radio user is 120 m. The CR nodes and the PU locations are randomly chosen.

The algorithm is compared with the greedy perimeter stateless routing protocol (GPSR) that has been extended for multichannel environment. This algorithm is oblivious to PUs and just tries to get the best path available. Two types of SEARCH algorithms are compared. The least latency does not include the route enhancements or the channel optimization, while the other does.

Figure 10.11 shows the end-to-end delay as the number of PUs increases from 10 to 20, with 10 channels. It can be seen that with an increase in the PUs increases the delay by a small amount with the optimized SEARCH algorithm. The GPSR algorithm had a large increase in the delay, because when a PU breaks its path, the protocol does not try and fix the route, but it has to make an entire new route. The least latency SEARCH algorithm is outperformed by the optimized SEARCH when there is a larger number of PUs as it only considers going around PUs; thus, there is an increased delay caused by the additional hops. Also the route enhancement in the optimized SEARCH will ensure that the route is always the best route.

Figure 10.12 shows that the optimized SEARCH algorithm outperforms the other algorithms in terms of packet delivery ratio. Figure 10.13 shows the impact of an increased number of CR users. It can be seen that the optimized SEARCH algorithm outperforms the other algorithms as well.

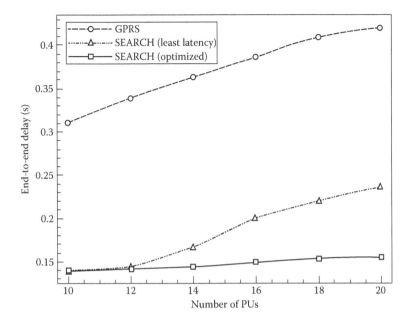

FIGURE 10.11 End-to-end delay versus number of PUs, 10 channels. (From Chowdhury, K. and Felice, M., *Comp. Commun.* (Elsevier), 32(18), 1983, 2009.)

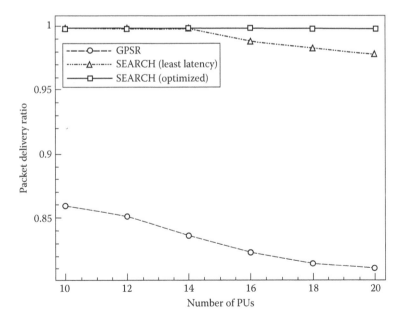

FIGURE 10.12 Packet delivery ratio versus number of PUs, 10 channels. (From Chowdhury, K. and Felice, M., *Comp. Commun.* (Elsevier), 32(18), 1983, 2009.)

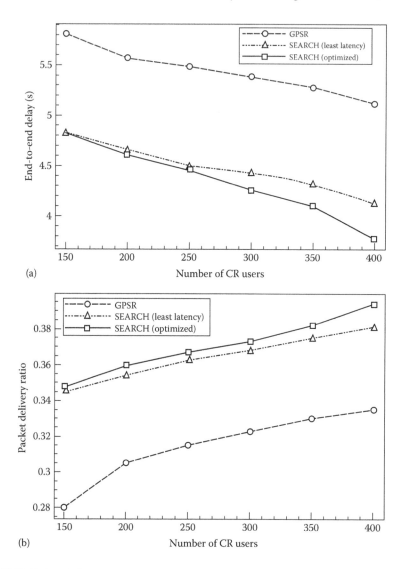

FIGURE 10.13 Influence of the number of CR users, 10 PU are considered: (a) end-to-end delay versus connection load and (b) packet delivery ratio. (From Chowdhury, K. and Felice, M., *Comp. Commun.* (Elsevier), 32(18), 1983, 2009.)

10.7 OPPORTUNISTIC COGNITIVE RADIO MULTIHOP PROTOCOL

Routing tables work better on networks that have static or moderately dynamic channel conditions. For networks that have highly dynamic channel conditions, once the route has been updated in the routing table, there is a high probability that the network has changed, and thus, the table may be outdated.

In [8], the authors show that the involved computations and communication overhead for rebuilding routing tables for all flows are nontrivial, especially when the channel status changes frequently.

This section presents the opportunistic CR multihop protocol designed for networks with PUs that have a highly dynamic mobility [9]. The proposed algorithm makes routing decisions on a per packet (or group of packets) basis and involves channel usage statistics in the discovery of the spectrum access opportunities (in order to improve the transmission performance of all SUs). The forwarding links are selected based on local spectrum access opportunities. A routing metric is also adopted, which tries to increase the performance while maintaining reasonable complexity.

10.7.1 PROTOCOL OVERVIEW

This protocol considers a multihop cognitive radio network with multiple PUs and SUs that share a set of orthogonal channels. All SUs communicate on a common control channel, and it is assumed that the devices have two radios, one for transmission and one for control messages. SUs use the on/off time of the PU to model channel usage statistics and help make future decisions during the routing process.

Whenever a user wants to forward a packet, the following two steps occur:

10.7.1.1 Channel Sensing

In the channel sensing step, the SU searches for a temporarily unoccupied channel in collaboration with its neighbors.

After selecting a channel, the sender broadcasts a short message in the CCC to inform its neighbors of its selected data channel and the location of the sender and destination. The other SUs in the network that receive this message set that data channel as *nonaccessible*. This is to prevent cochannel interference from current transmissions. Using the location information from the CCC message, the SUs determine whether they are eligible relay candidates (i.e., if the relay node is closer to the destination node than the current sender). The user will then initiate a handshake with the relay candidates.

10.7.1.2 Relay Selection

After a channel has been chosen, the sender selects the next hop from the potential relay candidates. This is done in two steps; the sender first broadcasts a routing request (RREQ) message to the relay candidates. When a relay candidate receives the RREQ and is eligible to be a relay, it creates a back-off timer until it sends a routing response RRSP message back to the sender. This back-off timer is based on the candidate's throughput, delay, and relay distance achieved. The better the specifications, the smaller the back-off timer. All eligible candidates listen to the channel to see if another user replies first, if so they stop their back-off timer and wait for the next RREQ.

The sender waits until it receives its first RRSP messages and chooses that node as its relay. Since the back-off timers are based on the specifications, the first RRSP should have the best specification. The sender makes a handshake with the chosen relay and begins to transmit data on the selected channel. If the sender does not hear any RRSP message, this means that there are no eligible candidates for relaying. It will then return to the channel sensing stage and try again.

10.7.2 OCR PERFORMANCE METRICS

The relay distance advancements and per-hop transmission delay are used as metrics for this protocol.

10.7.2.1 Relay Distance Advancement

The relay distance advancement achieved by a relay is determined by the difference between the distance of the sender to the destination and that of the relay node to the destination, as shown in the following equation:

$$Dist(S,R) = d(S,D) - d(R,D) \tag{10.11}$$

where
  S represents the sender location
  D represents the destination location
  R represents the relay node location

10.7.2.2 Per-Hop Delay

The OCR per-hop transmission delay comprises three parts (see Figure 10.14):

1. *Sensing delay*: The sensing delay includes the transmission time of the sensing invitation (SNSINV) and the energy detection time. It is denoted by T_{SNS}.
2. *Relay selection delay*: The *i*th relay candidate R_i sends an RRSP message that is received first only when the first $i - 1$ higher-priority candidates are

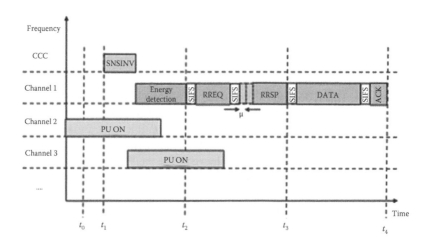

FIGURE 10.14 Timeline for the OCR protocol: Illustration of the one-hop case.

not eligible or have worse channel statistics. Thus, the delay can be represented in Equation 10.12, which was proposed in [2]:

$$T_{RS}(i) = T_{RREQ} + (i+1)\mu + T_{RRSP} + 2\,SIFS \qquad (10.12)$$

where

T_{RREQ} and T_{RRSP} are the transmission times of the two types of messages
μ is the duration of one minislot in the back-off period
$SIFS$ is the short interframe spacing

3. *Packet transmission delay*: Once the relay has been selected, the packet transmission delay is determined using Equation 10.13. This includes the delay of the packet transmission T_{DATA} and the ACK transmission times.

$$T_{DTX} = T_{DATA} + T_{ACK} + 2\,SIFS \qquad (10.13)$$

The total delay can be represented by the following equation:

$$T_{relay} = T_{SNS} + T_{RS} + T_{DTX} \qquad (10.14)$$

10.7.3 PROBABILISTIC FORMULATIONS

10.7.3.1 Channel Sensing

The probability that a secondary device will find an idle channel to transmit and the past channel statistics are used to help improve the channel selection.

Let $I_R^{c_j}$ denote the event that channel c_j is sensed to be idle by an SU R. A channel is determined to be idle if it was sensed idle at t_1 and remains idle till t_2. Given the channel status observed at earlier times, the current channel status can be estimated. Given that the on/off durations are modeled to follow an exponential distribution with parameters $1/E[T_{ON}^{c_j}]$ and $1/E[T_{OFF}^{c_j}]$, respectively, the following probability can be calculated using the following equation:

$$P_{OFF,R}^{c_j}\left(t_0,t_1\right) = \begin{cases} \rho_{c_j} + \left(1-\rho_{c_j}\right)e^{-\Delta_{c_j}(t_1-t_0)}, & \text{if } c_j \text{ is OFF at } t_0 \\ \rho_{c_j} - \rho_{c_j}e^{-\Delta_{c_j}(t_1-t_0)} & \text{if } c_j \text{ is ON at } t_0 \end{cases} \qquad (10.15)$$

where

$$\begin{cases} \rho_{c_j} = \dfrac{E[T_{OFF}^{c_j}]}{E\left[T_{OFF}^{c_j}\right] + E[T_{ON}^{c_j}]} \\[4mm] \Delta_{c_j} = \dfrac{1}{E\left[T_{OFF}^{c_j}\right]} + \dfrac{1}{E\left[T_{ON}^{c_j}\right]} \end{cases}$$

ρ_{c_j} represents the chance for an idle state in c_j.

The following expression describes the probability that a channel is in the idle state during the sensing period $[t_1, t_2]$:

$$P_R^{c_j}(t_1, t_2) = \int_{t_2 - t_1}^{\infty} \frac{\mathcal{F}_{\text{OFF}}^{c_j}(u)}{E\left[T_{\text{OFF}}^{c_j}\right]} du \tag{10.16}$$

Where $\mathcal{F}_{\text{OFF}}^{c_j}(u)/E\left[T_{\text{OFF}}^{c_j}\right]$ represents the probability density function of the residual time of an idle channel since the time in which it was sensed idle originally. Thus, we can say that the probability that R detects a channel c_j that is free is represented in the following equation:

$$\Pr\left\{I_R^{c_j}\right\} = P_{\text{OFF,R}}^{c_j}(t_0, t_1) \cdot P_R^{c_j}(t_1, t_2). \tag{10.17}$$

10.7.3.2   Relay Selection

After an idle channel has been found, the sender needs to select which node to forward its data to. In the OCR algorithm, the candidate with the highest relay priority will be chosen. However, a PU may interrupt this process and cause a failure in the relay selection. This case is quite rare, since it will only occur if a PU appears in the relay selection period, which is usually short. This probability can be represented using the following equation:

$$P_{\text{RSfail}}^{c_j} = \Pr\left\{I_S^{c_j}\right\} \cdot \Pr\left\{\bigcap_{R_i \in R_D^c} \overline{I_{R_i}^{c_j}} \mid I_S^{c_j}\right\} \tag{10.18}$$

where $\Pr\{I_S^{c_j}\}$ is the probability that the sender initiates the relay selection when it detects an idle channel. $\Pr\left\{\bigcap_{R_i \in R_D^c} \overline{I_{R_i}^{c_j}} \mid I_S^{c_j}\right\}$ is the probability that all SUs sense the channel busy in the previous sensing period, which is equivalent to the event that no relay candidate replies in the relay selection step.

10.7.3.3   Data Transmission

Once a relay has been selected, the data transmission on the link is successful if no active PU appears during the transmission period. This can be expressed by Equation 10.20

$$P_{\text{relay},R_i}^{c_j} = P_i^{c_j} \cdot P_{lSR_i}^{c_j}(t_3, t_4) \tag{10.19}$$

$$P_{\text{relay},R_i}^{c_j} = P_i^{c_j} \cdot P_S^{c_j}(t_3, t_4) \cdot P_i^{c_j} \cdot P_{R_i}^{c_j}(t_3, t_4)^{\left(1 - X_{SR_i}^{c_j}\right)} \tag{10.20}$$

where
  l_{SR_i} represents the data transmission link
  $X_{SR_i}^{c_j} = 1$ if the two SUs are affected by the same PU and 0 otherwise $P_i^{c_j}$ is the probability that the relay will be selected as the next hop.

10.7.4 FURTHER IMPROVEMENTS

To improve the performance of the OCR protocol, the selection of the channel and relay node can be made jointly. This is beneficial to the algorithm since its performance depends on the channel and the relay node; therefore, they both must be considered. To do this, a new metric is introduced to capture the effects of the joint channel and relay selection and apply it to a heuristic algorithm to select the best relay and channel. The metric is called the cognitive transport throughput (CTT) and characterizes the one-hop relay performance of the algorithm. CTT is the expected bit advancement per second for one hop of a packet with a payload L on channel c_j.

To choose between a number of channels/hops, protocols may employ an exhaustive search for all possible combinations. However, this restricts the scalability of the protocols because the more nodes and channels a network has, the longer the convergence time of the search. Thus, to improve the scalability of the OCR protocol, the authors proposed a heuristic algorithm to decrease the number of relays that need CCT computation. The optimization problem can be broken down into two phases due to the independent usage statistics for different channels. The first step is to find all the candidate relays in each channel and determine the relay with the highest CTT. The second step is to compare all CTT values from all available channels and then choose the channel with the highest CTT. The second step is usually limited.

10.7.5 SIMULATION RESULTS AND ILLUSTRATIONS

The simulations were conducted using a C++ event-driven simulator [9]. The PUs are modeled as an exponential process with parameters $1/E[T_{OFF}]$ and $1/E[T_{ON}]$ and an idle rate of $\rho = E[T_{OFF}]/(E[T_{OFF}] + E[T_{ON}])$. The network is set up with multiple PUs and SUs placed randomly in an area of 800×800 m². The source and destination are chosen 700 m away from each other, and a constant bit rate is used. The performance is evaluated in terms of end-to-end delay, packet delivery ratio, and hop count. The simulations show comparisons with SEARCH algorithm and the following algorithms:

OCR (CTT) is the OCR protocol where the relay and channel candidates are set jointly determined using the CTT metric and the heuristic algorithm.
OCR (OPT) is the OCR protocol where the channel and the relay are determined using an exhaustive search for the route that provides the largest CTT value.
GOR: A geographic opportunistic routing algorithm (GOR) in which the SU first selects the channel with the greatest success probability of packet transmission. If that channel is sensed idle, then the SU selects a relay on that channel based on its relaying capabilities.
GR: This is a general geographic routing protocol (GR). In this algorithm, the SU first selects a frequency band by sensing for idle channels. After a channel has been chosen, it selects the relays that provide the largest distance toward the destination.

TABLE 10.2

Simulation Parameters

Number of Channels	6
$\{\rho_{c1}, \rho_{c2}, \rho_{c3}, \rho_{c4}, \rho_{c5}, \rho_{c6}\}$	{0.3, 0.3, 0.5, 0.5, 0.7, 0.7}
Number of PUs per channel	11
PU coverage	250 m
$E[T_{OFF}]$	[100 ms, 600 ms]
Number of SUs	[100, 200]
SU transmission range	120 m
Source–destination distance	700 m
SU CCC rate	512 kbps
SU data channel rate	2 Mbps
CBR delay threshold	2 s
Minislot time	4 μs
Per channel sensing time	5 ms
Channel switching time	80 μs
PHY header	192 μs

Source: Lin, Y.L. and Shen, X., *IEEE J. Sel. Areas Commun.*, 30(10), 1958, November 2012.

The parameters used in the simulations can be found in Table 10.2.

Impact of PU activates: SEARCH and OCR are compared with different channel conditions in Figure 10.15. We can see that OCR greatly outperforms the SEARCH algorithm when the primary expected OFF value decreases. This is due to the fact that the SEARCH algorithm involves large overhead in updating the routing tables when PUs are active.

Impact of the number of SUs: The number of SUs is varied from 100 to 200. Table 10.3 gives the average number of neighbors based on the number of SUs. We can see from Figure 10.16 that the end-to-end delay decreases as the number of SUs increases. This is because with more SUs in the network, there are more potential routes. Therefore, there is a better chance to find one with a smaller delay. The OCR algorithm is significantly better than the GOR and GR. In GOR and GR, a channel is selected first, and then, the relay is selected. The coordination overhead increases with the number of existing relays. However, in OCR(CTT), the channel and relay are selected jointly, which reduces the overhead. We also can see that OCR(OPT) and OCR(CTT) are very similar in terms of performance. However, in terms of computational load, OCR (OPT) takes five times longer to converge, since it exhaustively checks all routes.

Effectiveness in Routing Metric: The end-to-end delay versus the flow rate is shown in Figure 10.17. We can see that as the flow rate increases, the GOR and GR delays increase much faster than in OCR. This is again because

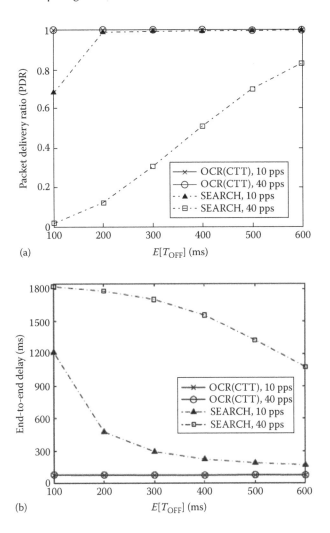

FIGURE 10.15 Performance comparison of SEARCH and OCR with varying PU activity: (a) packet delivery ratio and (b) end-to-end delay. (From Lin, Y.L. and Shen, X., *IEEE J. Sel. Areas Commun.*, 30(10), 1958, November 2012.)

the OCR jointly considers optimized channel and link selection, while the other two protocols select the channel and relays separately.

10.8   SUMMARY OF PROTOCOLS

Table 10.4 shows how important QoS requirements are for different applications [10].

Table 10.5 provides a summary of protocols described in this chapter. The table shows the evolution from simple centralized solutions to distributed opportunistic solutions.

TABLE 10.3

Average Neighbors Based on SUs

Number of Secondary Users	Average Number of Neighbors
100	7
120	8
140	9
160	11
180	12
200	14

Source: Lin, Y.L. and Shen, X., *IEEE J. Sel. Areas Commun.*, 30(10), 1958, November 2012.

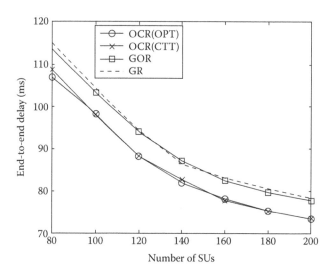

FIGURE 10.16 End-to-end delay versus number of SUs. (From Lin, Y.L. and Shen, X., *IEEE J. Sel. Areas Commun.*, 30(10), 1958, November 2012.)

The routing metric is usually specific to each protocol, as shown in Table 10.5. This makes it difficult to compare the algorithms since each one is based on specific metrics.

Table 10.6 summarizes the simulation results of the main protocols discussed in this chapter.

10.9   CONCLUSION

Multihop cognitive radio networks are among the most promising research areas in wireless communications. There are a variety of real-world multihop networks that would benefit the implementation of cognitive radios. This chapter focuses on

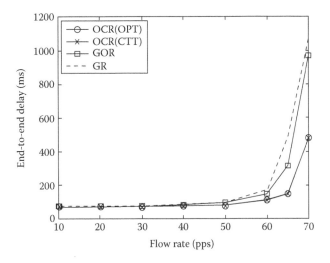

FIGURE 10.17 End-to-end versus flow rate. (From Lin, Y.L. and Shen, X., *IEEE J. Sel. Areas Commun.*, 30(10), 1958, November 2012.)

TABLE 10.4
Quality of Service Requirements

Application	Bandwidth	Delay	Jitter	Loss
Email	Low	Low	Low	Medium
File sharing	High	Low	Low	Medium
Web access	Medium	Medium	Low	Medium
Audio streaming	Low	Low	High	Low
Video streaming	High	Low	High	Low
VOIP	Low	High	High	Low
Videoconferencing	High	High	High	Low

Source: Tanenbaum, A. and Wetherall, D., *Computer Networks*, 5th edn., Pearson, Boston, MA, 2011.

routing protocols for multihop cognitive radio networks. A number of protocols are discussed showing the variety of approaches to address the challenges of cognitive radio networks. Each protocol tries to address a number of goals targeting a specific network. This implies that no single routing algorithm will work for all networks and that there are always trade-offs that occur when trying to improve certain aspects.

Future work will consider improvements in the scalability of centralized networks, reducing complexity and overhead in highly dynamic distributed networks, etc. Cross-layer design is another important area to explore where dynamic interaction between layers will play a key role [14].

TABLE 10.5

Summary of the Protocols Presented

Protocol	Routing Metric	Centralized or Distributed	Geographical	Applicable in Opportunistic Networks	Applicable in Dynamic Networks	Applicable in Static Networks	Route Maintenance	Routing Methodology	Scalability	Further Improvement
Coloring graph	Number of channels	Centralized	No	No	No	Yes	Updates graph periodically	Exhaustive search for best route	Poor	Ignore redundant routes when checking for best path, introduce weights for the colored edges to incorporate other specifications into the metric
Conflict graph	No-conflict route	Centralized	No	No	No	Yes	Updates graph periodically	Exhaustive search for best route	Poor	A better searching technique and route maintenance to reduce the overhead from generating new graphs
Layered graph	Minimize edge weight	Centralized	No	No	No	Yes	Updates graph periodically	Exhaustive search for best route	Poor	Introduce a technique to reduce the number of routes searched to reduce complexity

Protocol	Routing Metric	Centralized or Distributed	Geographical	Applicable in Opportunistic Networks	Applicable in Dynamic Networks	Applicable in Static Networks	Route Maintenance	Routing Methodology	Scalability	Further Improvement
Optimization approach	Maximize spectrum reuse factor	Centralized	No	No	No	Yes	Optimizes updated problem	Solving an optimization problem that models network	Poor	Formulate into a binary mixed integer program to reduce the convergence time
SEARCH	Shortest path (latency)	Distributed	Yes	No	Yes	Yes	Incorporate primary user awareness and CR user mobility into the algorithms metric	Generate routing tables	Medium	Reduce the number of routes checked to improve scalability. incorporate past primary user statistics to make better channel decisions
Spectrum-aware opportunistic routing algorithm	Relay distance advancement and per-hop transmission delay	Distributed	Yes	Yes	Yes	Yes	Periodically check route to ensure it is still viable	Per packet routing decision	High	Consider varying channel conditions. Protection measure for determining eligible candidates. Add congestion measure to OCR metric

TABLE 10.6

Simulation Summary of Presented Protocols

Protocol	Simulation Details	Results and Comments
Layered graph	• MATLAB® was used to simulate the network/protocol • Both 15- and 30-node networks were tested • Nodes placed in a 100×100 grid • Six available channels with bandwidth of 10 Mbps	• For 15 nodes with load varying from 10 to 60 packets per second, normalized throughput decreased from 98% to 75% • For 30 nodes with load varying from 10 to 60 packets per second, normalized throughput decreased from 90% to 50% • No delay simulations presented
SEARCH	• Ns-2 was used to simulate the network/protocol • On/Off application used to model users data flows • 400 Nodes used in a 1000×1000 m range • 10 PUs	• End-to-end delay with $E[T_{OFF}]$ ranging from 100 to 600 ms, delay varied from 1200 to 150 ms • End-to-end delay with load (kbps) ranging from 100 to 700 ms, delay varied from 1 to 3.8 s
Spectrum-aware opportunistic routing algorithm	• A C++ event-driven simulator • Node placed on a 800×800 range • Source and destination always 700 m away • Six channels with bandwidth of 2 Mbps	• End-to-end delay with $E[T_{OFF}]$ ranging from 100 to 600 ms, delay stayed constant at 120 ms • End-to-end delay with number of SUs varying from 80 to 120, delay decreased from 107 to 75 ms • End-to-end delay with flow rate (pps) varying from 10 to 70, delay increased from 50 to 425 ms

REFERENCES

1. S. Abdelaziz and M. ElNainay, Metric-based taxonomy of routing protocols for cognitive radio ad hoc networks, *Elsevier Journal of Network and Computer Applications*, 40 (4), 151–163. April 2014.
2. Y. Bi, L. X. Cai, X. Shen, and H. Zhao, Efficient and reliable broadcast in inter-vehicle communication networks: A cross level approach, *IEEE Transactions on Vehicular Technology*, 59 (5), 2404–2417, June 2010.
3. M. Cesana, F. Cuomo, and E. Ekici, *Routing in Cognative Radio Networks: Challenges and Solutions, Ad Hoc Networks*, 9 (3), 228–248, May 2011.
4. K. Chowdhury and M. Felice, Search: A routing protocol for mobile cognitive radio ad-hoc networks, *Computer Communications* (Elsevier), 32 (18), 1983–1997, 2009.
5. E. Buracchini, The software radio concept, *IEEE Communication Magazine*, 39 (9), 138–143, 2000.
6. F. S. Force, Report of spectrum efficieny working group, Retrieved from http://www.fcc.gov/sptf/files/SEWGFinalReport_1.pdf, November 2002. Last accessed April 2013.

7. Y. Hou, Y. Shi, and H. Sherali, Optimal spectrum sharing for mult-hop software defines radio networks, in *26th IEEE International Conference on Computer Communications*, 2007, pp. 1–9.

8. H. Khalife, N. Malouch, and S. Fdida, Multihop cognitive radio networks: To route or not to route, *IEEE Network*, 23 (4), 20–25, July/August 2009.

9. Y. L. Lin and X. Shen, Spectrum-aware opportunistic routing in multi-hop cognitive radio networks, *IEEE Journal on Selected Areas of Communications*, 30 (10), 1958–1968, November 2012.

10. A. Tanenbaum and D. Wetherall, *Computer Networks*, 5th edn., Pearson, Boston, MA, 2011.

11. J. Wang and Y. Huang, A cross-layer design of channel assignment and routing in cognitive radio networks, in *IEEE International Conference on Computer Science and Information Technology (ICCSIT)*, Chengdu, China, pp. 242–547, 2010.

12. Q. Wang and H. Zheng, Route and spectrum selection in dynamic spcetrum networks, in *Third IEEE Comsumer Communication and Networking Conference*, 2009, pp. 625–629.

13. C. Xin, B. Xie, and C.-C. Shen, A novel layered graph model for topology formation and routing in dynamic spectrum access networks, *in Proc. First IEEE International Symposium on New Frontiers in Dynamic Spectrum Access Networks (DYSPAN)*, Baltimore, MD, 2005, pp. 308–317.

14. J. Mack and M. Ibnkahla, Routing protocols in cognitive radio networks, Queen's University, Kingston, Ontario, Canada, WISIP Lab, Internal Report, April 2013.

11 Economics of Cognitive Radio

11.1 INTRODUCTION

The prime objective of cognitive radio technology is to improve spectrum utilization. This may involve any level of economic complexity from anonymously using an unlicensed channel at no cost, to temporarily leasing exclusive access rights from a wireless service provider (WSP), which would have leased the rights from the spectrum owner. Spectrum access rights are generally grouped into three categories:

1. Commons
2. Shared use
3. Exclusive use

The commons model describes a free, unlicensed system where all users share the spectrum. Although game theory is relevant to user behavior under this system, there is no potential economic framework to discuss.

The shared-use category can also be free, with the difference that there is a primary user (PU) that has priority to access the channel. Anonymous secondary users (SUs) may access the channel, so long as they do not interfere with the PU's transmissions. There is no economic framework to discuss for this case. However, if the PU's access rights are treated as property rights to the channel, then SUs can pay a fee to the PU for the privilege of sharing the channel. The time available for the SU to transmit is not necessarily constant or predictable. Therefore, the fee must be proportional to the quantity of time that is used by the SU for its transmissions. In this chapter, a cooperative model for spectrum sharing between one PU and multiple SUs is analyzed.

The exclusive-use category describes the case where spectrum owners lease their spectrum access rights exclusively to one user, or to a WSP that may then sublease the access rights to an end user. Due to the exclusive transmission privilege, channels leased for exclusive use are of higher value to transmitters. The drawback to this is that the spectrum owner cannot make use of its own channels while they are leased. The model is only effective when those who own the rights do not need them or only rarely need them.

This chapter analyzes the effectiveness of fixed-price markets, single auctions, and double auctions for access right distribution. In fixed-price markets, users do not have price control. Users have the opportunity to select from a set of sellers who must compete with one another for the users' business. For example, users may purchase access rights from a WSP. By contrast, single auctions have only one seller, but buyers are capable of bidding their desired price for access rights. Double auctions involve bidding by both buyers and sellers. In the double-auction mechanism, sellers compete to bid the lowest sale price, while buyers compete to bid the highest purchase price. This chapter investigates the models discussed earlier following the survey study in [1].

11.2   GAME THEORY

Prior to analyzing proposed models for the market types introduced earlier, a brief overview to game theory is needed to describe the methodology for modeling buyer and seller market behavior [2–11].

11.2.1   Strategic-Form Game Model

The strategic-form game model is a standard method for defining game parameters. It is denoted $(N,(A_i),(u_i))$, where N represents the number of players, (A_i) represents the action set available to player (i), and (u_i) represents the utility/payoff function of player (i).

For cognitive radio networks (CNRs), the pool of players includes both buyers and sellers. Their action sets constitute the options they may choose from to optimize their respective utility functions. For instance, in a fixed-price market, sellers' action sets are the prices they may choose to offer. They may take a particular action (select a particular price) because it results in the maximum payoff. The sellers also are aware that in order to achieve maximum payoff, they must determine the probability that another player (a buyer) will pay to lease their channel at a given price as well as the corresponding price to maximize revenue. Buyers consider the offered price with respect to their own utility and respond by taking an action (selecting a seller/price). The sellers then learn from the response and update their strategy/action accordingly. Player actions depend heavily on available information. It is prudent to make assumptions sparingly when modeling a real market.

11.2.2   Market Evolution and Equilibrium

When buyers and sellers can no longer benefit from changing their strategy, market equilibrium is achieved. Although not always the case, markets usually converge or become cyclical. When all players always select their optimal action in a noncooperative game, the resultant equilibrium is called the Nash equilibrium (NE). It is not necessarily economically efficient, as collusion tends to be more profitable than competition. Markets may have multiple NE, a unique NE, or no NE at all.

To demonstrate how a market approaches equilibrium, the method presented by [3] for predicting a WSP's channel occupancy is described next. This method was devised for a fixed-price market where WSPs compete for users to purchase access rights to their channel.

$N_c(t)$ is the number of links active on channel (c) at time (t). The traffic rate of a link is given by (λ) and follows a Poisson distribution. For a period of time, (Δt), the number of new transmitter–receiver pairs (links) in the network can be defined as $(N \times \lambda \times \Delta t)$. For the steady state where the number of links joining is equal to links departing, this also represents the number of departing links. To determine the number of active links on a given channel, it is necessary to determine P_c, the probability that an SU will select the channel from the complete set of available channels. This is given by

$$P_c = \text{Prob}\left\{c = \arg\max_{c^* \in C} \mu_i\left(c^*\right)\right\}, \forall c \in C \qquad (11.1)$$

For a particular user analyzing the utility of channel c, utility can be described simply as the difference between a function of the signal-to-noise ratio (SNR) and the price of using the channel

$$\mu_c = B \times \log_{10}\left(\frac{P_o g_c}{I_c + N_o}\right) - p_c \qquad (11.2)$$

where p_c is the price of the channel.

Sellers who accurately model buyers' utility functions can anticipate the ideal price to maximize P_c. Knowing P_c, the number of new links that occupy channel c can be computed as $(N \times \lambda \times \Delta t \times P_c)$. The next state of any channel can then be described as the current state, plus all new links occupying the channel and minus the links that depart from the channel:

$$N_c\left(t + \Delta t\right) = N_c\left(t\right) + \left(N \times \lambda \times \Delta t \times P_c\right) - N_c\left(t\right) \times \lambda \times \Delta t \qquad (11.3)$$

As the number of links in the channel is considered effectively constant, the rate of change of $N_c(t)$ with respect to time is zero. For this condition, the probability that an SU selects channel c becomes

$$\Pi_c\left(t\right) = P_c = \frac{N_c\left(t\right)}{N} \qquad (11.4)$$

This is termed also as the channel occupancy measure. It can be used to describe the market share of the channel.

Parameter P_c is dependent on the factors that maximize SU utility. These factors are

1. Interference, I_c, on the channel
2. Spectrum price, p_c
3. Total spectrum demand, ρ

As a change to any one of these parameters will affect the others, each SU's perceived utility changes on an ongoing basis. This results in a natural progression toward an equilibrium state.

11.3   COOPERATIVE TRADING MODEL FOR COGNITIVE RADIO

Shared-use spectrum markets are based on the concept that the PU owns property rights to a spectrum band and that those rights can be leased to secondary users if desired. To ensure that billing is fair, costs for secondary users must be proportional to the amount of time that the spectrum is used.

A shared-use model is proposed in [10] based on the cooperation between PUs and SUs. In this model, the PU makes use of SUs to acquire spatial diversity to reduce the effects of channel fading. A relay selection technique is incorporated to select the optimal set of SUs to relay the signal. In exchange, the PU allows noninterfering SUs to use the channel for their own purposes for a period of time. This process occurs as a cycle, with the SUs relaying signals for the PU and transmitting during their time slot. The PU selects a price per unit of time for the SUs, and the SUs select a power level for transmission to support the PU's transmission. Game theory can be used to model user behavior for the practical case where SUs do not know complete channel information.

The goal of the cooperative model is to maximize throughput of the SUs without compromising the ability of the PU to achieve its own goals. To achieve this, it is necessary to jointly optimize the primary and SU utility functions. A Stackelberg game was used in [10] to do so. In a Stackelberg game, a leader takes the first action and the followers respond. For a cooperative CR market, the PU is the leader and chooses the best possible price to maximize its utility. The SUs are the followers and purchase channel access time that optimizes the utility they receive from transmission.

It is notable that in the cooperative model, the SUs incur a trade-off not present in the noncooperative alternatives. In addition to the cost of using the channel, SUs must select their transmission power to support the PU. There is an added cost to increasing power use when transmission is not of benefit to the SU.

Figure 11.1 demonstrates the system behavior over three time slots that comprise one period.

The primary transmitter (PT) leases access rights for its channel to secondary transmitters (STs), which act as a set of relays, S. The number of active relays in the set is denoted by k. The PU sets a price, c, per unit of access time, t_i, for the ith SU. Cooperative transmission power, P_i, and channel power gain, $G_{i,p}$, are incorporated into the function for purchasing access time:

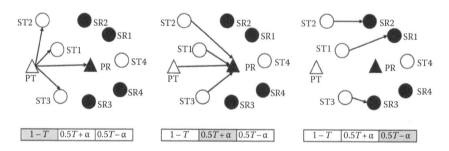

FIGURE 11.1   Three time slots for cooperative CR. (Adapted from Wang, X. et al., *IET Networks*, 1(3), 116, September 2012.)

$$ct_i = P_i G_{i,p}, \quad i \in S \tag{11.5}$$

The channel power gain $G_{i,p}$ is between ST_i and the primary receiver (PR), as the PU's objectives are to both profit and improve its own transmission.

In Figure 11.1, the first time slot $(1 - T)$ is the time during which the user transmits data to its selected relays as well as the PR. The second time slot $(0.5T + \alpha)$ is the fraction of the remaining time in the cycle that both the PT and the relays transmit directly to the PR. The final time slot $(0.5T - \alpha)$ is reserved for SU transmissions to their own receivers (SRs). Time-division multiplexing access (TDMA) is used for the SU transmissions.

In order to examine market behavior and equilibrium, it is necessary to derive the PU and SU utility functions. As described earlier, in the Stackelberg game, the PU is the leader and selects the price per unit of access time. The SUs are followers and select strategies that maximize their personal utility functions. A key factor in transmitter utility is the signal to noise ratio (SNR). The SNR for direct transmission from PT to PR is given by

$$\Gamma_{dir} = \frac{P_p G_p}{\sigma^2} \tag{11.6}$$

where σ^2 represents the noise power.

The SNR for the relay set is denoted similarly but is dominated by the lowest channel gain:

$$\Gamma_{PS}(k) = \frac{P_p \min_{i \in S} G_{p,i}}{\sigma^2}, \quad i \in S \tag{11.7}$$

These two equations represent the set of SNR values associated with broadcasting in the first time slot. In the second time slot, the combined transmission from the PT and its relays becomes

$$\Gamma_{SP}(k) = \frac{P_p G_p}{\sigma^2} + \sum_{i \in S} \frac{P_i G_{i,p}}{\sigma^2}, \quad i \in S \tag{11.8}$$

Using decode-and-forward (DF) cooperation protocol and space-time coding, the overall transmission SNR for the PU is

$$\Gamma_{coop}(c,k) = \min \left\{ (1 - T)\Gamma_{PS}(k), \left(\frac{T}{2} + \alpha\right)\Gamma_{SP}(k) \right\} \tag{11.9}$$

This can be used to define a simple utility function with a scaling parameter, w_p, to convert SNR into a useful utility value for the PU:

$$U_p = w_p \Gamma_{coop}(c,k) \tag{11.10}$$

Pricing is very important, as it is the only parameter that the PU can directly manipulate. It has the capability of influencing the number of SUs in set S as well as the quantity of time purchased by the SUs. The optimal price is the one that maximizes U_p. However, this model [10] does not include a separate term incorporating the PU's revenue in this equation. All users have some form of operating cost, and for that reason, revenue should always play some role in the utility function.

SUs derive all utility from their own transmissions. Their goal is to maximize their utility by determining the ideal quantity of access time for transmitting. Note that transmission rates for SU_i are given by

$$R_i = W \log_2\left(1 + \frac{P_s G_i}{\sigma^2}\right), \quad \forall i \in S \qquad (11.11)$$

Let w_1 represent the revenue generated for one unit of transmitted data, and let w_2 represent the cost of one unit of power. The SU utility function can be described as the revenue generated from data transmission minus the cost of transmitting both data for the SU and the PU:

$$U_i = w_1 R_i t_i - w_2 \left[P_s t_i + P_i \left(\frac{T}{2} + \alpha\right) \right], \quad \forall i \in S \qquad (11.12)$$

Each SU must decide the quantity of access time to purchase in order to maximize its utility.

Stackelberg equilibrium (SE) is the market equilibrium for which PU price and SU access times are both optimized, denoted $(c^*, \mathbf{t}^*)$. Note that $\mathbf{t}^*$ is a vector $[t_1, t_2, \ldots, t_k]^T$ corresponding to the set of active SUs. Within the cooperative game, SUs compete with one another in a noncooperative manner, as they are assumed to select access times that maximize their own utility. For this behavior, the SE matches the NE where players only select actions that improve their current utility the most. To select the optimal price, the PU is assumed to know the SUs' utility functions. It compiles the set of best responses $\mathbf{t}^*$ for a candidate price, c, and then selects $\mathbf{t}_c^*$ that maximizes U_p.

The noncooperative game for selecting access times t_i is formulated as a strategic game where S is the set of players (the SUs), $\{T_i\}$ is the strategy set $T_i = \left[t_i\right]_{i \in S} : 0 < t_i \leq 1$, and $\{U_i\}$ is the set of utility functions of the SUs:

$$G = \left[S, \{T_i\}, \{U_i\} \right] \qquad (11.13)$$

In order to determine the best response of the SU, the first derivative is taken with respect to t_i and is set to zero. When the rate of change of the SU's utility is zero with respect to changes in time purchased, utility has reached its maximum.

$$\frac{\partial U_i}{\partial t_i} = \frac{-2ct_i - c\displaystyle\sum_{j \in S, j \neq i} t_j - 2\alpha c}{G_{i,p}} + \frac{w_1}{w_2} R_i - P_s = 0 \qquad (11.14)$$

where θ_i is defined as follows:

$$\theta_i = \left(\frac{w_1}{w_2}\right) R_i G_{i,p} - P_s G_{i,p} > 0 \tag{11.15}$$

θ_i captures private information about the SU. A large θ_i means that the SU's own data transmission is efficient (i.e., a large channel power gain G_i) or it has a good relay link between ST_i and PR (a large channel gain $G_{i,p}$).

The NE is shown in [10] to be unique and obtainable by solving the derived access time equation for all SUs:

$$t_i^* = \frac{(1+k)\theta_i - \sum_{i \in S}\theta_i - 2\alpha c}{(1+k)c}, \quad \forall i \in S \tag{11.16}$$

It can be seen from this equation that the optimal user access time depends on the total number of users (k) cooperating with the same PU.

To obtain the minimum price necessary for the PU to justify cooperation, the maximum utility for the PT to PR must be computed for

$$U_{dir} = w_p \Gamma_{dir} = w_p \frac{P_p G_p}{\sigma^2} \tag{11.17}$$

This is the PU's utility function as introduced earlier, but without the SU's contributions to SNR. If the achievable utility from cooperation is less than this value, there is no incentive for a PU to participate in a shared-use model.

Overall, the model proposed in [10] is practical and simple to implement. However, it does not consider any competitive element between PUs. There are some similarities between this particular cooperative model and the fixed-price model [3] analyzed in this chapter. The probabilistic methodology from [3] could be used to incorporate competing prices between PUs.

11.4   FIXED-PRICE TRADING MODEL

Fixed-price trading models are exclusive-use economic models most commonly used by WSPs to distribute short-term leases for access rights to end users. As buyers are price takers, practical models tend to have fewer variables to consider than the more complex auction models. Buyers in fixed-price markets are nearly always end users, and a simpler system that does not require the end users to bid may have its advantages. For example, ease of entry encourages the development of new businesses. Revenue generated in fixed-price markets is not as high as auctions, since users are aware of prices. If prices are below the users' reservation price (the highest price the user is willing to pay), the user saves the difference. However, in most auctions, bidders must compete against others without knowing any of the bids. To maximize the probability of a winning bid, buyers bid their reservation price.

A trading model for a fixed-price market is described in [3]. It considers a duopoly where two WSPs compete to maximize their profits. It is shown that frequency propagation characteristics play a large role in user decisions when demand is low and that lower frequencies are typically preferable. When demand is high, cochannel interference can have a much greater impact on the SINR than noise. Due to this, price differences between channels become small for high cochannel interference. Further, the significance of cochannel interference indicates that the geographical distribution of users has an effect on pricing. WSPs can increase revenue by exploiting the geographical separation of SUs, as they can re-lease spectrum bands to buyers who do not interfere with one another. To communicate pricing, WSPs have two options: use a dedicated control channel for price broadcasting and provide a database query system for the users.

To model the market, three steps are developed to determine player behavior:

1. Examine the effect of other users' WSP subscriptions on buyer utility functions.
2. Determine the optimal WSP selection strategy for the users to maximize their utility functions.
3. From the WSP selection strategies, pricing strategies are developed for WSPs in a noncooperative duopoly.

Once player behavior has been modeled, it is possible to determine the NE for the market. It is assumed that two WSPs each have long-term access rights for a single licensed channel with a unique center frequency. The term *heterogeneous* is used to describe spectrum resources with distinctly different center frequencies and propagation profiles. Channels in a heterogeneous spectrum tend to have different transmission and interference ranges. User utility in [3] describes purchasing behavior based on band capacity and cost. All users are assumed to have the same utility function

$$\mu_i(c) = B \times \log_{10}\left(1 + \frac{P_o g_{c,i}}{I_{c,i} + N_o}\right) - p_c \tag{11.18}$$

where
  B is the channel bandwidth
  $g_{c,i}$ is the channel gain
  N_o is the noise power
  P_o is the transmission power

The bandwidth is constant for all channels. Receiver i experiences interference $I_{c,i}$ on channel c from other users. All users are assumed equidistant from the transmitter, causing the channel gain to depend only on channel frequency. Based on this, user i selects channel c_i such that utility is maximized.

The assumption that this utility function is consistent for all users is not necessarily realistic, but it demonstrates the effects of fixed-price heterogeneous spectrum markets on buyer behavior.

The user plane is considered to have an ad hoc secondary network with N transmitter–receiver pairs. The WSPs set fixed spectrum prices at which the users buy. Since users are assumed to have CR capabilities, there is a broad range of spectrum suitable for their purposes. This gives the users flexibility in choosing their WSP.

The market model has users randomly distributed across geographic areas following a Poisson point process. The distribution of the quantity of active links within the area is given by

$$n_A \sim \mathrm{Poisson}\left(n; \rho \,|\, A\,|\right) \tag{11.19}$$

where
   n is the number of links
   ρ is the average density
   A is the deployment area

Lower frequency bands have better propagation characteristics. For example, consider a simple spectrum propagation model

$$P_R = P_o \times g_c\left(r\right) = P_o \times \left(\frac{c_o}{f_c}\right)^{\alpha} \times r^{-\alpha} \tag{11.20}$$

where
   P_R is the received signal power
   P_o is the transmission power
   c_o is the speed of light
   f_c is the carrier frequency
   r is the distance between the transmitter and the receiver
   α is the path-loss exponent

It can be seen that as the carrier frequency decreases, received power increases.

In order to compute users' cochannel interferences, it is necessary to determine a reasonable interference range. This range is larger for lower frequencies:

$$R_I^c \triangleq \sup\left\{ r \in \mathbb{R} \,|\, P_o \times g_c\left(r\right) > \eta \right\} \tag{11.21}$$

The threshold η defines the interference range. It depends on the data rate, type of modulation, and other factors. Users who have subleases for the same channel are normally sufficiently far apart so as not to cause interference.

The normalized interference will be assumed following a Gaussian distribution $I_{in,c} \sim N\left(\mu_c, \sigma_c^2\right)$ and is denoted as $I_{in,c} = \displaystyle\sum_{S_c} g_c\left(r\right)$, where S_c is the set of cochannels that interfere with channel c, and $g_c(r)$ is the channel gain for distance r from the transmitter.

$$I_{in,c}(x) = \frac{1}{\sqrt{2\pi}} e^{\left(-\frac{(x-\mu_c)^2}{2\sigma_c^2}\right)}$$ (11.22)

The mean and variance of the normalized interference can then be modeled as [3]

$$m_k(\rho,c) = \frac{2\pi\rho_c}{(k\alpha-2)}\left(\frac{c_o}{f_c}\right)^{\alpha k}\left[\frac{1}{\varepsilon^{k\alpha-2}} - \frac{1}{\left(R_I^c\right)^{k\alpha-2}}\right]$$ (11.23)

where
   m_1 represents the mean
   m_2 represents the variance
   ε is the smallest distance between the transmitter and the receiver
   ρ_c is the density of SUs using the same channel
   c_o is the speed of light
   f_c is the carrier frequency

The total interference on channel c that is caused by users outside the interference region is given by

$$I_{out,c} = 2\pi P_o\left(\frac{c_o}{f_c}\right)^{\alpha}\frac{\rho_c\left(R_I^c\right)^{2-\alpha}}{(\alpha-2)}$$ (11.24)

Interference $I_c \sim N\left(\mu_c,\sigma_c^2\right)$ is then described by the following two equations:

$$\mu_c = E\left[I_{in,c}\right] + I_{out,c} = \left(\frac{c_o}{f_c}\right)^{\alpha}\frac{2\pi\rho_c P_o}{(\alpha-2)}\left(\frac{1}{\varepsilon^{\alpha-2}}\right)$$ (11.25)

$$\sigma_c^2 = \frac{\pi\rho_c P_o}{(\alpha-1)}\left(\frac{c_o}{f_c}\right)^{2\alpha}\left[\frac{1}{\varepsilon^{2\alpha-2}} - \frac{1}{\left(R_I^c\right)^{2\alpha-2}}\right]$$ (11.26)

These equations describe the interference characteristics of the user's signal to interference ratio (SIR) for a given channel.
   There are two key observations to be made:

1. The influence of spectrum heterogeneity can be seen on the channel, as interference is related to center frequency f_c and interference range R_I^c.
2. Changes to user density on a channel affect the channel's interference mean and variance.

These give insight into the effects that users have on one another's SIR as well as the significance of purchasing access rights to a frequency with desirable propagation characteristics.

A mean-field approach can be used to predict the development of the heterogeneous spectrum market as it converges to a steady state. In mean-field theory, the impact of a large group of small individuals on any particular individual is described by a single-averaged effect, reducing the complexity of the problem.

Recall that the probability that a user will select channel c from the complete set of available channels can be modeled as

$$P_c = \text{Prob}\left\{c = \arg\max_{c^* \in C} \mu_i\left(c^*\right)\right\}, \quad \forall c \in C \tag{11.27}$$

The equation for the user's utility is

$$\mu_c = B \times \log_{10}\left(\frac{P_o g_c}{I_c + N_o}\right) - p_c \tag{11.28}$$

As [3] considers a duopoly, there are only two WSPs to select from. One WSP is considered to have channel c available, and the other has channel a. The probability of selecting channel c can then be described by the following:

$$P_c = \text{Prob}\left(\mu_c - \mu_a > 0\right) \tag{11.29}$$

Rewriting this function to more clearly express the impact of interference and spectrum price yields

$$P_c = \text{Prob}\left(I_a + N_o - e^{p_c - p_a}\left(\frac{f_c}{f_a}\right)^\alpha (I_c - N_o) > 0\right) \tag{11.30}$$

where f_c and f_a are the center frequencies of the two channels. It is common to assume that in steady state, the utility of selecting either channel will be equal. However, due to heterogeneous propagation characteristics at different frequencies, this assumption is not always reliable.

In order to isolate the effects of heterogeneous spectrum characteristics on fixed-price market behavior, consider the case that the two competing WSPs offer identical prices. Their distinguishing feature becomes the characteristics of the channel they lease.

Let $I_{ca} \sim N\left(\mu_{ca}, \sigma_{ca}^2\right)$ be the difference between Gaussian random variables I_c and I_a. P_c can then be simplified to

$$P_c = \text{Prob}\left(I_{ca} > 0\right) = Q\left(\frac{-\mu_{ca}}{\sigma_{ca}}\right) \tag{11.31}$$

The mean and variance are given by

$$\mu_{ca} = \mu_a + N_o - e^{p_c - p_a}\left(\frac{f_c}{f_a}\right)^{\alpha}(\mu_c + N_o), \quad \sigma_{ca}^2 = \sigma_a^2 + \left(e^{p_c - p_a}\left(\frac{f_c}{f_a}\right)^{\alpha}\right)^2 \sigma_c^2 \quad (11.32)$$

Since a duopoly is considered $P_a = 1 - P_c$.

From the equations given earlier, it can be seen that the channel frequencies offered by competing WSPs will directly impact one another's interference means and variances.

It is notable that μ_c(respectively, μ_a) is proportional to ρ_c(respectively, ρ_a). A channel with low user density will have lower average interference, but will increase as users recognize and take advantage of this. For steady state, ρ_c(respectively, ρ_a) is constant.

The goal of each WSP is to lease their spectrum to as many users as possible at the highest possible price. As WSPs are competing with one another in a pricing game, their payoff function is modeled as

$$V_c(p_c, p_{-c}) = N_c(p_c, p_{-c}) \times p_c - b_c \quad (11.33)$$

where N_c is the number of SUs purchasing from the WSP at price p_c.

The cost to lease the spectrum from the owner, b_c, is considered in terms proportional to the duration of a user's sublease. It is assumed in [3] that this is a fixed cost. The competing WSP's price is denoted p_{-c}. As the goal is to maximize profit, p_c^* is selected as to maximize $V_c(p_c, p_{-c})$:

$$p_c^* = \arg\max_{p_c \in \mathbb{R}} V_c(p_c, p_{-c}) \quad (11.34)$$

It is shown in [3] that as average SU density approaches infinity, the probability of choosing channel c approaches $1/C$, where C is the number of unique channels. This assumes that prices are equal between all competitors. In other words, as the density of users increases, the significance of spectrum heterogeneity on WSP selection decreases, eventually becoming negligible. In order to determine how spectrum acquisition price b_c is set, auction markets for WSPs shall be analyzed.

11.5   AUCTION MODELS

Auctions can be carried out in a variety of ways. Models for single-sided and double-sided auctions will be considered in this chapter. A single-sided auction involves a single seller, typically a spectrum owner, and multiple buyers who compete for spectrum access rights by bidding against one another for bands in the auction. This has the effect of driving up the market price toward the buyers' actual valuations of the bands, provided demand exceeds supply. Where demand does not exceed supply, the buyers have no incentive to bid any more than they are required by auction rules. The result is that all buyers pay the minimum bid and the seller's revenue is minimized. The double-sided auction allows sellers to bid against one another, competing for the lowest asking price for their channels. This element of competition between sellers reduces the cost of leasing individual channels, compared to the monopolistic single-auction scenario. However, once again demand must exceed supply. Where there are more sellers in the market, supply is likely to be higher, and therefore, greater demand is required to justify the double auction. Additionally, a trusted third party such as the Federal Communication Commission (FCC) is required to facilitate the auction, since sellers are also bidding.

Auctions can be carried out as sequentially or concurrently. Sequential auctions have users bid for one available band at a time. Concurrent auctions have users bid for all available bands at once, and the highest bidders are each awarded a band. Sequential auctions were found in [12] to generate higher revenue when all other factors were equivalent. Cases where bidders could only be awarded a single band and where bidders could be awarded unlimited bands both shared this result.

It is possible to hold auctions where buyers request their own bandwidth, w_i, provided it is less than or equal to the total available spectrum, W, being auctioned:

$$w_i \leq W, \quad i = 1, \ldots, n \tag{11.35}$$

The sequential auction is constrained to seller-defined bandwidths, which reduces buyer's ability to optimize their utility. If a buyer can bid for their optimal bandwidth, they can bid with their highest achievable reservation price. The drawback of allowing buyer-defined bandwidths is that determining the combination of bids that maximizes revenue becomes more complex. It must be defined as a type of optimization problem known as the 0–1 knapsack problem. This is used when a set of quantities has greater sum than a known constraint (e.g., total demand exceeds supply). Each quantity has a value associated with it, and the maximum sum of values must be achieved without the total sum of the quantities exceeding the known constraint. The 0–1 represents the decision whether or not to select any given quantity from the set, using a δ-function. Bids selected from optimization will not necessarily make use of all available bandwidth.

For comparison, Figures 11.2 and 11.3 present an illustration of complete and incomplete allocation of available spectrum.

It is shown in [13] that in spite of this limitation, revenue from buyer-defined bandwidths is slightly higher when all other factors are equal (Figure 11.4). Average spectrum usage is improved as well. This result was compared to a second-price auction where winning bidders pay the second highest price bid. No comparison was provided for the case of a first-price auction.

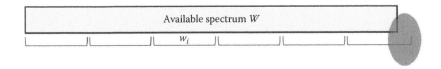

FIGURE 11.2   Illustration of an incomplete allocation of available spectrum.

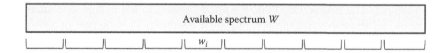

FIGURE 11.3   Illustration of a complete allocation of available spectrum.

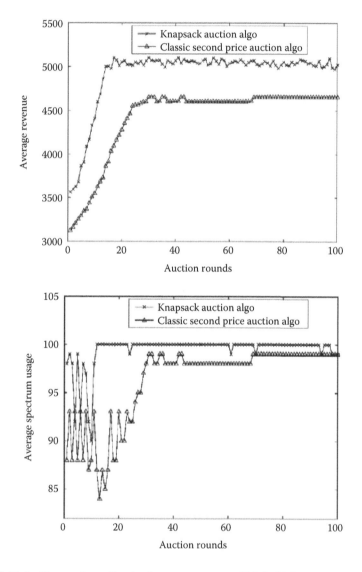

FIGURE 11.4   Knapsack auction is shown to generate slightly higher revenue and average spectrum usage than a second-price auction with seller-defined channel bandwidths. (From Sengupta, S. and Chatterjee, M., *IEEE/ACM Transactions on Networking*, 17(4), 1200, August 2009.)

11.5.1 SINGLE-SIDED AUCTIONS

A trading model formulated in [12] for single-sided auctions will be considered. First, consider the model where bidders may only win a single channel:

$$S = \{s_1, s_2, \ldots, s_m\}$$

$$N = \{N_1, N_2, \ldots, N_n\} \qquad\qquad (11.36)$$

$$B = \{b_1, b_2, \ldots, b_n\}$$

$$n > m$$

where

S represents the set of m substitutable frequency bands

N represents the set of n bidders, where the number of bidders is greater than the number of available bands

B represents the set of bids presented by the bidders

The available spectrum, S, for dynamic allocation is called the coordinated access band (CAB). Bidders cannot see one another's bids, and aim to maximize their utility (in this case, profit from providing services to end users).

Winners are awarded a lease for their corresponding band, s_i. When the lease expires, the band is returned to the spectrum owner. Losing bidders are assumed to increase their bids in subsequent auctions, and winning bidders are assumed to decrease their bids. This is because losing bids have undervalued the market worth of the bands, while the winning bids have matched or overvalued the market worth. In [12], the spectrum owner broadcasts the minimum bid after each auction, to identify the lower bound for the market. The intent of this is to encourage only those who have a higher reservation price to participate in future auction rounds.

WSPs use their revenues from services to end users in order to determine their valuation of the spectrum bands. Consequently, real-world systems depend heavily on trends in end-user behavior. Spectrum valuations vary between WSPs, as their revenues may be different for the same band.

Each bidder's payoff (profit, in this case) can be modeled as the difference between revenue generated from a frequency band and the cost to lease the band from the owner. V is used to represent the valuation/revenue:

$$V = \{V_1, V_2, \ldots, V_n\} \qquad\qquad (11.37)$$

$$\begin{cases} V_i - b_i & i\text{th bid, wins} \\ 0 & i\text{th bid, loses} \end{cases}$$

As there are only two possible outcomes, the opportunity cost for the losing bid is simply the payoff for the scenario in which it had the winning bid.

For sequential auctions, a winning bidder does not participate in bids on subsequent bands in the same auction. There are m rounds, with the number of bidders decreasing by one each round. Conversely, concurrent auctions have only a single round, with all bidders present.

For a sequential auction with uniformly distributed bidders, the probability density function of bid submissions is modeled as

$$f(b) = \frac{1}{V_{max} - b_{min}} \tag{11.38}$$

Before determining the probability of a bidder winning, a model is needed to determine the probability that one bid (i) is greater than another (j). This is given by

$$P\left(b_i > b_j \mid j \neq i;\, j \in (n-k-1)\text{bidders}\right) = \int_{b_{min}}^{b_i} f(b)\,db = \frac{b_i - b_{min}}{V_{max} - b_{min}} \tag{11.39}$$

Here, k represents the number of previous rounds in the auction.

If the highest bidder is always the winner, the probability of the ith bidder winning is then given by

$$P\left(b_i > b_j \; \forall j \neq i;\, j \in (n-k-1)\text{bidders}\right) = \prod_{j=1}^{n-k-1} P\left(b_i > b_j \mid j \neq i;\, j \in (n-k-1)\text{bidders}\right) \tag{11.40}$$

Therefore, the probability of winning the $(k+1)$th round of a sequential auction in which each bidder is awarded only one band is

$$P_{seq}\left(i\text{th bidder winning}\right) = \left(\frac{b_i - b_{min}}{V_{max} - b_{min}}\right)^{n-k-1} \tag{11.41}$$

This probability can then be multiplied by the payoff function to describe the expected payoff for the bidder. To maximize expected payoff, the derivative of the expected payoff function is set to zero. By doing so, the optimal bid to maximize expected payoff can be expressed as

$$b_{i_{seq}}^* = \frac{(n-k-1)V_i + b_{min}}{n-k} \tag{11.42}$$

As noted earlier, losing bidders are assumed to increase their bids, causing b_{min} to increase over the course of the auction. Over time, bids are expected to approach the true value of the spectrum bands. There is some risk in assuming that the value of spectrum bands remains constant or always increases. The perceived worth of transmitting data wirelessly can decrease. For instance, if a large quantity of bands became freely available to unlicensed users, the benefits of exclusive access rights for end users would have to outweigh the utility of free, shared spectrum access. The WSPs could potentially have to reduce prices to remain the preferable choice for their users.

For the concurrent case with substitutable bands, only one bid is submitted every auction. The highest m bidders are then awarded spectrum bands. Therefore, to win a band, the WSP must place a bid that is greater than $(n - m)$ bids. The probability of this can be modeled by

$$P_{con}\left(i\text{th bidder winning}\right) = \prod_{j=1}^{n-m} P\left(b_i > b_j \mid j \neq i;\ j \in \left(n - m\right)\text{bidders}\right) \quad (11.43)$$

$$P_{con}\left(i\text{th bidder winning}\right) = \left(\frac{b_i - b_{min}}{V_{max} - b_{min}}\right)^{n-m} \quad (11.44)$$

The optimal bid is then

$$b_{i_{con}}^* = \frac{\left(n - m\right)V_i + b_{min}}{n - m + 1} \quad (11.45)$$

Knowing the optimal bid in each case, it is possible to compare the two auction types to identify where one is preferable to the other, assuming bidders are rational.

Two cases have been examined in [12]: transient and steady state. In steady state (market equilibrium), bidders achieve no additional payoff for changing their bid, and are therefore fixed.

It is demonstrated that, prior to any bands auctioned, the optimal bid to win must be higher for sequential case. This difference grows as more bands are available for auction. It is also demonstrated that during both the transient states and steady state, the optimal bid for the sequential case remains higher. For the spectrum owner, the sequential auction is clearly preferable for the single-band award limit with fully substitutable bands.

Nonsubstitutable bands require additional information for the bidders. It is assumed that complete information is available for all m bands. Valuation prices must be described by a vector of vectors:

$$V = \left\{\left\{V_1\right\}, \left\{V_2\right\}, \dots, \left\{V_n\right\}\right\}$$
$$V_i = \left\{V_{i,1}, V_{i,2}, \dots, V_{i,n}\right\} \quad (11.46)$$

Similarly, each bidder has a unique reservation price for every spectrum band in the auction and a corresponding utility function:

$$R_i = \left\{r_{i,1}, r_{i,2}, \dots, r_{i,m}\right\}$$
$$U_i = V_i - r_{i,j}; \quad j \in m \quad (11.47)$$

With the constraint that every bidder may only win once, bidders will identify the spectrum band with the highest utility, with the intent of maximizing profit. There may be multiple bidders competing for a single optimal band. If bidders compete in a concurrent auction, there is potential for some of the bands to receive no bids (zero revenue for the seller). However, it was found that all bands were always sold in the sequential system. Unsurprisingly, it was found that sequential auctions provided higher returns than concurrent auctions for the spectrum seller/auctioneer.

The multiple unit grant scenario is a relaxation of the auction systems previously analyzed. In auctions that allow for the purchase of multiple units, the optimal bid is defined by a 0–1 knapsack problem. The complete set of spectrum bands in the CAB represent the capacity of the sack, W, with the WSPs' bids representing the valuations for the spectrum band(s) they request. This is proposed as a *dynamic spectrum allocator knapsack auction*. In this auction, there are n bidders, who do not know one another's bids.

Unlike the single unit grant system, where n bidders compete for m available bands ($n > m$), multiple unit grants compare supply and demand as a function of bandwidth. W is the total available bandwidth for auction, and w_i is the ith service provider's bandwidth request

$$\sum_{i=1}^{n} w_i > W \qquad (11.48)$$

Let x_i denote the bidding price for spectrum quantity w_i. The objective of the spectrum seller is then to maximize the sum of all x_i.

The authors in [12] analyze two types of auction for this scenario: asynchronous and synchronous.

Synchronous auctions perform allocation and de-allocation at fixed intervals, with bids taken simultaneously. Asynchronous auctions do not have fixed intervals for allocation and de-allocation, and bids do not need to be submitted simultaneously.

In an asynchronous auction, the amount of time T_i (that a spectrum band is requested) is considered in the bidding strategy, q_i^a, of the ith bidder:

$$q_i^a = \{w_i, x_i, T_i\} \qquad (11.49)$$

The decision to allocate the requested spectrum band is then made immediately, since there are no constraints at the time of allocation. If the bid meets a required minimum and the spectrum band is available, it is granted. This method does not necessarily maximize revenue and is difficult to model. For instance, if q_i provides x_i for a spectrum band, q_j may offer x_j shortly after, where ($x_j > x_i$). As q_i was already granted the band, q_j is declined and the spectrum seller profits less overall.

This issue is not present in the synchronous auctions, since requests are submitted simultaneously and bands are relinquished after a fixed time.

It was shown that the average revenue generated by asynchronous auctions is less than that generated by synchronous auctions.

Bidding strategies are dependent on the nature of the cost to the bidder. In a first-price auction, the top bidder pays their bid, but in a second-price auction, the top bidder pays the second highest bid. It was proven that in second-price bidding, the best strategy for the bidder is to bid their reservation price. Essentially, this guarantees that the bidder will pay at most the maximum they are willing to bid, with the opportunity to potentially pay less. It was also shown that for first-price bidding, the upper bound of any bid is the reservation price.

11.5.2 DOUBLE-SIDED AUCTIONS

Here, the model formulated in [14] for double-sided auctions is considered. The proposed mechanism is titled the "double truthful spectrum auction" (DOTA), and allows for multiple requests for spectrum bands from a single user. *Truthfulness*, in the context of an auction, is the property that ensures buyers are bidding based on their actual valuation of the good they are purchasing, independently of others. Without truthfulness, the auction process becomes an inefficient system due to the potential for collusion and the need for buyers to anticipate untruthful behaviors of other buyers. In order to ensure that the double auctions are carried out fairly and truthfully, a third party is required to facilitate the auctions. Their cost is a commission on each lease sold, and must be sufficient to cover the cost of operation. A spectrum broker acts as the trusted third party, auctioning the spectrum sellers' spectrum bands to spectrum buyers for short periods. The sellers specify asking prices for their bands, while the buyers specify their bids.

The DOTA method aims to minimize channel overhead by bundling multiple requests from buyers into a single bid. Simulations found that spectrum utilization could be increased as much as 61%, in addition to generating extra revenue from the auction.

In the double auction, a set of sellers, S, submit their asking prices as bids, B. Using both these bids and those of the buyers, the third party generates *market-clearing* prices to maximize revenues. In other words, this form of auction has winning buyers as well as winning sellers. The seller's utility function is the difference between the price paid for the spectrum band and the seller's valuation of the band. Conversely, the buyer's utility function is the difference between the buyer's valuation of the band and the price paid for it. As per the strategic-form game model, both buyers and sellers aim to maximize their utility functions. The common parameter is the price paid, but an increase in one function will result in a decrease in the other (i.e., a better buyer payoff comes at the expense of the seller).

If two users do not interfere with one another, they may both lease a channel simultaneously. If two SUs do not interfere with one another, they may both use a channel simultaneously. For m sellers, the jth seller provides k_j channels. For n buyers, the ith buyer requests d_i channels. It is assumed that all requested channels have uniform characteristics. Conflict conditions are those where two buyers cannot use the same channel due to interference.

The auction is a concurrent, synchronous auction with one auctioneer. It is assumed that all bids are submitted independently and that utility of nonwinning bids is zero for both buyers and sellers.

DOTA uses four steps:

1. Translate-into-one-Bid
   a. To reduce the cost due to multiple smaller transactions, a single larger transaction is given by

$$B_i^* = B_i^b + \frac{\left(k \times C(1) - C(k)\right)}{k} \tag{11.50}$$

   This combines the complete set of bids that a buyer makes on (k) channels at cost $C(k)$.
2. Buyer Grouping Rule
   a. It is assumed that all sellers' channels are available to all buyers.
   b. To maximize channel reuse, nonconflicting buyers are grouped together. Each group is then treated as a single buyer, whose per-channel price, π_1, is the minimum bid from the set of buyers in the group for λ_1 channels.
3. Clearing Rule
   a. Buyer and seller bids are sorted in decreasing order in a single set. If a buyer's bid matches a seller, the buyer's bid receives a higher rank in the set.
   b. The auctioneer considers bids as single unit requests. For instance, group buyers or sellers bidding k channels are considered k bidders, each placing a 1-channel bid. Each of the k bids has the minimum bid as selected by the group.
      i. Range requests are fulfilled as a range from 0 to d_i
      ii. Strict requests can be fulfilled as all or nothing (0 or d_i)
4. Pricing Rule
   a. When the bidder of b_{k+1} from the combined set is a buyer, b_{k+1} is the clearing price. Accordingly, DOTA charges the same price, b_{k+1}, to all buyers in the winning group for each channel. Winning sellers are then paid the clearing price for each channel sold.
   b. When the bidder of b_{k+1} from the combined set is a seller, b_k is the clearing price. This price is multiplied by the number of channels allocated for *range requests* and by the number of channels requested for *strict requests*. For every channel sold, sellers are paid the market-clearing price, b_k.

A simple example was provided by [14] to demonstrate how a conventional double auction works (Figure 11.5). This does not incorporate the first two steps in the DOTA model.

Bidding strategies are similar to those described for the double auction. DOTA is a relatively straightforward extension of the conventional double-sided auction, simply adding elements that allow for reauctioning of frequency access rights to noninterfering buyers. For instance, if two WSPs are sufficiently far apart geographically, they may lease access rights to the same channel for their respective areas.

Bidders	Channels	Bids (per channel)
Seller A	2	$6
Seller B	1	$3
Seller C	1	$2
Buyer D	1	$5
Buyer E	2	$4

- Sorted in descending order: $\{6_A, 6_A, 5_D, 4_E, 3_B, 2_C\}$
- $K = 4$ (channels for lease)
- $(K+1)_{th}$ bid: 4_E
- Winning sellers: 3_B, 2_C
- Winning buyers: 4_E, 5_D
- Bid $(K+1)$, 4_E is charged to all winning buyers and paid to all winning sellers

FIGURE 11.5 Example of a conventional double-sided auction. (From Wang, Q. et al., DOTA: A double truthful auction for spectrum allocation in dynamic spectrum access, in *IEEE Wireless Communications and Networking Conference: MAC and Cross-Layer Design*, Orlando, FL, 2012.)

11.6 SIMULATIONS AND ILLUSTRATIONS

Here, simulation examples and discussions are presented for each of the strategies.

11.6.1 FIXED-PRICE MARKETS

A duopolistic fixed-price market was considered in [3] where WSPs *a* and *c* competed for user market share (Figure 11.6). Each WSP has one channel for lease, with an extreme frequency difference of 250 MHz between the two. This was clearly intended to highlight the effects of differences in propagation characteristics due to frequency, but it does not necessarily reflect a true competitive scenario. A smaller difference, in the range of 10–50 MHz, may have provided a fairer representation of a competitive market.

For $f_a = 500$ MHz and $f_c = 750$ MHz, it was found that the probability of selecting the higher frequency rose with user density, approaching a limit at 0.5.

At $\rho = 10$ users/km^2, the probability of a user selecting the 750 MHz frequency was 0.10.

At $\rho = 200$ users/km^2, the probability of a user selecting the 750 MHz frequency was approximately 0.48.

A more extreme case was then considered, where $f_c = 1$ GHz. It was found that the rate at which probability of selection approached 0.5 was much slower. The two channels were then assigned equal frequencies, resulting in equal probability of selection ($P_c = P_a = 0.5$) for all user densities.

To evaluate the impact of spectrum price on profits, it is necessary to set all other parameters constant (Figure 11.7). To retain heterogeneity, $f_c = 1$ GHz and $f_a = 500$ MHz. SU density was set to $\rho = 50$ users/km^2. With these frequencies, it was found that the lower frequency WSP achieved a greater profit for all price combinations in the duopoly [3].

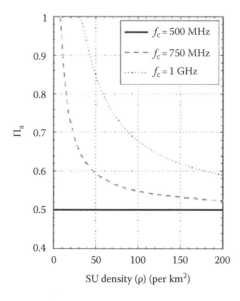

FIGURE 11.6   Market share of WSP_a versus user density. Channel $f_a = 500$ MHz, Π_a is the channel occupancy measure of channel a at market equilibrium. (From Min, A.W. et al., *IEEE Trans. Mobile Comput.*, 11(12), 2020, December 2012.).

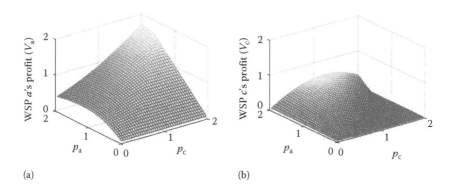

FIGURE 11.7   WSP profitability for price ratios between zero and two. WSP_a leased channel $f_a = 500$ MHz and WSP_c leased channel $f_c = 1$ GHz. Profit of WSP (a) a and (b) c. (From Min, A.W. et al., *IEEE Trans. Mobile Comput.*, 11(12), 2020, December 2012.)

The optimal price for this duopolistic competition is dependent on the price ratio between the two competitors. As user density increases and the property of heterogeneity becomes less significant, the optimal ratio approaches 1. The WSP providing the higher frequency is found to need much lower pricing (as low as 30% for $\rho = 20$ users/km²) in order to achieve maximum profit.

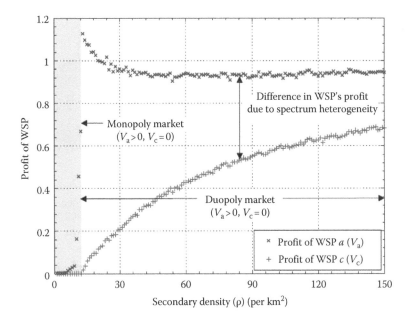

FIGURE 11.8 WSP profits versus user density. (From Min, A.W. et al., *IEEE Trans. Mobile Comput.*, 11(12), 2020, December 2012.)

An iterative search algorithm is used [3] to step through an evolving market, with one competitor setting an optimal price to maximize profit and the other competitor responding. Depending on user density, it is found that there may exist no NE, one NE (unique NE), or multiple NE. It is found that as the low frequency provider's price increases, the best response of the higher frequency provider also increases. Price differences reduce as user density increases, but the two prices remain at minimum 20% different in simulation. Where user demand is not high enough ($\rho < 10/\text{km}^2$), the WSPs cannot justify their business because pricing could not be high enough to profit. Once user density breaks the profitability threshold, the WSP leasing the lower frequency band monopolizes the market. Further increases in user density convert the market into a true duopoly. A high enough user density causes the heterogeneous qualities distinguishing the two WSPs to become insignificant due to the dominance of cochannel interference over noise, resulting in equal market share. See Figure 11.8 for a graphical representation.

11.6.2 Single-Sided Auctions

In [12], two cases were used to compare sequential and concurrent single-sided auctions. In the first, only one subband could be awarded to each bidder. In the second, multiple subbands could be awarded. These were denoted *single unit grant* and *multiple unit grant*. In both cases, sequential auctions were more profitable for the auctioneer (Figure 11.9).

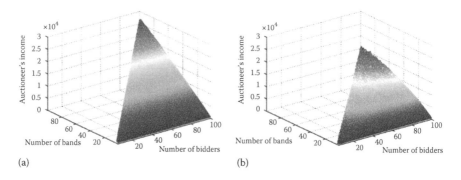

(a)   (b)

FIGURE 11.9 Auctioneer income for single-sided sequential (a) and concurrent (b) auctions. (From Sengupta, S. and Chatterjee, M., *Mobile Network Appl.*, 13, 498, June 2008.)

11.6.2.1 Single Unit Grant

Reservation prices for bidders followed a uniform distribution across a known range of 250–300 units, with bids uniformly distributed between 100 and 300 units.

Sequential auctions were shown to generate more revenue for sellers than concurrent auctions, for the substitutable spectrum band case (homogeneous characteristics for all subbands). As the number of auction periods grows (dynamic spectrum access [DSA] periods in Figure 11.10), seller income grows too. A higher number of bands available correlated to significantly higher seller revenue as well as a faster market settling time (Figure 11.10).

Simulation of nonsubstitutable spectrum band auctions was performed with valuations uniformly distributed between 450 and 500 units [12]. Once again, the sequential auction produced higher revenue for the auctioneer, but this time significantly more so. Market settling times remained unchanged.

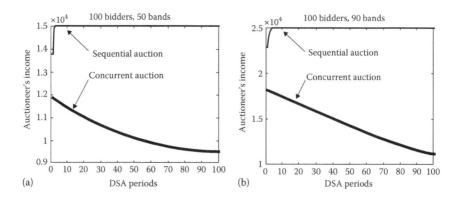

(a)   (b)

FIGURE 11.10 Auctioneer income with substitutable bands with 100 bidders: (a) 50 bands; (b) 90 bands. (From Sengupta, S. and Chatterjee, M., *Mobile Network Appl.*, 13, 498, June 2008.)

11.6.2.2   Multiple Unit Grant

The DSA knapsack auction model was simulated in [12], and results indicated that synchronous allocation performed better than asynchronous allocation for bidders who are granted multiple nonsubstitutable bands (Figure 11.11).

In addition to the constraint on available bandwidth in the CAB, minimum and maximum constraints were placed on bandwidth and time requests. A minimum bid constraint was placed as well. For the synchronous system, spectrum lease time was fixed to one unit.

Second-price bidding was selected over first-price bidding, simply because it is less predictable.

Synchronous knapsack auctions both generated higher revenue and more consistent revenue than asynchronous knapsack auctions. Asynchronous auctions experienced occasional, random drops in seller revenue by more than 15%.

As expected, any increase in CAB capacity was accompanied by a linear increase in revenue. As capacity increased, the gap between synchronous and asynchronous revenues grew slightly.

11.6.2.3   Double-Sided Auctions

In [14], simulations were conducted for 10 sellers and 100 buyers, with all bids greater than zero but less than or equal to one (Figures 11.12 through 11.14). Only one channel was sold per seller, but buyers had a random number of channel requests between one and three. Five seeds were used, and each was run for 100 rounds.

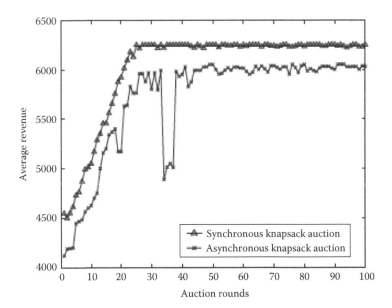

FIGURE 11.11   Revenue generated for synchronous and asynchronous auctions. (From Sengupta, S. and Chatterjee, M., *Mobile Network Appl.*, 13, 498, June 2008.)

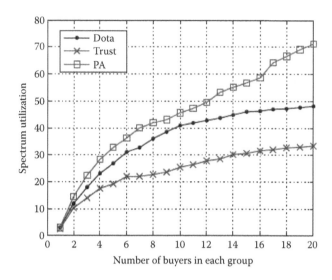

FIGURE 11.12 Spectrum utilization results for DOTA simulation. Spectrum utilization is the total number of winning buyers. PA prioritizes optimal spectrum utilization above bid prices, and does not represent a practical economic model. (From Wang, Q. et al., DOTA: A double truthful auction for spectrum allocation in dynamic spectrum access, in *IEEE Wireless Communications and Networking Conference: MAC and Cross-Layer Design*, Orlando, FL, 2012.)

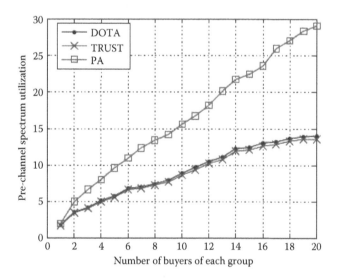

FIGURE 11.13 Total number of buyers who obtain the same channel. (From Wang, Q. et al., DOTA: A double truthful auction for spectrum allocation in dynamic spectrum access, in *IEEE Wireless Communications and Networking Conference: MAC and Cross-Layer Design*, Orlando, FL, 2012.)

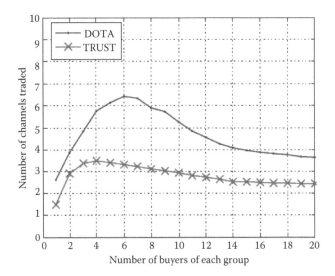

FIGURE 11.14 Total number of channels allocated to buyers. (From Wang, Q. et al., DOTA: A double truthful auction for spectrum allocation in dynamic spectrum access, in *IEEE Wireless Communications and Networking Conference: MAC and Cross-Layer Design*, Orlando, FL, 2012.)

Spectrum utilization, per-channel utilization, and number of traded channels were the observed metrics for this simulation. Three auction methods were compared: DOTA, TRUST (truthful double spectrum auction), and pure allocation (PA).

In TRUST, buyers only request one channel. In PA, the group with the largest size is selected in order to maximize spectrum utilization, regardless of the best economic interests of the seller. The simulation did not address pricing or revenue. PA consistently achieved better spectrum utilization than DOTA and TRUST by all metrics, but this is due to the seller's disregard for the highest bidder in the PA model. DOTA maintained equivalent per-channel spectrum utilization to TRUST, but with a greater number of channels traded. Additionally, spectrum utilization for DOTA was greater as a whole.

DOTA's strength over TRUST is seen to stem from the flexible bidding system that allows users to bid on multiple channels simultaneously. This allows for an increased number of traded channels, as bidder i may lose a bid with group j, but win their bid with group k. By comparison, under the TRUST mechanism when bidder i loses a bid with group j, they have no further opportunities to win during that round.

11.7 AREAS FOR FUTURE RESEARCH

1. *Study of the impact of changes to market equilibrium with the entrance of new sellers.*

   New WSPs emerge and others disappear from time to time, and there is a resultant effect on spectrum lease pricing. Similarly, these WSPs typically start out as small players with limited resources for bidding. The impact of a dichotomy between large, incumbent WSPs and small, new competing WSPs should be studied for auction scenarios. Regulations to protect small players from larger ones exploiting market advantage can be built on findings from such research.

2. *SUs may be capable of facilitating their own cooperative TDMA framework in order to optimize their transmissions in fixed-price markets.*

   When multiple SUs cannot justify the purchase of exclusive rights to a channel, they may purchase a channel as a collective, with each SU receiving a portion of total time allocated to their own transmission purposes. Further, it may be possible to use distributed relay techniques similar to those described in [10] to improve the overall SNR of SUs using the channel.

3. *It is not apparent whether spectrum owners who auction access rights to WSPs have greater profits than the WSPs, who lease the rights to end users.*

   The relative payoffs of spectrum owners and WSPs should be analyzed for such two-market scenarios. In the event that the WSP generates a higher profit, the spectrum owners may be inclined to lease directly to end users, or possibly a hybrid model. For instance, WSPs may purchase bulk spectrum rights for greater lease periods (either fixed price or auction, depending on competition), while end users with shorter leases have the option to choose between leases from WSPs and spectrum owners.

11.8   CONCLUSIONS

As cognitive radio technology is emerging as a popular new technology for dynamic spectrum access, standards, protocols, and other infrastructure must be introduced to facilitate its practical uses. Currently, spectrum owners have exclusive access rights to a large part of the available spectrum. These rights are either purchased from or allocated by a government agency and are tantamount to property rights. For this reason, economic models have been proposed to allow spectrum sharing and leasing between spectrum owners and other wireless transmitters. The models describe the trading relationship (not limited to monetary exchanges) between spectrum owners and the transmitters that wish to use their channel. Decisions made by stakeholders in each model are analyzed using game theory, and it is assumed that all stakeholders take actions to optimize their own objectives. This chapter reviews the nature of four economic models for spectrum leasing: cooperative shared use, fixed-price exclusive use, single auctions, and double auctions. The cooperative shared-use model is found to be the most suitable model for spectrum owners who wish to continue using their channel as they please. Unsurprisingly, it is also the model that depends least on demand from CR users and has the highest network flexibility. Fixed-price exclusive-use models are the best of the four when demand does not necessarily exceed supply, and is a typical market model for service providers leasing access rights to end users. Fixed-price DSA markets are the simplest to model and implement. Single auctions generate the highest revenue for spectrum owners, since they lease exclusive-use rights in a monopolistic bidding environment. Double auctions are the most complex to implement and simulate, due to competition between sellers in addition to competition between buyers. A trusted third party is required to facilitate double auctions and must be compensated, typically with a commission. Although buyers in the market benefit from the sellers' competition, auctions are only sustainable if demand exceeds supply. Double auctions rely on higher demand than single auctions due to the higher supply and lower price of available channels. For all economic models discussed in this chapter, there is an element of noncooperation. Even in the cooperative case, SUs seek to optimize their own utility functions without regard for the impact of their decisions on other SUs. It is not practical to create purely cooperative models due to the limited information available for SUs to make their decisions.

Of the four trading models discussed in this chapter, each presents strengths and weaknesses.

The cooperative shared-use model provides the greatest network flexibility, as payments for channel use are based on the amount of time that the channel is accessed for personal transmission. Other models involve short-term leasing of spectrum access rights for exclusive use, which restricts the channel mobility of users. The cooperative shared-use model also has the benefit of functioning independently of user demand, whereas auction-based mechanisms are only profitable if demand exceeds supply. Therefore, it is useful in a wider variety of circumstances. The PU retains their right to transmit, and the capacity for SU demand is very high due to TDMA. TDMA enables multiple users in the same geographic area to access the same PU's channel. The primary drawbacks to the cooperative model are that SUs pay the added cost of relaying the PU's transmission, and there is substantially less time available for SUs to utilize the spectrum due to TDMA. PUs benefit both from the additional revenue for channel sharing and the enhanced SNR due to spatially distributed transmission.

Compared to shared use, exclusive-use models have much higher revenues due to their granting exclusive access rights to the lease owner. The trade-off is that these models are only suitable for spectrum owners who do not need to transmit using their spectrum, or transmit *very* infrequently. Exclusive-use models have the advantage of allowing for higher prices, as buyers derive higher utility from their exclusive privileges. Further, competitive WSP and user demand drives up prices for access rights, benefit sellers.

Fixed-price exclusive-use models function well when demand does not necessarily exceed supply, since buyers are price takers and can only influence pricing by selecting their optimal seller. Where supply exceeds demand, prices tend to decrease as buyers trend toward the lowest-priced seller. However, cases of low demand experience greater effects due to spectrum heterogeneity. The lowest price may not guarantee a buyer if propagation characteristics are unfavorable. The model is the simplest of all discussed.

Auction-based exclusive-use models are only practical where user demand exceeds spectrum availability. When demand is less than supply, the result is an auction where all buyers bid the minimum possible price, as they are guaranteed to receive access rights for a band.

Double auctions require sellers to compete with one another, with motivations to bid a low sale price to ensure they win a buyer's bid. This is somewhat offset by the buyers' motivation to bid higher prices than competing buyers, to ensure they win their bid. The motivation stems from the market-clearing price. The market-clearing price is the value of the lowest winning buyer's bid, and is the price paid by all winning buyers to all winning sellers. The effect of selecting the lowest winning buyer's bid is that all winning buyers pay less than or equal to what they bid, while all winning sellers receive more than or equal to their bid. Double auctions promote competitive selling, which is preferable to monopolistic single auctions. However, a larger buyer's base is required for a double auction, due to the larger available set of channels for lease. Additionally, the double auction is more complex and requires a trusted third party, such as the FCC, to perform the role of auctioneer. The third party must be paid a commission to maintain its role.

Single auctions generate higher revenue for individual sellers, as the seller does not need to compete. Market behavior is easier to model, and heterogeneous propagation characteristics are less likely to influence buyer behavior. It was shown in [13] that knapsack synchronous auctions perform better than classical second-price synchronous auctions both in terms of maximizing seller revenue and increasing spectrum utilization. It was shown in [12] that synchronous auctions generate higher and more consistent revenue than asynchronous auctions. Also shown in [12], sequential auctions, where a separate subband was bid for every round, generated higher revenue than concurrent auctions. From this, it is clear that homogeneous spectrum bands are best auctioned in sequential, synchronous auctions, and heterogeneous bands are best auctioned in synchronous knapsack auctions. Single auctions have the drawback of being somewhat monopolistic in nature, due to the individual seller.

These findings are summarized in Table 11.1 [1].

TABLE 11.1
Summary of Market Types

	Cooperative Markets	Auction Markets		Fixed-Price Markets
		Single	Double	
Sellers	PUs	Spectrum owners	Spectrum owners	WSPs
Buyers	SUs	WSPs	WSPs	End users
Access rights	Shared	Exclusive	Exclusive	Exclusive
Seller revenue	Lowest	Highest		
Complexity			Highest	Lowest
Market flexibility	Highest	Lowest		
Requirement for third party	No	No	Yes	No

Note: Strengths and weaknesses between distinct market types depend heavily on the spectrum owner's need for spectrum access and the quantity of user demand.

REFERENCES

1. A. Bloor and M. Ibnkahla, Economics of spectrum access, Internal Report, Queen's University, WISIP Laboratory, Kingston, Ontario, Canada, April 2013.
2. C. W. Chen et al., Optimal power allocation for hybrid overlay/underlay spectrum sharing in multiband cognitive radio networks, *IEEE Transactions on Vehicular Technology*, 62 (4), 1827–1837, May 2013.
3. A. W. Min, X. Zhang, J. Choi, and K. G. Shin, Exploiting spectrum heterogeneity in dynamic spectrum market, *IEEE Transactions on Mobile Computing*, 11 (12), 2020–2032, December 2012.
4. Y. Xing et al., Price dynamics in competitive agile spectrum access markets, *IEEE Journal on Selected Areas in Communications*, 25 (3), April 2007.
5. J. Neel, J. Reed, and R. Gilles, The role of game theory in the analysis of software radio networks, in *SDR Forum Technical Conference*, San Diego, CA, November 2002.
6. J. Sairamesh and J. O. Kephart, Price dynamics of vertically differentiated information markets, IBM Thomas J. Watson Research Center, Ossining, NY, 1998.
7. B. Wang et al., Game theory for cognitive radio networks: An overview, *Computer Networks*, 54, 2537–2561, October 2010.
8. D. Niyato and E. Hossain, Competitive pricing for spectrum sharing in cognitive radio networks: Dynamic game, inefficiency of Nash equilibrium, and collusion, *IEEE Journal on Selected Areas in Communications*, 26 (1), 192–202, January 2008.
9. J. M. Chapin, The path to market success for dynamic spectrum access technology, *IEEE Communications Magazine*, 45 (5), 96–103, May 2007.
10. X. Wang et al., Pricing-based spectrum leasing in cognitive radio networks, *IET Networks*, 1 (3), 116–125, September 2012.
11. D. Niyato et al., Dynamic spectrum access in IEEE 802.22-based cognitive wireless networks a game theoretic model for competitive spectrum bidding and pricing, *IEEE Wireless Communications*, 16 (2), 16–23, April 2009.
12. S. Sengupta and M. Chatterjee, Designing auction mechanisms for dynamic spectrum access, *Mobile Networks and Applications*, 13, 498–515, June 2008.
13. S. Sengupta and M. Chatterjee, An economic framework for dynamic spectrum access and service pricing, *IEEE/ACM Transactions on Networking*, 17 (4), 1200–1213, August 2009.
14. Q. Wang et al., DOTA: A double truthful auction for spectrum allocation in dynamic spectrum access, in *IEEE Wireless Communications and Networking Conference: MAC and Cross-Layer Design*, Orlando, FL, 2012.

12 Security Concerns in Cognitive Radio Networks

12.1 INTRODUCTION

Security is a major challenge in cognitive radio networks (CRNs) [1] and needs to be addressed in an effective way prior to real deployment [2–10]. For example, in primary user (PU) cooperative scenarios where the PU provides the secondary user (SU) with all the information about the usage of the spectrum, a malicious user could pretend to be a PU and give false information about idle spectrum to an SU, such as by saying that the spectrum is free when it is actually occupied by a PU. In this scenario, the SU would cause interference to the real PU. At other times, the malicious user might say that the spectrum band is occupied when it is in reality free. Now the spectrum is unused by either a PU or an SU, which reduces the intended throughput. In noncooperative scenarios, the SU needs to sense the spectrum and find an unused spectrum band for its usage, without causing interference to the PU. In this case, a malicious user can pretend to be a PU and selfishly occupy the spectrum, thus denying spectrum access to other users.

This chapter is based on the survey study presented in [11]. It is organized as follows. Section 12.2 gives an overview of the security problem in CRNs. Section 12.3 covers trust as the metric used to determine if users in the network are good or bad. Then three selected types of attacks are presented. Section 12.4 investigates route disruption attacks. Section 12.5 studies jamming attacks. Section 12.6 presents PU emulation (PUE) attacks. The chapter provides an in-depth discussion on how these attacks affect the network and what measures can be taken to mitigate the risks.

12.2 SECURITY PROPERTIES IN COGNITIVE RADIO NETWORKS

This section follows the discussion provided in [2], which categorizes CRN users as good users and bad users. Good users make the network successful and run as designed, by sending and forwarding correct and truthful information. On the other hand, bad nodes try and break down the network; they can do this in a multitude of ways. For example, they can be misbehaving users, selfish users, cheating users, and malicious users. Misbehaving users are SUs that do not obey the rules set in place by the system. Selfish users are SUs that want to always have the most bandwidth and do not care about fairness rules or system equilibrium. A cheating user is an SU that cheats with other users on purpose to increase its own gains. Finally, malicious users are SUs that intentionally and harmfully interfere with the service of others.

There are many functions of the network that could be compromised because of the presence of bad users. For example, concerning availability, a spectrum band belonging to a PU should always be available to them; however, a selfish behavior can compromise that. Another potential problem is that an SU looking for an idle band could get wrong information from a cheating user. Moreover, in a CRN, there is an agreement in place between PUs and SUs (e.g., respecting an interference threshold) and a malicious or selfish user may break this agreement. Another potential risk arises when some users in the network are not trustworthy. Users that enter the system need to be authenticated, and once they are admitted, the system will treat them as good users. However, cheating users can take advantage of this. Finally, we want to ensure stability in the network, such that if anything should happen, the system can return to a state of equilibrium after being interrupted by a physical disturbance. Cheating and malicious users can compromise the stability of the system. Table 12.1 lists a number of network attributes and the behavior that could affect them.

Bad users in a CRN can harm the system in many ways. Here, we briefly describe the following potential attacks: routing disruption attacks, jamming attacks, PUE attacks, biased utility attacks, false feedback attacks, common control channel attacks, key depletion attacks, lion attacks, and jellyfish attacks [3].

Route disruption attacks: The network layer routing table provides the path that a packet travels from source to destination. An attack can be launched anywhere on this path and can disrupt the whole network.

Jamming attacks: An attacker can jam the network by sending arbitrary messages into the network. For example, a user intentionally transmits high power signals so that he can use licensed bands continuously uninterrupted. In overlapping unlicensed jamming, a user can send random malicious packets in order to interrupt service across the network.

PU emulation (PUE) attacks: In a CRN, when a PU uses the spectrum, an SU is not allowed to use it. This leads to the thought that if an SU can emulate the characteristics of a PU, it can have that spectrum band all to itself. This means that the spectrum is not being used efficiently and effectively anymore.

Biased utility attacks: This kind of attack occurs when an SU is selfish and alters its function parameters to get access to more bandwidth; it also decreases the bandwidth/throughput of other SUs nearby.

TABLE 12.1
How Bad Behavior Can Affect Attributes of Cognitive Radio Networks

Cognitive Radio Attribute	Behavior That Could Affect It
Availability	Selfish
Reliability	Cheating
Nonrepudiation	Malicious and selfish
Authentication	Cheating
Stability	Physical circumstances

TABLE 12.2
Attacks and Targeted Layers

Attack	Layer
Route disruption	Network
Jamming	Physical
PU emulation	Physical
Biased utility	Data link
False feedback	Data link
Common control channel	Data link
Key depletion	Transport
Lion	Transport
Jellyfish	Transport

False feedback attacks: This attack arises when a malicious user hides the existence of a PU; thus, SUs are unable to make an accurate decision about the PU's presence.

Common control channel (CCC) attacks: In CCC attacks, the bad user can take control over the CCC channel and inhibit the channel negotiation and allocation process. This may lead, for example, to excessive denial of service (DoS) to SUs.

Key depletion attacks: For this kind of attack, the unwanted user breaks the encryption system with the help of the probability of session key repetitions. This will allow for unauthenticated users to enter into the system.

Lion attacks: Lion attacks target frequency handoffs. They are launched by an external entity with the goal of decreasing throughput.

Jellyfish attacks: Jellyfish attacks use closed loop flow to diminish the throughput of TCP. It also interferes with end-to-end protocols.

Table 12.2 shows the list of attacks and the targeted layers in the protocol stack. It is clear that attacks can occur at all levels of the protocol stack. Therefore, CRNs need to implement effective security measures overall layers in order to minimize or eliminate attacks.

12.3   TRUST FOR COGNITIVE RADIO SECURITY

Liu and Wang provide a comprehensive mathematical foundation for trust in CRNs [3]. Here, we present their work and provide some illustrations based on their findings.

12.3.1   Foundation of Trust Evaluation

Trust is a metric for communication networks, which originates from trust in human culture. It is easier for computers to come up with an expectation of trust. In a communication network, a user has a trusting belief, which can be on (1) the information the user is getting and (2) the system as a whole, that is, a trust that the system is operating as the user expects. In this system, trust can be viewed as a belief. For example, a user believes that the system (or another user) will act in a certain way.

Trust is established between two users in the network who want to perform a specific action. In this part of the chapter, user one will be called the subject and user two will be called the agent. The trust relationship will be denoted as {subject: agent, action} [3].

Now that we have established that trust acts between two users, we need to come up with a way to evaluate trust. Since it has been established that trust is a belief, it is presented that trust should be the uncertainty in the user's belief. Now, we will examine three different cases of trust. In the first case, the subject believes that the agent will certainly perform the action. In this case, the subject fully trusts the agent and there is no uncertainty. In the second instance, the subject believes the agent will definitely not perform the action. In this case, the subject fully distrusts the agent and once again, there is no uncertainty. In the third case, the subject has no idea about the agent at all. Here, the subject has the maximum amount of uncertainty and the subject has no trust in the agent.

The entropy has been used in [3] to express uncertainty in the trust metric. It is needed so that the first case yields 1 (full trust), the second case yields −1 (full distrust), and the third case yields 0 (maximum uncertainty). Let T{subject, agent, action} denote the value of a trust relationship and P{subject, agent, action} denote the probability that the agent will perform that action from the point of view of the subject. The trust function is therefore defined as [3]

$$T = \begin{cases} 1 - H(p), & \text{for } 0.5 \le p \le 1 \\ H(p) - 1, & \text{for } 0 \le p \le 0.5 \end{cases} \tag{12.1}$$

where
$T = T${subject, agent, action}
$p = P${subject, agent, action}
$H(p) = -p\log_2 p - (1-p)\log_2(1-p)$ is the entropy function

This definition for trust covers both trust and distrust, where trust is a positive value and takes place when $p > 0.5$, that is, more likely to perform the action, and is a negative value when $p < 0.5$, that is, the agent is less likely to perform the action. As we can see from the entropy function, $H(p)$, that the value of trust is not linear. With that said gaining more probability in an action you already think is likely going to happen, yields a higher gain in trusts, than gaining the same probability in a different action you are uncertain about. To compare values mathematically, the trust value is determined by the mean value of the estimated probability, and the confidence in that trust value is calculated by the variance of the estimation.

12.3.2 FUNDAMENTAL AXIOMS OF TRUST

A trust relationship can be established either through direct observations or by recommendations. If direct observations are available, the subject can estimate the probability and then calculate the associated trust value. If the subject does not have direct contact with the agent, it can establish trust through propagation. Some funda-mental axioms are needed for the process of trust propagation.

Let us consider two users A and B [3]. A and B have established {A: B, action$_r$}, and B has established {B: C, action} with a third user C. We can say {A: C, action} can be established if the following two conditions are satisfied. First, action$_r$ is to make recommendations of other users about performing action, and second, {A: B, action$_r$} has a positive trust value.

The rationale for the first condition is due to the fact that users that perform the action are not necessarily making accurate recommendations. The second con-dition is essential since recommendations of untrustworthy users could be com-pletely uncorrelated to the truth. To put it conceptually, an enemies' enemy is not necessarily a friend. Therefore, we do not take recommendations from untrust-worthy users.

In the following, three axioms are discussed [3]:

Axiom 1 [3]: Concatenation propagation of trust does not increase trust. This axiom states that when a subject establishes a trust relationship with the agent with the recommendation of a third party, the trust value between the subject and the agent should not be greater than the trust between the subject and recommender as well as the trust value between the recommender and the agent. This is given by the following equation:

$$|T_{AC}| \leq \min\left(|R_{AB}|, |T_{BC}|\right) \tag{12.2}$$

where
$T_{AC} = \{A: C, action\}$
$R_{AB} = \{A: B, action_r\}$
$T_{BC} = \{B: C, action\}$

Axiom 1 is similar to the data processing theorem in information theory, as entropy cannot be reduced via data processing.

Figure 12.1 shows the actions along the path where dashed lines indicate *recom-mendations* and solid lines indicate *performing the action*.

Axiom 2 [3]: Multipath propagation does not reduce trust. This axiom states that if a subject receives the same recommendation for an agent from multiple sources, the result should not be less than the case if it were to receive fewer recommendations (Figure 12.2). A_1 establishes trust with C_1 through one concat-enated path.

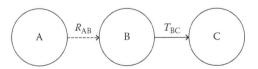

FIGURE 12.1 Transit of trust along a chain. (From Liu, K.R. and Wang, B., *Cognitive Radio Networking and Security*, Cambridge University Press, New York, 2011.)

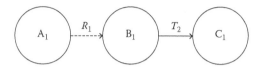

FIGURE 12.2 Combining trust recommendations. (From Liu, K.R. and Wang, B., *Cognitive Radio Networking and Security*, Cambridge University Press, New York, 2011.)

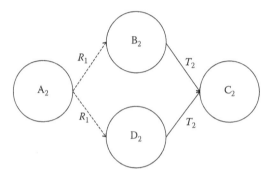

FIGURE 12.3 Multipath trust propagation. (From Liu, K.R. and Wang, B., *Cognitive Radio Networking and Security*, Cambridge University Press, New York, 2011.)

In Figure 12.3, A_2 establishes trust with C_2 through two equivalent trust paths. It should be noted that the number scheme is added to help identify which path we are examining; both figures actually represent the same scenario of A sending information to C. Similar to earlier, we will let $T_{AC1} = T\{A_1: C_1, \text{action}\}$ and $T_{AC2} = T\{A_2: C_2, \text{action}\}$. To put axiom 2 in mathematical terms, we have

$$
\begin{aligned}
T_{AC2} \geq T_{AC1} \geq 0, \quad &\text{for } R_1 > 0 \text{ and } T_2 \geq 0 \\
T_{AC2} \leq T_{AC1} \leq 0, \quad &\text{for } R_1 > 0 \text{ and } T_2 < 0
\end{aligned}
\tag{12.3}
$$

where
$R_1 = T\{A_2: B_2, \text{making recommendation}\} = T\{A_2: D_2, \text{making recommendation}\}$
$T_2 = \{B_2: C_2, \text{action}\} = \{D_2: C_2, \text{action}\}.$

Axiom 2 states that if a subject gets another opinion about the agent, its opinion of the agent will be at least the same as its original opinion. It should be noted that axiom 2 only holds when multiple sources make the same recommendations. The collective combination of different recommendations can generate different trust values according to different trust models.

Axiom 3 [3]: Trust based on multiple recommendations from a single source should not be higher than that derived from independent sources. When a trust relationship is established through concatenation and multipath propagation, it is possible to get multiple recommendations from a single source. This can be seen in Figure 12.4.

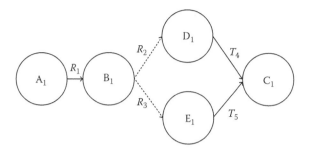

FIGURE 12.4 Multiple recommendations from a single source. (From Liu, K.R. and Wang, B., *Cognitive Radio Networking and Security*, Cambridge University Press, New York, 2011.)

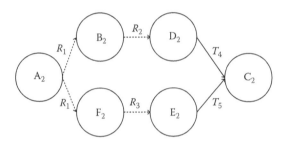

FIGURE 12.5 Recommendations over multiple paths.

Here, we will let $T_{AC1} = T\{A_1: C_1,$ action$\}$ be the trust from Figure 12.4 and $T_{AC2} = T\{A_2: C_2,$ action$\}$ be the trust from Figure 12.5.

For the two scenarios shown in Figures 12.4 and 12.5, axiom 3 states that

$$\begin{aligned} T_{AC2} \geq T_{AC1} \geq 0, \quad &\text{if } T_{AC1} \geq 0 \\ T_{AC2} \leq T_{AC1} \leq 0, \quad &\text{if } T_{AC1} < 0 \end{aligned} \quad (12.4)$$

where all the recommendations R_1, R_2, and R_3 are positive.

Axiom 3 implies that recommendations from independent sources can reduce uncertainty more successfully than recommendations from associated sources.

12.3.3 TRUST MODELS

With the notion of trust defined and with the fundamental axioms developed, it is possible to define methods for calculating trust via concatenation and multipath propagation. These are called trust models. Here, we will look at two main trust models, the entropy-based trust model and the probability-based trust model, as presented in [3].

12.3.3.1 Entropy-Based Trust Model

In the entropy-based trust model, the trust values are used as defined by Equation 12.1 as the input [3]. It should be noted that this model only considers the value of the trust, and not the confidence in that trust. From Figure 12.1, it can be seen that user B observes the behavior of user C and then makes a recommendation to user A. This value is represented by $T_{BC} = \{B: C, action\}$. User A then has a trust on user B given by $T\{A: B, making\ recommendation\} = R_{AB}$. These models need to satisfy the fundamental axioms. Therefore, in order to satisfy axiom 1, $T_{ABC} = T\{A: C, action\}$ is calculated as [3]

$$T_{ABC} = R_{AB}T_{BC} \qquad (12.5)$$

From this equation, we can see that if user B has no idea about user C or if user A has no idea about user B, then the trust between user A and C is zero.

For multipath trust propagation, we have $R_{AB} = T\{A: B, making\ recommenda-tion\}$, $R_{AD} = T\{A: D, making\ recommendation\}$, $T_{BC} = \{B: C, action\}$, and $T_{DC} = \{D: C, action\}$. Therefore, A can establish trust with C through two different paths, $A \rightarrow B \rightarrow C$ or $A \rightarrow D \rightarrow C$. This can be seen in Figure 12.3. To combine the trust established through multiple paths, maximal ratio combining is used, which is given as [3]

$$T\{A: C, action\} = w_1\left(R_{AB}T_{BC}\right) + w_2(R_{AD}T_{DC}) \qquad (12.6)$$

with w_1 and w_2 defined as

$$w_1 = \frac{R_{AB}}{R_{AB} + R_{AD}} \quad and \quad w_2 = \frac{R_{AD}}{R_{AB} + R_{AD}} \qquad (12.7)$$

We can see that if any path in this model has a trust value of 0, it will not affect the final results. We can also see from Equations 12.5 and 12.6 that this model satisfies the three fundamental axioms of trust.

12.3.3.2 Probability-Based Trust Model

In the probability-based trust model, concatenation propagation and multipath trust propagation are calculated using the probability values of the trust relationship. These probabilities can then easily be turned back into trust values by using Equation 12.1. Unlike the entropy-based model, this model uses both the trust and confidence in that trust (i.e., the mean and variance of the trust value). Here, the concatenation propagation probability–based model is presented [3]. We will look at the concatenation of trust propagation in Figure 12.1. The following definitions and notations are used as in [3].

P is a random variable that C *will perform the action*. In A's opinion, the trust value $T\{A: C, action\}$ is determined by $E(P)$ and the confidence is determined by $Var(P)$.

We have a second random variable X, which is binary. If $X = 1$, then B provided an honest recommendation; if it did not, then $X = 0$.

A third random variable Q is the probability that $X = 1$. This can be given by the equation $P(X = 1 \mid Q = q) = q$ and $P(X = 0 \mid Q = q) = 1 - q$. In user A's opinion $P\{A: B, making\ recommendation\} = p_{AB} = E(q)$, and $Var(q) = \sigma_{AB}$.

B provides a recommendation about C according to $P\{B: C, \text{action}\}$ is p_{BC} and the variance of $P\{B: C, \text{action}\}$ is σ_{BC}.

In order to get $E(P)$ and $\text{Var}(P)$, the probability density function (PDF) of P needs to be derived.

A's opinion about C depends solely on whether user B makes honest recommendations. It has been shown in [3] that

$$E(p) = p_{AB} \cdot p_{BC} + (1 - p_{AB}) \cdot (1 - p_{BC}) \tag{12.8}$$

and

$$\text{Var}(P) = p_{AB}\sigma_{BC} + (1 - p_{AB})\sigma_{C|X=0}$$
$$+ p_{AB}(1 - p_{AB})(p_{BC} - p_{C|X=0})^2 \tag{12.9}$$

where
$\sigma_{C|X=0} = \text{Var}(P \mid X = 0)$
$p_{C|X=0} = 1 - p_{BC}$

The choice of $\sigma_{C|X=0}$ depends on specific application scenarios. As an example, if P is uniformly distributed on the interval $[0, 1]$, we can choose $\sigma_{C|X=0}$ to be the maximum possible variance, which is $1/12$. On the other hand, if we assume that the PDF of P is a beta function with mean $m = p_{C|X=0}$, the authors in [3] have shown that we can choose $\sigma_{C|X=0}$ by

$$\sigma_{C|X=0} = \begin{cases} m(1-m)^2 / (2-m) & \text{for } m \geq 0.5 \\ m^2(1-m)/(1+m) & \text{for } m < 0.5 \end{cases} \tag{12.10}$$

The expression in Equation 12.10 is the maximum variance for a given mean m in a beta distribution.

Therefore, we can conclude that the concatenation propagation probability–based model can be expressed by Equations 12.8 and 12.9.

12.3.4 EFFECT OF TRUST MANAGEMENT

In Figure 12.6, three scenarios are examined [3]. The first is a baseline system, where there is no trust management and no attackers. The second is a system with five attackers who randomly drop 90% of the packets that are routed through them. The third scenario is a system that has those same five attackers from the second scenario only with trust management in place. The probability-based trust model has been used for the system. As can be seen in Figure 12.6, the network throughput can be drastically diminished by malicious attackers. Another conclusion that can be drawn is that trust management allows the performance of the network to improve as untrustworthy routes can be avoided. Finally, it

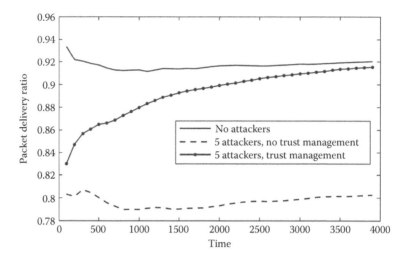

FIGURE 12.6 Network throughput with and without trust management. (From Liu, K.R. and Wang, B., *Cognitive Radio Networking and Security*, Cambridge University Press, New York, 2011.)

can be seen that as the simulation time increases, the trust management system approaches the no attacker's case. This shows that more accurate trust records are built over time.

12.4 ROUTE DISRUPTION ATTACKS

In this section, we present the mechanisms that have been proposed by Liu and Wang and provide some of their illustrations [3].

12.4.1 INTRODUCTION

This section considers CR mobile ad hoc networks as in [3], where security is of paramount importance. Since wireless links in ad hoc networks are fragile, there is a high risk of broken links; users lacking sufficient physical protection can be captured, compromised, and hijacked. The sporadic nature of connectivity and the dynamically changing topology may cause frequent route updates, and a lack of centralized monitoring or management points can cause further deterioration of the situation. Attackers can easily launch attacks ranging from passive eavesdropping to active interference.

To work properly, routing protocols in ad hoc networks need trustful working environments, which are hard to come by, since malicious or selfish users or users who are compromised by outside attackers might exist in the network. In the security schemes, which have been considered in classical approaches, most focus has been on preventing attackers from entering the network. This is done through secure key distribution/authentication and secure neighbor discovery. However, these schemes are not effective in situations where malicious users have entered the network, or users in the network that have been compromised.

This section focuses on the scenario where all users in the network belong to the same authority and pursue common goals. We will present a set of integrated mechanisms to defend against routing disruption attacks launched by inside attackers. In these scenarios, users are sorted into two classes: good and malicious nodes. Good users are fully cooperative, and they properly forward packets for others. Malicious users may manipulate routing messages and drop data packets of other good users. Their goal is to degrade the network performance and consume network resources.

Honesty, adaptivity, diversity, observer, and friendship (HADOF) [3] are the set of mechanisms used to defend against routing disruption attacks. Each node launches a traffic route observer to monitor the behavior of each valid route in its cache; it also gathers packet forwarding statistics from the nodes on those routes. Each node also keeps a cheating record to find malicious nodes. This database indicates which nodes are dishonest or thought to be dishonest. If a node is detected as cheating, it is excluded from future routes; also, nodes will not forward the packets coming from that node. If malicious nodes are smart, it is hard to prove their cheating; therefore, in order to improve malicious node detection, each node can also build friendship with nodes that it trusts.

Since there are multiple routes from the source to the destination, the best path can be determined adaptively. Due to the dynamics of the system, a route that was good in a previous time may not necessarily be good in the present time. Therefore, adaptive route discovery is important to implement in routing protocols. Adaptive route discovery dynamically determines when new route discoveries should be initiated.

The authors in [3] used an on-demand source routing (DSR) protocol as underlying routing protocol. On-demand routing here means that the source only finds routes when it wishes to send a packet (i.e., routes are not necessarily known in advance). DSR has two basic operations: (1) route discovery and (2) route maintenance.

Two types of attacks are commonly used to attack the network layer of ad hoc networks: resource consumption and routing disruption. In resource consuming attacks, the attacker injects extra packets into the network to consume valuable network resources. In routing disruption attacks, the attackers attempt to cause data packets to be routed in a way that causes packets to be dropped or extra network resources to be used. For example, once an attacker is on a certain route, it can create a black hole by dropping all the packets through it. It can also create a gray hole by selectively dropping some of the packets. Attackers can apply rushing attack to disseminate ROUTE REQUEST quickly through the network [3].

12.4.2   LIU–WANG SECURITY MECHANISMS

Liu and Wang have proposed the following security mechanisms for route disruption attacks.

12.4.2.1   Route Traffic Observers

Each node launches a route traffic observer (RTO) to periodically collect traffic statistics of each valid route in its route cache.

12.4.2.2   Cheating Records and Honesty Scores

When the RTO of a sender S collects packet forwarding statistics, false reports may be submitted by malicious nodes. For example, a malicious node might send a smaller RN (the number of packets originated by the source and received by the node) and larger FN (the number of packets originated by the source and relayed by the node) values to cheat the source and frame its neighbors. To counter this, each source keeps a cheating record database (ChR), which keeps track of whether some nodes have ever submitted or have been accused of submitting false reports to it. S will then mark these nodes as malicious if it has enough evidence to confidently say that a specific node is submitting false reports.

12.4.2.3   Friendship

Even when the ChR database is activated, the activity of malicious nodes can only be suspected, but it cannot be proved. For example, a malicious node knows the source S and destination D of the route it is on. It will only frame those nodes that are neither S nor D, to avoid being caught. This can be mitigated by taking advantage of the trustworthy relationship. Each node has a private list of nodes that it trusts and considers to be honest. For example, assume that B submits a false report to S in order to frame A. In this case, if S trusts A, then S will immediately detect B and will set B's honest score to 0.

12.4.2.4   Route Diversity

In most scenarios, there exist multiple paths from source to destination. Therefore, it is usually advantageous to find multiple routes from the source to the destination. Route diversity can protect against route-disrupting attacks. In dynamic source routing, discovering multiple paths from a source to a destination is straightforward [12].

12.4.2.5   Adaptive Route Discovery

Due to mobility and dynamic traffic patterns, some routes may no longer be valid, and their quality might change as well. In a basic protocol, route discovery takes place when no routes from a sender S to a destination D exist on S's routing table. However, this will not be efficient in dynamic networks. Therefore, adaptive route discovery is adopted [3]. Here, each time S has a packet to send to D, a new route discovery procedure is initiated if there are no routes that fulfill some QoS conditions.

12.4.3   Illustrations of Liu–Wang's Security Mechanisms

The following simulations compare HADOF to the basic watchdog mechanism [3]. The idea of watchdog is that a node will report another node to the source if it does not forward a certain number of packets. In the simulations, this number is set to 5.

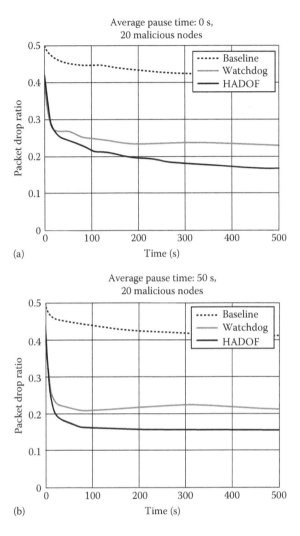

FIGURE 12.7 Packet drop ratio comparisons under gray hole attacks. (From Liu, K.R. and Wang, B., *Cognitive Radio Networking and Security*, Cambridge University Press, New York, 2011.)

The first scenario considers only gray hole attacks (Figure 12.7). In gray hole attacks, an attacker selectively drops packets passing through it [3]. In this scenario, nodes do not submit false reports. In the simulations, the gray holes drop half of the packets that go through them. It can be seen in Figure 12.7 that HADOF outperforms watchdog. For example, under the configuration with a pause time of 50 s and 20 malicious users, the baseline packet drop ratio is 40%, the watchdog mechanism can reduce this number to 22%, and HADOF can reduce it even further to 16%.

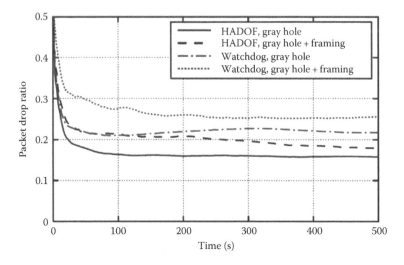

FIGURE 12.8 The effects of framing attacks. (From Liu, K.R. and Wang, B., *Cognitive Radio Networking and Security*, Cambridge University Press, New York, 2011.)

The next scenario considers gray hole and framing attacks. In a framing (or frame-up) attack, a malicious user makes a good node look like a malicious node. In HADOF, the only way for a malicious node to frame a good node is to let the source suspect the good node is cheating [3]. This is done by reporting a smaller RN than the actual value to frame the node ahead of it on the path, and/or report a larger FN to frame the node following it on the path. In HADOF, the malicious node can never make the source believe a good node is cheating, as it cannot create solid evidence. In the watchdog scenario, there are multiple ways for a malicious node to frame good nodes.

Simulation results are shown in Figure 12.8 [3] for 20 malicious users. It can be seen that HADOF outperforms watchdog.

In the previous simulations, friendship was not used, and the source only trusted the destination. The authors in [3] discuss how friendship helps against framing attacks. In this scenario, each source has 20 friends, which are randomly chosen from the good nodes in the network. Figure 12.9 shows HADOF results with 50 s pause time and 20 malicious nodes. Half of the malicious nodes launches gray hole with framing attacks, and the other half launches gray hole attacks. From the simulation results, it can be seen that friendships can improve the system's performance against framing attacks. Given a long enough time with 20 friends, the packet drop ratio becomes 15%, which is even lower than the case of gray hole attacks at 16% packet drop ratio.

12.5 JAMMING ATTACKS

This section presents Wu–Wang–Liu's mechanism for jamming attacks [4], which is based on a Markov decision process (MDP).

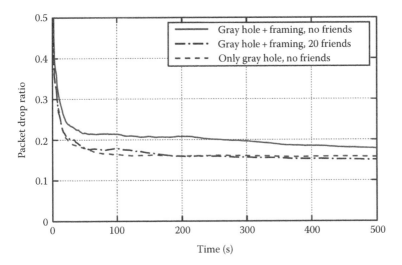

FIGURE 12.9 Effects of friendship. (From Liu, K.R. and Wang, B., *Cognitive Radio Networking and Security*, Cambridge University Press, New York, 2011.)

12.5.1 INTRODUCTION

Jamming attacks are among the major threats to CRNs. Jamming attacks occur when multiple malicious users try to interrupt the communications of SUs, by injecting interference into their signal. In this section, we look at the scenario where an SU could hop across multiple bands of spectrum to avoid the possibility of being jammed. An optimal defense strategy has been proposed in [4] using the Markov decision process. The strategy makes a balance between the cost of hopping bands and the cost of damages that would be caused by a jammer. Here the scenario where the SU has perfect knowledge of the system is considered.

12.5.2 SYSTEM MODEL

The system consists of M licensed bands. Malicious attackers aim at jamming the SUs' communications.

Each licensed band is assumed to be divided into time slots. The access of PUs in each time slot can be characterized by an ON–OFF model. As can be seen in Figure 12.10, during each time slot, the PU can have a busy (ON) or idle (OFF) state. There can also be the transition from ON to OFF with probability α, and the probability of switching from OFF to ON is given by β.

Since SUs must avoid interfering with PUs, SUs must be synchronized with PUs to detect their presence at the beginning of each time slot. In this system, it is assumed that each SU is equipped with only one radio, and can thus only sense one of the M spectrum bands in any time slot. In the case that a PU is not using that band, the SU can take advantage of that band, which will lead to a communications gain R; if not, the SU must tune their radio to a different band of spectrum, and determine the

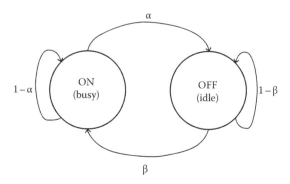

FIGURE 12.10   ON–OFF model for PUs' spectrum usage.

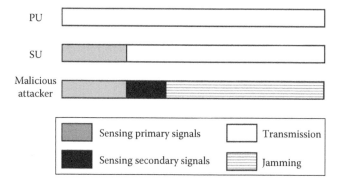

FIGURE 12.11   Time slot structure as proposed by Wu et al. (From Wu, Y. et al., Optimal defense against jamming attacks in cognitive radio networks using the Markov decision process approach, *IEEE GLOBECOM*, Miami, FL, 2010.)

availability of that band in the upcoming time slot. The cost that is associated with this hopping of spectrum is denoted in Wu–Wang–Liu's model by C.

Wu–Wang–Liu's model assumes that there are m (where m is at least one) malicious users who are attempting to jam the communication link of the SU. The *m* malicious attackers are assumed to have a single radio and can therefore only be in one band at a time. The scenario is illustrated in Figure 12.11. PUs' usage of their bands of spectrum is enforced by their ownership of the spectrum, so malicious users do not want to interfere with PUs either. An attacker tunes its radio to a specific band at the start of a time slot to sense the presence of a PU. If a PU is absent, the attacker continues to detect whether an SU is using that band. If an attacker finds an SU in the band, it will immediately begin to inject jamming power into the SU's signal, which will make the SU fail to decode the incoming data packets. Note that all attackers can work together to maximize the damage. Wu–Wang–Liu's model assumes that an SU suffers from a significant loss when it is affected by jamming. Jamming causes interruption to SU communication, and an immense effort is required to reestablish the link. This loss is denoted by L.

When there are no attackers, the SU should always stay in an idle band until an SU comes, since hopping will incur a cost C. However, when there are attackers in a system, the longer an SU stays in a band, the higher the risk to be exposed to attackers in future time slots. So, proactive hopping might actually be beneficial, in order to help hide from malicious users.

Wu–Wang–Liu's model is represented by a multistage game, where the players are the SUs and there are m malicious users. At the end of each time slot, the SU either decides to stay in the same band or hop to a different band. This decision is made based on observations from the current and past time slots. The SU receives an immediate payoff of $U(n)$ in the nth time slot, which is given by the gain minus cost and damage [4]:

$$U(n) = R \times 1(\text{Successful transmission}) - L \times 1(\text{Jammed}) - C \times 1(\text{Hopping}) \quad (12.11)$$

where 1(.) is an indicator function, which returns 1 when the statement in the brackets is true and a 0 otherwise.

The SU wants to maximize the average payoff. However, the attackers' objective is to minimize $\bar{U}$, which is a function of the SU's payoff. Wu–Wang–Liu's model defines $\bar{U}$ as a weighted sum of the SU's immediate payoffs [4]:

$$\bar{U} = \sum_{n=1}^{\infty} \delta^n U(n) \quad (12.12)$$

where $\delta(0 < \delta < 1)$ is the discount factor that measures how much the future payoff is valued over the current.

12.5.3 OPTIMAL STRATEGY WITH PERFECT KNOWLEDGE

In this section, the main goal is to derive the optimal strategy an SU should use when perfect information is available. In order to catch an SU as fast as possible, the attackers should coordinate and tune each of their radios randomly to m undetected bands in each time slot. This will happen until the process starts over when either all bands have been sensed or the SU has been found and then jammed.

Assuming the fixed attack strategy, the jamming game is reduced in [4] to a Markov decision process. This is due to the fact that only the defense strategy needs to be taken into place.

At the end of the nth time slot, the SU observes the state of the current time slot and chooses an action $a(n)$. This action is either to tune the radio to a new band or to stay in the current band, and this action will take place at the beginning of the next time slot. If the PU occupied the band in the nth time slot, $S(n)=P$, or the SU was jammed in the nth time slot, $S(n)=J$, the SU will have to hop to a new band, meaning $a(n)=h$. If neither of the previously mentioned events happened, the SU has successfully transmitted a packet in that time slot and has the choice of hopping to a new band ($a(n) = h$), or staying in the current band ($a(n) = s$). If the SU is on the Kth consecutive time slot with a successful transmission in the same band, the current state is given by $S(n) = K$. Since there is minimal ambiguity forthcoming, we will drop the index n for convenience.

According to Equation 12.11, the immediate payoff depends on the state and the action:

$$U(S,a) = \begin{cases} R, & \text{when } S \in \{1,2,3,....\}, \ a = s \\ R - C & \text{when } S \in \{1,2,3,....\}, \ a = h \\ -L - C & \text{when } S = J \\ -C & \text{when } S = P \end{cases} \qquad (12.13)$$

The transition of states can be described by Markov chains. The transition probabilities depend on the action taken. Wu–Wang–Liu's model uses $p(S' \mid S, h)$ and $p(S' \mid S, s)$ to represent the transition probability from an old state S to a new state S' when the SU either hops (h) or stays (s).

If the SU hops to a new state, the transition probabilities remain the same, as they do not depend on the previous state. Also, the only new possible states are P (the new band is occupied by a PU), J (transmission in the new band is detected by an attacker and they will jam them), and 1 (a successful transmission takes place in the new band). When the total number of bands M is large, it is safe to assume that the probability of a PU in a new band is equal to the steady-state probability of the ON–OFF model in Figure 12.10. Neglecting the case when an SU hops back to some band in very short time, we have

$$p(P \mid S,h) = \frac{\beta}{\alpha + \beta} \triangleq \gamma, \quad \text{for } S \in \{P, J, 1, 2, 3, ...\} \qquad (12.14)$$

Provided that a new band is available, the SU will be jammed with probability m/M, since each attacker will be in one band without overlapping. This yields the following transition probabilities

$$p(J \mid S,h) = (1-\gamma)\frac{m}{M}, \quad \text{for } S \in \{P, J, 1, 2, 3, ...\} \qquad (12.15)$$

$$p(1 \mid S,h) = (1-\gamma)\frac{M-m}{M}, \quad \text{for } S \in \{P, J, 1, 2, 3, ...\} \qquad (12.16)$$

In the scenario where the SU stays in the same band, the PU may come back and use the band. This happens with probability β, given in the ON–OFF model. If the band stays idle (no PU), then the state will transition to J if the SU gets jammed, and it will increase by 1 otherwise. It should be noted that when the state is either J or P, it is not practical to stay. At state K, only max(M-Km, 0) bands have not been detected by an attacker, and another m bands will be detected in the next time slot; therefore, the probability of jamming given that a PU is absent is given by [4]

$$f_J(K) = \begin{cases} \dfrac{m}{M\text{-}Km} & \text{if } K < \dfrac{M}{m} - 1 \\ 1, & \text{otherwise} \end{cases} \qquad (12.17)$$

This gives the transition probabilities associated with s as: $\forall K \in \{1, 2, 3, \dots ,\}$

$$p(P \mid K,s) = \beta \qquad (12.18)$$

$$p(J \mid K,s) = (1-\beta)f_J(K) \qquad (12.19)$$

$$p(K+1 \mid K,s) = (1-\beta)(1 - f_J(K)) \qquad (12.20)$$

12.5.3.1 Markov Decision Process

In Wu–Wang–Liu's model, if an SU stays in the same band for a long period of time, it will eventually be found by an attacker. It can be seen from Equations 12.17 and 12.20, that $p(K+1 \mid K, s)=0$ if $K > M/m - 1$. With that known, the state S can be limited to a finite set $\{P, J, 1, 2, 3, \dots, \bar{K}\}$, where $\bar{K} = \left\lfloor \dfrac{M}{m} - 1 \right\rfloor$, and the floor function $\lfloor x \rfloor$ returns the largest integer that is not greater than x.

For the Markov decision process, a policy is defined by a mapping from a state to an action as $\pi : S(n) \to a(n)$. This says that Wu–Wang–Liu's policy π specifies an action $\pi(S)$, which we shall take when we are in a certain state S. Out of all possible policies, the optimal policy is the one which maximizes the discounted payoff. Wu–Wang–Liu's model defines the value of state S as the highest expected payoff given that the MDP started in state S:

$$v^*(S) = \max_\pi \left(\sum_{n=1}^{\infty} \delta^n U(n) \mid \text{initial state is } S \right) \qquad (12.21)$$

Since the goal here is to maximize the expected payoff, the optimal policy is the optimal defense strategy for the SU. Another important concept is that after a first move, the remaining part of an optimal policy should still be optimal. This means that the first move should maximize the sum of the immediate payoff, and the expected payoff conditioned on the current action. This is known as the Bellman equation, which here is given by [4]

$$v^*(S) = \max_{a \in \{h,s\}} \left(U(S,a) + \delta \sum_{S'} p(S' \mid S,a) v^*(S') \right) \qquad (12.22)$$

The optimal policy $\pi^*(S)$ is the one that maximizes the Bellman equation.

As stated earlier, the probability of being jammed will get larger as the SU stays in the same band for longer periods of time. Therefore, we can expect that there is a critical state K^* ($K^* \leq \bar{K}$), after which the damage overwhelms the hopping cost. If the SU stays in the same band for a period of time less than K^* time slots, they should stay to exploit the band more. If it has been more than K^* time slots, the SU should proactively hop to another band, since the risk for getting jammed has become substantial. K^* can be found by solving the MDP, and thus, the optimal strategy in the Wu–Wang–Liu's model becomes

$$a^* = \pi^*(S) = \begin{cases} s, & \text{if } 1 \leq S \leq K^* \\ h, & \text{otherwise} \end{cases} \tag{12.23}$$

12.5.4   ILLUSTRATIONS FOR WU–WANG–LIU'S MODEL

Here, we present the results provided in [4] using Wu–Wang–Liu's model to combat jamming. In the simulations, the different parameters are fixed as follows: $R=5$, $C=1$, $M=60$ bands, $\delta=0.95$ as discount factor, $\beta=0.01$, and $\gamma=0.1$.

Figure 12.12 shows the value of the critical state K^* obtained through the MDP versus the value of the damage of jamming L, and the number of attackers m.

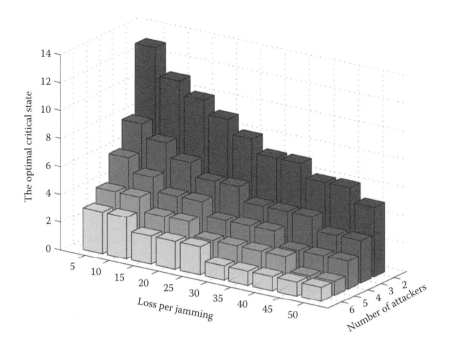

FIGURE 12.12   The optimal state K^* for different L and m. (From Wu, Y. et al., Optimal defense against jamming attacks in cognitive radio networks using the Markov decision process approach, *IEEE GLOBECOM*, Miami, FL, 2010.)

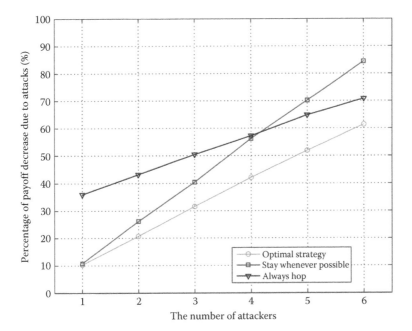

FIGURE 12.13 Percentage of payoff decrease due to jamming attacks with different number of attackers. (From Wu, Y. et al., Optimal defense against jamming attacks in cognitive radio networks using the Markov decision process approach, *IEEE GLOBECOM*, Miami, FL, 2010.)

The simulation is run under the assumption that the SU has perfect knowledge of the environment. As explained in [4], it can be seen that if the damage from jamming is fixed, say for $L = 10$, then the critical state K^* decreases from 11 to 3 as the number of attackers m increases from 2 to 6. It can also be noted that if the number of attackers is fixed, the critical state also decreased as L increases. Intuitively, this makes sense as the user should proactively hop more often as the threat of being jammed increases.

The percentage of payoff loss is studied in [4] where the damage L is set to 20. The optimal strategy is compared to two naive approaches. The first is to always hop, and the second is to hop only when necessary. It can be understood intuitively that when there are few attackers, it is better to stay than to hop, but once the number of attackers gets large, it is better to hop than to stay. However, it can be seen in Figure 12.13 that the optimal hopping strategy outperforms both of these naive strategies over the entire range of attackers. It can also be noted that as the system gets more attackers, more damage is caused to SUs.

12.6 PU EMULATION ATTACKS

As seen in the previous chapters, one of the major challenges in spectrum sensing is to be able to accurately tell whether a signal belongs to a PU or an SU. For example, as stated in [5], usually spectrum sensing schemes that are based on energy detection, implicitly assume a *naive* transmitter verification scheme to distinguish between primary and secondary signals. When using energy detection, an SU can recognize signals coming from other SUs but not signals from PUs. This means that when an SU senses a signal and recognizes it, it assumes that the signal is from an SU; and if it does not recognize the signal, it assumes that it belongs to a PU. This opens the door for malicious users to exploit this simplistic approach. Now, a malicious user can disguise as a PU by transmitting a signal that is unrecognized by other SUs in one of the licensed bands. In this case, SUs will not enter the band, believing that a PU is using the band. This kind of attack is called a PUE attack.

In PUE attacks, the attackers only transmit in idle bands, as the malicious users need to avoid interfering with legitimate PUs. Therefore, the malicious attacker prevents other SUs from using that band by emulating the signal characteristics of a legitimate PU.

Chen et al. [5] classify PUE attacks into two different kinds of attacks depending on the goal of the attacker. The first type is a selfish PUE attack. Here, the goal of the attacker is to maximize its own spectrum use. If selfish PUE attackers find an idle band, they will prevent other SUs from entering the band by emulating PU signal characteristics. This type of attack is likely to be implemented by two users who want to establish a dedicated link. The second type of attack is a malicious PUE attack. The overall goal of this attack is to prevent genuine SUs from detecting and then using unused licensed spectrum bands, causing DoS. The difference between this type of attack and selfish attack is that here the malicious attacker does not necessarily use the band of spectrum it is blocking for its own use. It is possible for an attacker to obstruct the dynamic spectrum access (DSA) process in multiple bands by exploiting two DSA mechanisms in every SU. The first mechanism requires an SU to wait a specified amount of time before it can transmit in an idle band to make sure it is indeed vacant, and this delay is not negligible. The second mechanism requires an SU to sense and immediately leave the band in case of a PU presence. By launching a PUE attack in multiple bands randomly, an attacker could make it difficult for a genuine SU to find an idle band.

Both PUE attacks will be very disruptive if they occur on CRNs. Figure 12.14 illustrates the impact of malicious PUE attackers on the available bandwidth. It can be seen that the available bandwidth decreases significantly as the number of attackers increases. In order to be able to fight these attacks, it is important to accurately detect them. The goal of this section is to show the concept and feasibility of using location-based techniques to detect attackers. The Chen–Park–Reed transmitter verification and localization schemes are described [4]. The schemes can be integrated into the spectrum sensing procedure to detect PUE attacks.

12.6.1   Transmitter Verification Scheme for Spectrum Sensing

In Chen–Park–Reed's model, PUs belong to a network of TV broadcast towers with known locations. It is also assumed that SUs form a mobile ad hoc network. Each SU is assumed to have a self-localization function. Finally, an attacker is assumed to be equipped with a CR device that is capable of changing its modulation mode, frequency, and transmission power.

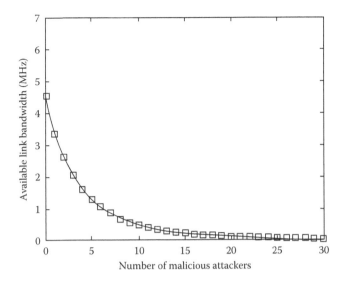

FIGURE 12.14 Effect of malicious PUE attackers on the available link bandwidth. (From Chen, R. et al., *IEEE J. Sel. Areas Comm.*, 26(1), 25, 2008.)

The Chen–Park–Reed's transmitter verification scheme has been designed for hostile environments. With the known location of the TV towers, if a primary signal characteristic is sensed, but their location is different from that of the TV tower, it can be found that the detected signal is a PUE attack. If the PUE attacker positions itself near a TV tower, it could potentially render the location-based detection ineffective. In this case, the signal's energy level, in conjunction with its location, will serve as the way to detect a PUE attack. It would be infeasible for an attacker to impersonate both the PU's location and energy level, since the TV tower has a much larger transmission power than a handheld CR device. Once the PUE attacker has been detected, its estimated signal level can be used to estimate its location.

The transmitter verification scheme includes three phases. The first phase verifies the signal characteristics; this is followed by measuring the received signal's energy levels (second phase), and finally localization of the signal source (third phase). The technical problems related to the first two phases have been covered in previous chapters. As in [4], we emphasize on the problem of transmitter localization. The primary signal transmitter (PST) localization problem is more challenging because no modification should be made to PUs to accommodate the DSA of the licensed spectrum. Due to this requirement, it is not appropriate to include the location about a PU in its signal. This requirement also excludes the possibility of using a localization protocol since it would involve PUs and localization devices interacting. Therefore, the PST localization problem becomes a noninteractive localization problem [5]. Moreover, the transmitter needs to be localized, but not the receiver. When a receiver is localized, it does not need to take into account the existence of other receivers. On the other hand, if there are multiple transmitters, it could be more difficult to do transmitter localization.

12.6.2 Chen–Park–Reed's Noninteractive Localization of Primary Signal Transmitters

Traditional localization approaches include time of arrival (TOA), time difference of arrival (TDOA), angle of arrival (AOA), and radio signal strength (RSS) techniques.

TOA technique would need significant improvements to be applied to the PST localization problem [5]. TDOA and AOA can both be used for transmitter localization, and both have fairly high precision in their estimates. In order to apply them in the PST localization problem, they must be equipped with the ability to handle multiple transmitters (including the attacker) with directional antennas. The major disadvantage of both techniques is the need for expensive hardware.

The Chen–Park–Reed localization system uses the premise that the magnitude of an RSS value decreases exponentially with the distance between the receiver and the transmitter. Therefore, if we can collect a sufficient number of RSS measurements from a group of receivers, the location with the peak RSS value is most likely the location of the transmitter. The main advantage of this technique is that it supports localizing multiple transmitters simultaneously [5].

A wireless sensor network [12] is implemented throughout the CRN to collect RSS measurements across the network. The measured RSS distribution will then be used to solve the PST localization problem. However, the path loss may change over time and a PUE attacker may be in motion and changing its location or its transmission power to avoid being located. This would cause RSS measurements to change over time. Therefore, RSS measurements made throughout the sensor network need to be synchronized. Moreover, RSS regularly varies over short distances due to random shadowing. This makes it difficult to decide the location of a PU based on the raw RSS measurements. A smoothing technique is then needed in this case. In the Chen–Park–Reed scheme, data smoothing is used in order to capture the important patterns present in the raw RSS measurements while eliminating noise.

12.6.3 Simulation Results of the Chen–Park–Reed Scheme

The Chen–Park–Reed scheme is first investigated for three 500 mW PU transmitters denoted by T_1, T_2, and T_3 with coordinates (1000 m, 1000 m), (1000 m, 50 m), and (50 m, 50 m), respectively [5]. The localization errors of the localization scheme are shown in Figure 12.15. When a PST is found to be away from any known location of PUs more than the localization error, the transmitter is considered a PUE attacker. Once the PUE attacker is detected, the localization error will define the area range for which the attacker will be tracked. The computation time is the simulation time to run the localization algorithm but does not include the wireless sensor network delay resulting from RSS data collection.

The results show the system to be effective. However, the wireless sensor network overhead is very high [5]. Note here the high density of sensor nodes, for example, in the 10,000 sensor network scenario; the distance between two adjacent sensor nodes is 20 m, which is close to the localization error of 21.9 m for T_1.

Next, we look at the case when the attacker uses a directional antenna in an attempt to evade localization (Figure 12.16) [5]. It is assumed that the attacker

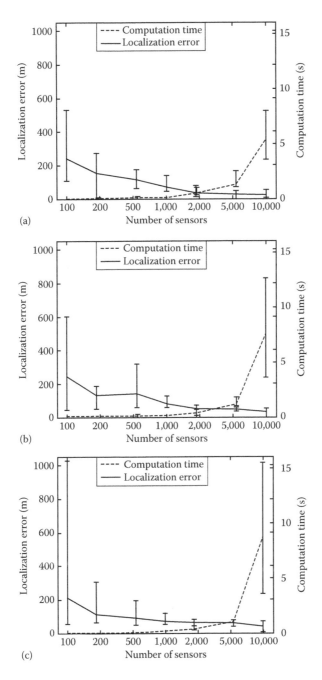

FIGURE 12.15 The localization error of Chen–Park–Reed's localization system. (a) T_1(1000 m, 1000 m). (b) T_2(1000 m, 50 m). (c) T_3(50 m, 50 m). (From Chen, R., et al., *IEEE Journal on Selected Areas in Communications*, 26(1), 25, 2008.)

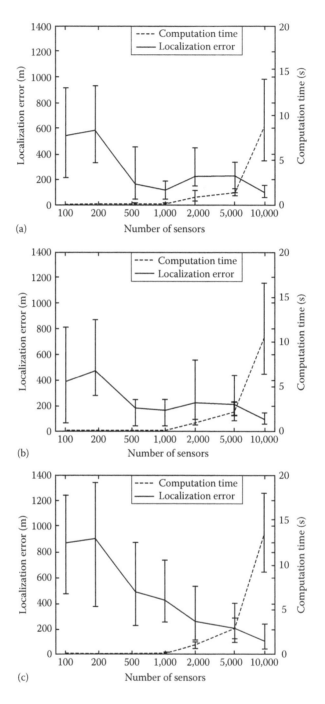

FIGURE 12.16 Localization error when the attacker uses a 10-element Yagi antenna. (a) T_1(1000 m, 1000 m). (b) T_2(1000 m, 50 m). (c) T_3(50 m, 50 m). (From Chen, R., et al., *IEEE Journal on Selected Areas in Communications*, 26(1), 25, 2008.)

uses a 10-element Yagi-Uda antenna. In the simulation, the major lobe in the antenna's radiation pattern is pointed toward the increasing direction of the x-axis. The directional antenna has increased the localization error and computation time. Here, the directional antennas caused fewer sensors to detect the transmitted signal. Therefore, increasing the density of the sensors leads to performance improvement. However, the performance remains weak compared to classical localization techniques. Moreover, the number of sensors remains unreasonably high (hundreds or a few thousands)! Another limitation of this technique is that the locations of PUs are considered fixed, and the technique cannot be applied in the context of mobile PUs, or PUs with unknown locations. It should be noted though that most future handheld devices will certainly be equipped with efficient localization techniques, which can be used in future implementations of location-based security algorithms.

12.7 CONCLUSION

This chapter addresses the security issue in CRNs and provides some methods to enhance security in CRNs. A number of security threats are briefly presented in this chapter. The chapter introduces the notion of trust with some mathematical trust models. Three types of threats are examined thoroughly based on [3–5]. The first threat is route-disrupting attacks. HADOF technique is described as a good defense mechanism against these attacks. The second threat is jamming attacks. An optimal defense strategy based on Markov decision process has been discussed. Here, SUs proactively hop after staying in a certain band for a given period of time to hide from malicious attackers. The third threat concerns PUE attacks. Here, the transmitter's signal characteristics, transmitted signal energy level, and its location are necessary to mitigate this type of attack. These parameters allow an SU to distinguish between a genuine PU and an attacker.

REFERENCES

1. S. Haykin, Cognitive radio: Brain-empowered wirless communications, *IEEE Journal on Selected Areas in Communications*, 23 (2), 201–220, 2005.
2. S. Alrabaee, A. Agarwal, D. Anand, and M. Khasawneh, Game theory for security in cognitive radio networks, in *International Conference on Advances in Mobile Network, Communication and Its Applications*, Bangalore, India, 2012.
3. K. R. Liu and B. Wang, *Cognitive Radio Networking and Security*, Cambridge University Press, New York, 2011.
4. Y. Wu, B. Wang, and K. R. Liu, Optimal defense against jamming attacks in cognitive radio networks using the Markov decision process approach, *IEEE GLOBECOM*, Miami, FL, 2010.
5. R. Chen, J.-M. Park, and J. H. Reed, Defense against primary user emulation attacks in cognitive radio networks, *IEEE Journal on Selected Areas in Communications*, 26 (1), 25–37, 2008.
6. X. Zhang and C. Li, The security in cognitive radio networks: A survey.
7. W. Weifang, Denial of service attacks in cognitive radio networks, in *Conference on Environmental Science and Information Application Technology*, Wuhan, China, 2010.

8. D. Hao and K. Sakurai, A differential game approach to mitigating primary user emulation attacks in cognitive radio network, in *IEEE International Conference on Advanced Information Networking and Applications*, Victoria, British Columbia, Canada, 2012.

9. T. C. Clancy and N. Goergen, Security in cognitive radio networks: Threats and mitigation.

10. J. Shrestha, A. Sunkara, and B. Thirunavukkarasu, Security in cognitive radio, San Jose State University, San Jose, CA, 2010.

11. J. Spencer and M. Ibnkahla, Security in cognitive radio networks, Internal report, Queen's University, WISIP Lab., April 2014.

12. M. Ibnkahla, *Wireless Sensor Networks: A Cognitive Perspective*, CRC Press—Taylor and Francis, Boca Raton, FL, 2012.

Index